Incorporating Uncertainty into Fishery Models

Edited by

Jim M. Berkson
Department of Fisheries and Wildlife Sciences, Virginia Polytechnic Institute and State University, Blacksburg, Virginia

Lisa L. Kline
Atlantic States Marine Fisheries Commission, 1444 Eye Street, NW, 6th Floor, Washington, D.C.

Donald J. Orth
Department of Fisheries and Wildlife Sciences, Virginia Polytechnic Institute and State University, Blacksburg, Virginia

American Fisheries Society Symposium 27

Proceedings of the Workshop
Incorporating Uncertainty into Fishery Models
Held at Jekyll Island, Georgia, USA
28–29 June 1999

American Fisheries Society
Bethesda, Maryland
2002

The American Fisheries Society Symposium series is a registered serial. Suggested citation formats follow.

Entire book

Berkson, J. M., L. L. Kline, and D. J. Orth, editors. 2002. Incorporating uncertainty into fishery models. American Fisheries Society, Symposium 27, Bethesda, Maryland.

Chapter within the book

Mace, P. M., and M. P. Sissenwine. 2002. Coping with uncertainty: evolution of the relationship between science and management. Pages 9–28 *in* J. M. Berkson, L. L. Kline, and D. J. Orth, editors. Incorporating uncertainty into fishery models. American Fisheries Society, Symposium 27, Bethesda, Maryland.

Printed in the United States of America on acid-free paper.

Cover design by Debra Tyler

Library of Congress Control Number 2002101267
ISBN 1-888569-31-X
ISSN 0892-2284

American Fisheries Society
5410 Grosvenor Lane, Suite 110
Bethesda, Maryland 20814-2199
USA

It's ok to be uncertain. It's ok to be ignorant.
But it's never ok to be certain when you are, in fact, uncertain or ignorant.
Dr. Donald J. Orth

Contents

About the Editors vii
Preface ix
Symbols and Abbreviations xi

Evolving Methodologies: From Creation to Application
Jim Berkson, Lisa L. Kline, and Donald J. Orth 1

Coping with Uncertainty: Evolution of the Relationship between Science and Management
Pamela M. Mace and Michael P. Sissenwine 9

Management in the Face of Uncertainty: A Case Study with Amberjack
Gregg T. Waugh and Susan Shipman 29

Mixed Monte Carlo/Bootstrap Approach to Assessing King and Spanish Mackerel in the Atlantic and Gulf of Mexico: Its Evolution and Impact
Christopher M. Legault, Joseph E. Powers, and Victor R. Restrepo 37

Incorporating Uncertainty into Stock Assessments: A Case Study of Atlantic Striped Bass
Gary R. Shepherd 45

A Stochastic Decision-Based Approach to Assessing the Status of the Delaware Bay Blue Crab Stock
Thomas E. Helser, Alexei Sharov, and Desmond M. Kahn 63

Consideration of Uncertainty in Stock Assessment of Atlantic Menhaden
D. S. Vaughan, M. H. Prager, and J. W. Smith 83

Bayesian Approaches to the Analysis of Uncertainty in the Stock Assessment of Pacific Halibut
Ana M. Parma 113

Bayesian Analysis of Stock Survival and Recovery of Spring and Summer Chinook of the Snake River Basin
Richard B. Deriso 137

Incorporating Uncertainty into Marine Mammal Management
Debra Palka 157

Uncertainty, Risk, and Decision: The PVA Example
Daniel Goodman 171

About the Editors

Dr. Jim Berkson is an assistant professor in the Department of Fisheries and Wildlife Sciences at Virginia Polytechnic Institute and State University (Virginia Tech), specializing in population dynamics and resource modeling. He received his bachelors degree under Dr. Michael Gilpin at the University of California at San Diego, his masters degree under Dr. Carl Walters and Dr. Donald Ludwig at the University of British Columbia, and his doctoral degree under Dr. Daniel Goodman at Montana State University. Prior to coming to Virginia Tech in 1998, Dr. Berkson spent 15 years working for state, federal, and tribal resource agencies, developing and applying population dynamics models to wildlife, marine mammal, and fish populations. Dr. Berkson has worked on a variety of technical committees, including 10 years on the Pacific Salmon Commission's Chinook Technical Committee. He is currently the vice-chairman of the Science and Statistics Committee of the South Atlantic Fishery Management Council. Dr. Berkson has developed a course titled, "Introduction to Resource Modeling," to be taught at the National Conservation Training Center in Shepherdstown, West Virginia. Dr. Berkson's current research projects include developing a risk-assessment model for introducing a nonnative triploid oyster into the Chesapeake Bay, designing a benthic trawl survey for horseshoe crab stock assessment, and evaluating the effects of a closed fishery on the rebuilding of the red porgy stock.

Dr. Lisa Kline is Director of Research and Statistics for the Atlantic States Marine Fisheries Commission. She received her bachelors degree under Dr. Samuel Ha at Millersville University of Pennsylvania and her doctoral degree under Dr. James Colvocoresses and Dr. Herbert Austin at the Virginia Institute of Marine Science, College of William and Mary. Prior to her appointment as Director of Research and Statistics, Dr. Kline worked for the commission as the recreational fisheries statistics coordinator and for the Maryland Department of Natural Resources as a fisheries specialist. During her employment with the commission, Dr. Kline has been involved in coordination of Atlantic Coast Research and Data-Collection Activities. Dr. Kline has been involved in the development and implementation of the Atlantic Coast Cooperative Statistics Program (ACCSP), the Northeast Area Monitoring and Assessment Program (NEAMAP), and the Southeast Area Monitoring and Assessment Program (SEAMAP) for coordination of fisheries-dependent and independent data. Dr. Kline has also been involved in developing and conducting stock-assessment training workshops and seminars for state biologists and managers, coordinating external peer reviews of commission stock assessments, and coordinating research issues such as aquaculture, power plant and multispecies assessments, tagging programs, and socioeconomics.

Dr. Donald Orth is a professor and department head of the Department of Fisheries and Wildlife Sciences at Virginia Polytechnic Institute and State University (Virginia Tech), where he teaches fisheries management, modeling, fish population dynamics, and stream habitat analysis. He also teaches continuing-education courses in fish stock assessment. Currently, his academic interests focus on bioethics, leadership, and restoration of stream habitats in highly altered systems. He is a Fellow of the American Institute of Fisheries Research Biologists and the Virginia Natural Resources Leadership Institute.

Preface

The Atlantic States Marine Fisheries Commission (ASMFC) was established by an interstate compact, entered into by the fifteen Atlantic coast states (Maine through Florida). The ASMFC provides a forum for discussion and resolution of common problems and assists the states in developing joint research and management programs. Since about 1995, the ASMFC has focused attention on educating state fisheries and stock-assessment biologists on current assessment principles and models. Several levels of workshops have been sponsored, including an advanced workshop conducted jointly with Virginia Polytechnic Institute and State University (Virginia Tech). This workshop was developed to provide advanced training on how to incorporate uncertainty into fishery models and to develop recommendations for the ASMFC and Atlantic Coast states on methods to more fully incorporate uncertainty into state and interjurisdictional stock assessments.

Given that there is no standardized method for incorporating uncertainty into fishery models, the workshop was designed to expose state biologists to a range of approaches. The workshop focused on the presentation of case studies from assessment and management of U.S. marine fisheries, as opposed to the presentation of theoretical modeling approaches. Development of this workshop was coordinated by a Scientific Advisory Panel, which was specifically tasked with the following:

1. Ensure that the invited workshop presenters were experts in the field of stock assessment and that a full range of methods to incorporate uncertainty were included in the workshop and book;
2. Ensure that the format of the workshop and book would be appropriate for the dissemination of information to other parties interested in this topic; and
3. Ensure that the reviewers of all chapters had the appropriate expertise.

The Scientific Advisory Panel was composed of three groups:

1. The organizers of the workshop and book;
2. Recognized experts in the field of incorporating uncertainty into fishery models; and
3. Stock-assessment biologists from state agencies representing the needs of the state agencies.

The Scientific Advisory Panel was composed of the following members:

Jim Berkson, Virginia Tech;
John Boreman, National Marine Fisheries Service;
Ray Hilborn, University of Washington;
Lisa Kline, Atlantic States Marine Fisheries Commission;
Najih Lazar, Rhode Island Division of Fish and Wildlife;
Steve Murawski, National Marine Fisheries Service;
Michael Murphy, Florida Department of Environmental Protection;
Donald Orth, Virginia Tech;

David Simpson, Connecticut Department of Environmental Protection; and
Michael Sissenwine, National Marine Fisheries Service.

Speakers/authors were selected by the Scientific Advisory Panel and invited to contribute to both the workshop and book.

The ASMFC and Virginia Tech conducted the workshop on "Incorporating Uncertainty into Fishery Models" 28–29 June 1999 in Jekyll Island, Georgia. Invited attendees consisted of approximately 40 state, federal, ASMFC, and university scientists. Each of the workshop speakers contributed a chapter to this book. Each chapter was reviewed by two experts selected by the Scientific Advisory Panel. The editors accepted the chapter for publication when the reviewers' comments had been properly addressed.

The editors, authors, and sponsors of this book hope that you will find this book informative and that it will increase your familiarity and comfort with methods for incorporating uncertainty into fishery models. The trend of fishery stock assessment and fishery management is clearly moving in the direction of not only incorporating uncertainty, but also embracing it.

We anticipate that ongoing education and training programs will continue to focus more time and resources on this topic. These methods are in the transitional stage of moving from their development by a few to their application by many. Though the methods may initially seem foreboding, many are actually quite straightforward. We hope that this book will take away many of the mysteries and misgivings you may feel about their application.

The editors would like to thank the following individuals for their time and effort spent ensuring the success of the workshop and book: the members of the Scientific Advisory Panel for assistance in planning the workshop; Geoffrey White and Jeffrey Brust from the ASMFC staff for assistance with conducting the workshop; David Waterman and his staff at Virginia Tech for assistance in planning the workshop; the presenters and authors for their commitment to completing this book; the chapter reviewers for improving the final chapter manuscripts; Vanessa Jones of the ASMFC staff for assistance in formatting; Deborah Tyler for the cover artwork; and Robert Rand, Debby Lehman, and the staff at the American Fisheries Society for their assistance in publishing this book.

Partial funding for the workshop on Incorporating Uncertainty into Fishery Models was provided by the ASMFC pursuant to U.S. Department of Commerce, National Oceanic and Atmospheric Administration Award No. NA97 FGO 0034. Additional funding was provided by Virginia Tech through a University Reach Out Grant.

Jim Berkson
Lisa L. Kline
Donald J. Orth

Symbols and Abbreviations

The following symbols and abbreviations may be found in this book without definition. Also undefined are standard mathematical and statistical symbols given in most dictionaries.

A	ampere
AC	alternating current
Bq	becquerel
C	coulomb
°C	degrees Celsius
cal	calorie
cd	candela
cm	centimeter
Co.	Company
Corp.	Corporation
cov	covariance
DC	direct current; District of Columbia
D	dextro (as a prefix)
d	day
d	dextrorotatory
df	degrees of freedom
dL	deciliter
E	east
E	expected value
e	base of natural logarithm (2.71828 . . .)
e.g.	(exempli gratia) for example
eq	equivalent
et al.	(et alii) and others
etc.	et cetera
eV	electron volt
F	filial generation; Farad
°F	degrees Fahrenheit
fc	footcandle (0.0929 lx)
ft	foot (30.5 cm)
ft^3/s	cubic feet per second ($0.0283\ m^3/s$)
g	gram
G	giga (10^9, as a prefix)
gal	gallon (3.79 L)
Gy	gray
h	hour
ha	hectare (2.47 acres)
hp	horsepower (746 W)
Hz	hertz
in	inch (2.54 cm)
Inc.	Incorporated
i.e.	(id est) that is
IU	international unit
J	joule
K	Kelvin (degrees above absolute zero)
k	kilo (10^3, as a prefix)
kg	kilogram
km	kilometer
l	levorotatory
L	levo (as a prefix)
L	liter (0.264 gal, 1.06 qt)
lb	pound (0.454 kg, 454 g)
lm	lumen
log	logarithm
Ltd.	Limited
M	mega (10^6, as a prefix); molar (as a suffix or by itself)
m	meter (as a suffix or by itself); milli (10^{-3}, as a prefix)
mi	mile (1.61 km)
min	minute
mol	mole
N	normal (for chemistry); north (for geography); newton
N	sample size
NS	not significant
n	ploidy; nanno (10^{-9}, as a prefix)
o	ortho (as a chemical prefix)
oz	ounce (28.4 g)
P	probability
p	para (as a chemical prefix)
p	pico (10^{-12}, as a prefix)

Pa	pascal
pH	negative log of hydrogen ion activity
ppm	parts per million
qt	quart (0.946 L)
R	multiple correlation or regression coefficient
r	simple correlation or regression coefficient
rad	radian
S	siemens (for electrical conductance); south (for geography)
SD	standard deviation
SE	standard error
s	second
T	tesla
tris	tris(hydroxymethyl)-aminomethane (a buffer)
UK	United Kingdom
U.S.	United States (adjective)
USA	United States of America (noun)
V	volt
V,	Var variance (population)
var	variance (sample)
W	watt (for power); west (for geography)
Wb	weber
yd	yard (0.914 m, 91.4 cm)
α	probability of type I error (false rejection of null hypothesis)
β	probability of type II error (false acceptance of null hypothesis)
Ω	ohm
m	micro (10^{-6}, as a prefix)
′	minute (angular)
″	second (angular)
°	degree (temperature as a prefix, angular as a suffix)
%	per cent (per hundred)
‰	per mille (per thousand)

American Fisheries Society Symposium 27:1–8, 2002

Evolving Methodologies: From Creation to Application

JIM BERKSON
Department of Fisheries and Wildlife Sciences, Virginia Polytechnic Institute and State University Blacksburg, Virginia 24061, USA

LISA L. KLINE
Atlantic States Marine Fisheries Commission, 1444 Eye Street, NW, 6th Floor Washington, D.C. 20005, USA

DONALD J. ORTH
Department of Fisheries and Wildlife Sciences, Virginia Polytechnic Institute and State University Blacksburg, Virginia 24061, USA

Introduction

There are many challenges facing fisheries modelers and stock-assessment biologists today. The systems being modeled are complex, and yet our understanding of them is limited. Our workload is increasing, while our manpower and resources remain constant at best. The tools of our trade, the technology and methodology, continue to evolve and expand, while the raw materials of our trade, our data, are limited and of poor quality. Often times, the results of our work are not welcomed by the decision makers for whom they are intended. Much of the time, our work is misunderstood by stakeholders in the management process. As a result of some combination of limited understanding, limited resources, and political realities, we often fall short of our goal to help shape policy by providing the likely consequences of alternative management actions.

Fisheries modeling is entering a new era that could increase the effectiveness of both fisheries models and fisheries modelers. Our modeling methodologies are undergoing a major evolution, where the uncertainty of our systems is being acknowledged and dealt with. At the same time, the political landscape is changing, where decision-makers are acknowledging the relationship between uncertainty and risk in the development of management policies. Given these events, fisheries modelers will be in a position to play a more substantive role by providing information that incorporates uncertainty to policymakers who will be more likely to welcome it. Of course, this depends on whether the fisheries modelers are aware of the changes taking place and whether they are able to incorporate the changes into their daily work. The purpose of this book is to disseminate this information to the stock-assessment biologists and fisheries modelers who will benefit from it.

Innovators and Assessors

As a group, fishery modelers and stock-assessment biologists come from a wide range of backgrounds and accept a wide range of jobs. To illustrate the process of information dissemination, we will generalize and break the broader group of modelers and assessment biologists into two groups. The first group, which we'll refer to as the Innovators, is composed of a small number of academic and federal fishery biologists whose job it is to develop new stock-assessment and modeling methodologies and approaches. These individuals have extremely strong, mathematical skills. In fact, many could be more accurately described as biological mathematicians, rather than mathematical biologists. Their job is not focused on the quantity of stock assessments produced or the range or number of species assessed, but rather, on pushing the envelope when it comes to methodology.

The second group, which we will refer to as the Assessors, is composed of a large number of federal and state fishery biologists whose job it is to apply existing stock-assessment and modeling methodologies, within regulatory timelines, to what is usually a long list of species. These individuals often have good to excellent mathematics backgrounds, but they do not consider themselves mathematicians. Instead of working through proofs and theory, they primarily work with existing data sets and software packages. These individuals often serve on interagency, species-specific technical committees responsible for putting together stock assessments on short notice with limited data sets. It is likely that the same 5–10 state fishery biologists will be on all of the species-specific technical committees in their region, merely changing their binders and data sets before each new meeting. The Innovators create the majority of the methods. The Assessors conduct the majority of the assessments. While there are obviously a large number of stock-assessment biologists and modelers between these two extremes, the two extremes are useful for illustrating the information dissemination process.

It is the nature of fisheries modeling that new methods are developed by the Innovators and eventually disseminated to the Assessors. There is usually a significant time lag involved. There are many causes for this problem. Much of the work produced by the Innovators, such as journal articles, is often directed to other Innovators, rather than to Assessors. Many Assessors have little to no time to keep track of innovations, and many are not comfortable dealing with the language of the Innovators.

Because of this inevitably slow dissemination process, agencies such as the Atlantic States Marine Fisheries Commission (ASMFC) place a high priority on providing training to its member states' biologists. By shortening the time lag from Innovator to Assessor through its education workshops, ASMFC has improved the effectiveness of its stock-assessment biologists and its stock assessments.

The Evolution of Methodology for Incorporating Uncertainty

There are many definitions and categories of uncertainty. In this book, we primarily deal with the case where we know that uncertainty exists; we know it follows some probability distribution; and we know it affects our fishery stock assessments. Another more complex form of uncertainty is where the uncertainty is not yet apparent and, therefore, cannot be acknowledged. We focus on recent attempts to acknowledge and incorporate uncertainty where we know it exists. We see this uncertainty in our data and in our fishery models. We may call it by a number of names, including process error and structural error. We can either knowingly choose to ignore it or knowingly choose to deal with it. Acknowledging our level of uncertainty permits managers to make more informed decisions regarding the risks of overfishing. This book provides the basics for those who choose to incorporate this uncertainty into their models.

Historically, most fisheries modelers did not acknowledge or deal with the uncertainty related to their fisheries models. Their work was based on the belief that there are some biological and ecological relationships critically important to proper management that are so basic that they simply must be occurring in the system being studied. Functional relationships, such as stock-recruitment relationships and growth curves, were fit to data, producing biological reference points. Management recommendations were based on point estimates, masking the true nature of uncertainty in the data and the analysis.

On occasion, modelers might attempt to present uncertainty to fishery managers, often in the form of confidence intervals. More often than not, it was met with one of two responses: 1) managers would ask the modelers to redo their reports without the confidence intervals, because the managers did not understand them; or 2) managers would use whichever extreme in the confidence interval that allowed the highest catch for their stakeholders, thereby taking a risk-prone strategy. Modelers found neither response very rewarding and, as a result, would often return to presenting point estimates.

Eventually, both managers and modelers began to realize that the point-estimate way of doing things, the method that presents results with complete certainty, was not working well. It became apparent that the standard stock-assessment approach described above was often not sufficient to produce acceptable management policies. The reason for this is that the uncertainty in our systems is too important to ignore. If we consider the parts of the fisheries systems we understand as the signal and the part of the system we do not understand in the form of functional relationships the noise, then the noise is too often greater than the signal. Historically, we were building fisheries models to see how alternate policies worked under average conditions when conditions were anything but average. Now, with the aid of more powerful computers and more advanced methodologies, we can assess how alternate policies work under a distribution of observed conditions, rather than just average conditions.

To do this, a variety of Monte Carlo bootstrapping models were developed to model policies against a backdrop of both signal and noise. Monte Carlo models have been used to test a range of uncertainty types across a variety of species. As you will see in the chapters in this book, there is not one standardized, accepted approach to dealing with uncertainty among Monte Carlo modelers. Meanwhile, a number of Innovators, primarily from the West Coast, were trying a different approach, incorporating uncertainty through the use of Bayesian statistics. The move to Bayesian techniques represented not only a shift in methodology, but also a shift in underlying philosophy.

The process of disseminating new methods from Innovators to Assessors has always been a slow one, for reasons previously discussed. The fact that there is no standardized, accepted way to incorporate uncertainty into fisheries models adds to the complexity of conveying the information. There isn't one new methodology to learn; there's a range of philosophies, approaches, methods, and specialized software packages to learn, choose among, and apply. For the majority of Assessors who already have more work than they can handle with methods they understand, the thought of delving into these new methods and approaches is often too time consuming.

To ease this task, fisheries modelers and stock assessment biologists would benefit by

1. Reading the history and context of incorporating uncertainty into fishery models and seeing some examples of how it has and hasn't worked;
2. Seeing a range of new methods applied in actual case studies; and
3. Seeing how new processes can be constructed to connect the new methods directly to the management process.

This book has been designed and organized to cover all three of these areas.

Timing

The move towards incorporating uncertainty into fisheries models has been going on for years. Why is this book needed now? We believe the book is needed at the present time for the following reasons:

First, the public's mindset, as reflected in legislation and policy, is now taking uncertainty into account, as Mace and Sissenwine (Chapter 2) point out. The adoption of the precautionary principle in key fisheries legislation requires the incorporation of uncertainty into policy development. Sustainability and conservation are moving beyond their long-term status as buzzwords to actual principles for the basis of management in policy forums.

Second, the incorporation of uncertainty has moved from the development of theoretical equations to the application of methods to actual stock assessments. Investigators are applying these methods to real data, producing results, and assessing how the various methods work. Many of these methods have also become more easily accessible in recent years.

Third, we are seeing the development of methods that directly link fisheries models and assessments to the setting of policies. Previously, managers could use, ignore, or misuse information on uncertainty as they wished. With the new methods being developed, which combine decision theory, control rules, and risk assessment, our incorporation of uncertainty can lead directly to setting policy. We will cover all three aspects in the following sections and chapters.

Fourth, a variety of software packages are currently available to Assessors that can be easily misused without sufficient knowledge of the methodologies.

Fifth, students at major universities are being introduced solely to statistical inference or classic frequentist approaches; yet in the real world of decision making, other approaches to analytic science are essential.

Contents of the Book

The book is divided into four sections containing a total of eleven chapters. In this chapter (Chapter 1), we provide the context for, and introduction to, the book.

Section 1 deals with the historical context of incorporating uncertainty into fisheries models and management. How has the omission of uncertainty affected management policy in the past, and how and why is that changing? How have the issues regarding data, methods, and policy interacted in the past? In Section 1, we'll quickly see that the problems of incorporating uncertainty into fishery models have not been solely a problem of methodologies.

In Chapter 2, Mace and Sissenwine provide a historical review of the incorporation of uncertainty into decision making. With regard to restrictive management policies, the burden of proof was, historically, often placed on the biologists, requiring the biologist to prove likely harm before any new regulations became enacted. As a result, the incorporation of uncertainty has traditionally worked against the biologist. The introduction of the precautionary principle into fisheries management has changed this. The precautionary principle states that the greater the uncertainty concerning the status of the resource, the more conservatively the resource should be managed. This is causing a shift in the burden of proof, albeit a slow shift. Mace and Sissenwine review the different types of uncertainty and provide a brief review of ways of presenting and dealing with uncertainty. They suggest that a shift in public opinion away from inexhaustibility and towards conservation also adds to the likelihood of effective management. Mace and Sissenwine also write that we have been slow to apply many of the methods developed in the 1970s and 1980s.

Chapter 3, by Waugh and Shipman, is the first of many case studies in this book. In this chapter, the authors explore the effects of uncertainty on the assessment and management of amberjack *Seriola dumerili* by the South Atlantic Fishery Management Council. Amberjack is an aggressive, top-order piscivore, occurring in temperate and tropical waters from Massachusetts to the Caribbean and into the Gulf of Mexico. Amberjack are caught in both commercial and recreational fisheries, and exploitation increased when regulations on red snapper resulted in unexpected consequences. Waugh and Shipman present a case study that is in many ways typical of how uncertainty has influenced fisheries assessment and management throughout the country. Lack of data is a frequent and ongoing problem that hampers both the biologists and the managers. With too little data, our ignorance of the system may be far more important than our uncertainty. Waugh and Shipman detail the interrelationships of data, analysis, and policy, illustrating the frequent disconnect between biologists' recommendations and the actual policies adopted.

In Chapter 4, Legault, Powers, and Restrepo's review of the evolution of the assessment and management of king mackeral *Scomberomorous cavalla* and Spanish mackerel *S. maculates* is in sharp contrast to the review of amberjack in the preceding chapter. King and Spanish mackerel are migratory coastal pelagic fish found from Maine to Texas in the Atlantic Ocean and the Gulf of Mexico. The history of the assessments and databases in this fishery has been shaped by the recreational sector. Mackerel are fished in a larger area than amberjack and are considered more valuable than amberjack. As a result, data collection and stock assessment of mackerel has been a high priority for management agencies. Legault, Powers, and Restrepo detail an assessment method that incorporates uncertainty in tuning indices, catch at age, and natural mortality rates. Their methods combine virtual population analysis (VPA) with Monte Carlo bootstrap methods. Legault, Powers, and Restrepo quantify how levels of uncertainty observed in the data related to levels of uncertainty observed in the management advice given.

Section 2 provides actual case studies demonstrating methods of incorporating uncertainty into stock assessments and fishery models for several Atlantic Coast species. The case studies presented in this section represent state-of-the-art methodology, and the results have all been presented in management forums. The case studies involve a range of species, a range of assessment and management goals, a range of uncertainties, and a range of assessment and modeling techniques. Given that no standard methodology exists for incorporating uncertainty, Section 2 presents a cross section of approaches in actual use on the east coast of the United States.

In Chapter 5, Shepherd presents a case study on Atlantic striped bass *Morone saxatilis*, a coastal migratory stock that originates in the Chesapeake Bay and its tributaries, the Hudson River and the Delaware River. Shepherd tests the effects of uncertainty in model input and model configuration in a virtual population analysis. A stochastic projection model was used to help understand the implications of the uncertainty. Shepherd emphasizes the need to understand the uncertainty within our data sources. He also points out that stock-assessment biologists must regularly make both objective and subjective decisions regarding their models.

Chapter 6 provides an interesting change of pace, an invertebrate case study. Helser, Sharov, and Kahn present a case study on the Delaware Bay blue crab *Callinectes sapidus*. Many current methods of dealing with uncertainty require biological reference points as input. Although there is obviously uncertainty associated with those reference points, that uncertainty is often ignored. As a result, point estimates are used as if they were true values in methods claiming to investigate the effects of uncertainty. To deal with this, Helser, Sharov, and Kahn introduce a two-tiered, decision-based framework that incorporates the uncertainty in biological reference points and the uncertainty in the assessment results. Their approach uses a Monte Carlo bootstrap procedure.

We return to fish, Atlantic menhaden *Brevoortia tyrannus* to be precise, in Chapter 7. Atlantic menhaden is a euryhaline clupeid found in coastal waters from Florida to Nova Scotia. Vaughan, Prager, and Smith take a different approach to investigating how uncertainty affects VPA methodology. They examine the development and coherence of the catch-at-age matrix; investigate historical, retrospective analyses; and test the method's sensitivity to age-varying, natural mortality. In addition, they consider stochastic analyses, including a bootstrapped application of a surplus-production model. In comparing virtual population analyses and surplus-production models, they examine the effects of model structure. They conclude by pointing out that the greatest uncertainty in their stock assessment is not a product of the model, but rather a consequence of our lack of understanding of a stock's capacity, under current conditions, to rebound.

Section 3 provides case studies on how Bayesian methods were applied to specific case studies of two important species from the West Coast of the United States. Both chapters explore uncertainty in model structure and provide an alternative method of addressing uncertainty from the examples presented in Section 2.

The Pacific halibut *Hippoglossus stenolepis* management forum has been one of the most sophisticated in its approach to and acceptance of methods to incorporate uncertainty. In Chapter 8, Parma presents a Bayesian approach to the analysis of uncertainty. Pacific halibut are flatfish that have been fished in a longline, commercial fishery for more than 100 years. Parma specifically investigates uncertainty in the specification of the model structure, a major source of assessment errors in closely monitored fisheries. Parma points out that more and more sophisticated assessment techniques are being developed, but most assume that a particular model structure is correct. A historical review of Pacific halibut assessments provides an excellent example of this. Parma's methods combine uncertainty in model choice with standard parameter uncertainty into risk computations. Posterior distributions of parameters are approximated using Markov Chain, Monte Carlo methods. This approach allows alternative models to be incorporated into the analysis, eliminating the need to select only one.

We continue with another Bayesian approach in Chapter 9, where Deriso presents a method used to assess Snake River spring and summer chinook *Oncorhynchus tshawytscha*. Snake River spring and summer chinook are an anadromous species whose major source of human-caused mortality comes from passage through the Columbia and Snake rivers' hydropower systems, rather than from fishing. This Bayesian approach tests the effects of changes in input parameters and model structure. In particular, this approach is useful when there is a great deal of uncertainty concerning future events. This paper is also unique in that it deals with an endangered species. Stock-assessment biologists are finding it increasingly common to have to combine a stock-assessment analysis with a population viability analysis for threatened and endangered species. Deriso provides an example in this chapter.

Historically, policymakers have had the ability to arbitrarily use the results of models incorporating uncertainty. Now that we are getting a whole range of new methods to incorporate uncertainty, how will the results be used? Will the results be used?

Does the use of these methods require the voluntary interpretation of our results by managers and politicians on a case-by-case basis? Does the job of the assessment biologist necessarily stop when the results of the assessment are presented? Must there be a disconnect between the assessment and the management decision, or should the results of the assessment lead directly into a management process where uncertainty directly affects the decision being made?

There must be a way to combine the results of these analyses incorporating uncertainty directly into management policy, leaving out the potential for subjectivity. The precautionary principle states that the more uncertainty in the system, the more conservative the policy. That appears to be an open invitation to connect the analysis results directly to policy, which is an important direction for research, and we preview it in Section 4. Section 4 provides several case studies that explore the linkage of uncertainty analyses results with the policy, decision-making process.

In Chapter 10, Palka describes two management processes, currently in use, directly connecting analyses incorporating uncertainty to policy. Both of these processes manage marine mammals, one by the International Whaling Commission (IWC) and one by the U.S. National Marine Fisheries Service (NMFS) operating under the Marine Mammal Protection Act. The methods used by both the IWC and NMFS investigate the robustness of proposed control laws to different types of uncertainty. Simulations are used where model assumptions are varied, incorporating the types of uncertainty encountered during the assessments. As a result, management-decision rules have been developed that take into account this uncertainty. Palka describes how earlier management processes relied on estimating biological parameters, but often the data available did not allow for them to be accurately estimated. As a result, no management actions were taken in some cases. Palka provides an example of how both the development and application of marine mammal assessments have often been ahead of the development and application of similar approaches in fisheries assessments.

In Chapter 11, Goodman gives us a glimpse of the future of incorporating uncertainty analyses directly into the policy process. Unlike the majority of our chapters, this chapter is highly theoretical, providing a template for future applications. Goodman argues that the decision-making process can be an exact science despite uncertainty, through the combination of cost-benefit analyses, decision theory, risk assessment, and Bayesian statistics. This process is developed using a population viability analysis, a standard modeling technique for endangered species, as a case study. The resulting risk-assessment procedure makes what must be uncertain predictions in an objective, rigorous, and reproducible way. Using a PVA to develop this approach, Goodman provides a template not only for fisheries biologists, but also for wildlife biologists, conservation biologists, and the resource management field as a whole.

Workshop Recommendations

The ASMFC has focused a great deal of attention and effort on improving the stock assessment capabilities of its member states from Maine through Florida. The two-day workshop sponsored by ASMFC and Virginia Tech provided an opportunity for an exchange of information and methods between stock-assessment biologists from both the east and west coasts of the United States. Given that there is no standard method for incorporating uncertainty into fishery models used by all U.S. stock-assessment biologists, an important conclusion to this workshop was the development of some basic recommendations for improving this process. Many of the recommendations developed through this workshop are applicable to all fisheries management agencies. A summary of these broad recommendations follows, in the hopes that they can be used to improve the capabilities of individual modelers to address uncertainty from a technical perspective, to improve the use of uncertainty modeling results in the decision-making process, and to improve the agency infrastructure and processes needed to more fully address these issues.

Methodology

- An evaluation of various management options through simulation and analysis is recommended because of its utility to fisheries managers in determining the most effective management strategy.
- A risk-analysis framework should be used in order to test the likely effects of alternative management actions. This requires decision makers to identify the specific management objectives prior to initiating an assessment.
- Biological reference points and/or control rules should be used to judge the status of the stocks, in conjunction with long-term projections of various management options.
- Decision analysis may provide the basis for the development of better control rules.
- Quantifiable goals and management objectives, in conjunction with well-defined control rules, will provide a framework for simulations and analyses of various management strategies. These control rules and analyses should then dictate the actions of fisheries managers.
- Long-term assessments and projections should be used to phase out control rules over the long term, in order to provide more stable and effective management strategies. Management decisions that extend for a period of time were believed to be much better for long-term consistency and more acceptable to the fishing industry.
- Agencies should increase the availability of stock-assessment tools, to allow more scientists to contribute to the process of conducting stock assessments.

Output

- Stock assessment biologists should present the likely consequences of management actions over a long-term time scale, rather than a one- or two-year time scale.
- Confidence intervals or posterior distributions should be included in all stock assessments, rather than point estimates alone. This would reduce the uneasiness that many scientists and managers have regarding long-term projections.

Communication

- Stock-assessment biologists must understand how to present the information effectively to fishery managers and stakeholders who will often have a limited, technical background. It must be straightforward and understandable.
- Fisheries managers need to be exposed to technical presentations involving risk-assessment on a more regular basis. They need clearer presentations to assist them in understanding the risk involved in making various management decisions.
- There is a need for greater buy-in from constituents and stakeholders. It was suggested that better

communication is needed not only with managers and decision makers, but also among all constituents.

Education

- Agencies should sponsor training workshops to teach state-of-the-art stock-assessment techniques to fisheries biologists. This will increase the cadre of stock assessment biologists available to meet assessment workloads.
- Higher-level training courses in the best stock-assessment techniques should be developed.
- Due to the general lack of understanding as to what is implied by decision analysis, it was suggested that workshops on decision theory be conducted to evaluate how to apply this method towards improving risk-assessment.
- A disconnect between science and management may exist, in that many fisheries managers may not have the knowledge to clearly understand the technical information being presented. Training for fisheries managers is recommended as a method to improve the knowledge of fisheries managers in interpreting assessment results and uncertainties.

Data

- Lack of good data is a major constraint to conducting quality stock assessments on Atlantic Coast species. Agencies should strive to implement the recommendations of the National Research Council to improve data collection.
- Agencies should strive to implement standard data-collection programs across jurisdictions and should increase resources for full implementation of these programs (i.e., the Atlantic Coastal Cooperative Statistics Program on the Atlantic Coast and the Fisheries Information Network on the Gulf and Pacific coasts).

Human Resources

The ability to incorporate uncertainty into fishery models and apply the results is dependent on having sufficient resources to do the job. The lack of needed resources will likely limit an agency's ability to implement the recommendations to incorporate uncertainty into fishery models on a consistent basis. Many agencies are currently limited by the large number of stock assessments required and by the limited number of qualified assessment biologists employed to maintain this level of work.

The number of species on which stock assessments must be conducted continues to increase; yet, the number of stock-assessment biologists does not appear to be growing. Postings for quantitative fishery biologists are common; yet, academic fishery biologists bemoan the lack of well-trained quantitative biologists applying for graduate school. With the upcoming loss of a large cohort of stock-assessment biologists to retirement, fishery managers and decision makers should be very concerned about the lack of qualified individuals needed to replace those leaving, much less increasing the workforce to the needed levels.

Meanwhile, the productivity of our existing stock-assessment biologists continues to decrease. Increased litigation against state and federal agencies has forced the agencies to increase the written documentation required for use in potential lawsuits. In addition, with so few stock-assessment biologists available, the same individuals are routinely tasked to work on multiple species.

No specific recommendations to address concerns with limitations on human resources were developed. However, it is important that all fisheries management agencies interested in improving their assessment capabilities be aware of the potential need to increase their workforce in order to keep up with the increasing demands on their current staff of stock-assessment biologists.

In Closing

It is important that all fisheries management agencies attempt to incorporate uncertainty into all of their stock assessments in order to allow managers to make more informed decisions. As pointed out in the recommendations from the ASMFC/Virginia Tech workshop, this may be difficult to accomplish due to several major impediments. These impediments include the need to quickly disseminate innovative modeling approaches for application in the fisheries management decision-making process, the need for accurate and timely fisheries data to support these modeling approaches, and the need for fisheries agencies to have an adequate level of staff to support the increasing demands for accurate stock assessments.

The chapters included in this book will assist stock-assessment biologists in becoming familiar with state-of-the-art modeling approaches for incorporating uncertainty into fishery models. The technical training workshops conducted by the ASMFC and other institutions will also assist in disseminating stock-assessment modeling information from the Innovators to the Assessors. There is also a growing need to do a better job of disseminating information from the technical people responsible for conducting assessments to fisheries managers who must use the information in making informed decisions. Agencies should strive to provide educational seminars for fisheries managers to improve the level of understanding by those using the information.

Improvements in data collection programs have been initiated in the United States through several state-federal cooperative programs. It is important that funding be provided to support these data-collection initiatives, in order to improve both the quantity and quality of data used in our assessments. Improvements in modeling approaches and a broad dissemination of these new innovative approaches will only provide limited returns if the data are not available to fully support these models. The old axiom that "the model is only as good as its data" is still true, in spite of the addition of Bayesian statistics or Monte Carlo bootstrap methods.

Finally, it is becoming more and more obvious that the limited number of stock-assessment biologists available cannot continue to complete an exponentially growing workload. Increased training opportunities will assist in providing higher-level training to those already familiar with population dynamics and stock-assessment principles. However, even with increased training and better dissemination of information, there is an overwhelming need to hire more people to conduct assessments in order to keep up with increasing demands.

As our modeling methodologies evolve, so must our methods to disseminate information from Innovators to Assessors and from Assessors to Managers. Given that workloads are increasing while manpower is not, and methodologies are getting increasingly complex and diverse, the time required to disseminate and put into practice new methods is likely too slow. We must work diligently to make these methods and approaches accessible through a wide range of approaches, from books such as this one to advanced training as offered by ASMFC. Where once we may have thought that improved management was dependent on the development of improved methods, now we realize that improved management depends on improved communication among the Innovators, Assessors, Managers, and Stakeholders.

American Fisheries Society Symposium 27:9–28, 2002

Coping with Uncertainty: Evolution of the Relationship between Science and Management

PAMELA M. MACE AND MICHAEL P. SISSENWINE

*National Marine Fisheries Service, National Oceanic and Atmospheric Administration
166 Water Street, Woods Hole, Massachusetts 02543, USA*

Abstract.—Scientists, managers and fishers have always known that uncertainty is an integral part of nature. Early assessment models dealt with uncertainty as a signal-noise problem, developing deterministic, equilibrium relationships to explain cause and effect. Other models coped with uncertainty by normalizing model outputs against inputs that were known to be highly variable (for example, yield per recruit analysis, which avoids consideration of recruitment variability by performing calculations on a per recruit basis). Although deterministic approaches are still widely used, it is also true that methods for quantifying and presenting uncertainty have been taught, developed, and applied for several decades. The question has been less one of how to present uncertainty, rather than whether to present it. What purpose would it serve? Experience showed that managers or politicians presented with a range of estimates frequently chose to set quotas from the risk-prone end of the range. In short, uncertainty was used as a reason for avoiding conservation measures. Development of the precautionary approach to fisheries management has initiated a transition in the perception of uncertainty from that of a reason to avoid action to that of a reason to exercise caution. Increased public awareness of the limits to the productivity of natural marine resources and increasing numbers of economically devastating stock collapses have been instrumental in initiating the transition, more so than scientific advances in the modeling and representation of uncertainty. This transition is, however, more active on paper than in actual implementation. Today, one challenge is to integrate science and management into a rigorous, risk-averse, decision analysis framework, and another is to further promote responsible management and the application of the precautionary approach in the face of uncertainty. Presentation of uncertainty as risk profiles or Bayesian posterior distributions rather than confidence intervals around point estimates has provided a more objective basis for evaluating the consequences of alternative management decisions, but there is a need for better comprehension of the full consequences of the large risks managers, fishers and other stakeholders (which includes the public in general) often appear to be willing to take.

Introduction

In the late 19th and early 20th centuries, fisheries scientists were primarily concerned with sorting the "signal" from the "noise" by creating rules or laws in the form of deterministic relationships between variables. In other words, they tried to create order in a world of enormous variability and uncertainty. Although the limitations of deterministic, equilibrium approaches are widely recognized, they still provide useful benchmarks in that they are essentially approximations to the mean response of a variable. In addition, they probably represent an essential step towards an understanding of complex biological and fishery interactions, without which there may never have been much progress in the relatively new discipline of fisheries science. It may have been necessary to assume constancy in certain processes in order to be able to investigate the nature and consequences of other processes. One tongue-in-cheek example of trying to create order (in this case, constancy) out of uncertainty is given by John Pope's portrayal of the evolution of the specification of natural mortality, M, which starts with $M = ?$ and gradually transforms the question mark to become $M = 0.2$, the value commonly assumed in stock assessments for temperate water finfish (Figure 1; Pope 1975).

In this paper, we first categorize the various types and sources of uncertainty and provide a brief overview of early responses to each type. Next, we discuss how uncertainty has in the past been misused to forestall effective management decisions. We then describe the paradigms under which fisheries have operated until recently and the transition in paradigms that is currently in progress. Such paradigm shifts include changes in the perception of maximum sustainable yield (MSY), development of precautionary approaches to fisheries management, and development of target and limit reference points and harvest control rules. Finally, we provide examples of modern methods for presenting uncertainty, and discuss current developments that will make the treatment of uncertainty more comprehensive, although they may also complicate its interpretation.

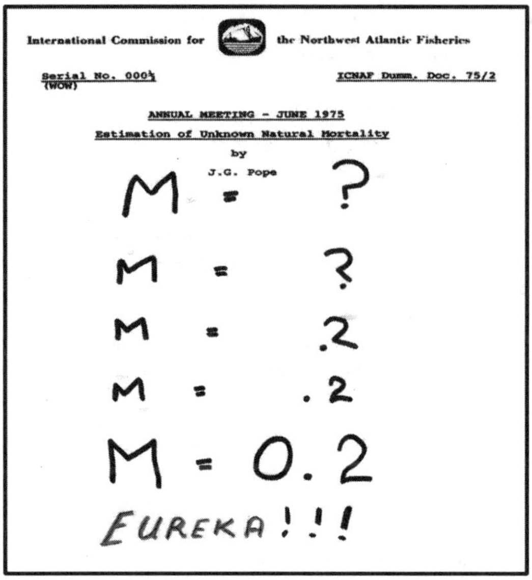

FIGURE 1. The evolution of the specification of natural mortality by Pope (1975).

Types of Uncertainty

Many papers and books have discussed the uncertain environment in which fishery scientists and managers operate, and have attempted to categorize the various sources of uncertainty (e.g., Sissenwine 1984; Hilborn 1987; Hilborn and Walters 1992; Fogarty et al. 1996; Quinn and Deriso 1999), including some chapters in the current book (Helser, Parma, Shepherd, and Vaughan). In this paper, we consider five categories of uncertainty of relevance to fisheries science and management:

(1) Natural variation, which to a large degree is irreducible,
(2) Observation errors in input data,
(3) Model misspecification,
(4) Translation of scientific advice into management actions designed to achieve stated objectives, and
(5) Imperfect implementation of management strategies.

There is no perfect or uniquely correct classification of the sources of uncertainty. However, the particular classification given here follows a logical sequence in terms of chronology: uncertainties are generated by nature and by fishing activities as natural variability, these are then augmented in the form of observation errors made by field biologists, further amplified in models constructed by mathematical biologists, possibly misconstrued or compromised by being combined with political considerations by decision makers, and exacerbated by imperfect implementation of management decisions. Early responses to each of these sources of uncertainty are briefly summarized in the next section.

Early Responses to Uncertainty

(1) Natural variation, which to a large degree is irreducible

Natural variation includes variation in the physical, chemical and biological environment, as well as variation associated with human activities. Examples of the former include variations in water temperature, salinity and currents, and variation in growth, survival and reproductive rates. Examples of the latter include changes in the distribution of fishing effort and changes in catchability.

One of the most fundamental sources of natural variability in marine resources is in the relationship between spawning stock size and recruitment. The magnitude of the variation is often large enough to obscure any underlying relationship, even though the nature of the relationship is obvious at the extremes (no spawners, no recruits). Because variability and other factors obscure the relationship, there was historically a tendency to treat recruitment as independent of stock size, and instead to focus on yield per recruit analyses and predictions, with F_{max}, and to a lesser extent $F_{0.1}$, being the most commonly used biological reference points. However, using F_{max} as the basis for management is essentially equivalent to assuming infinitely strong compensatory density dependence such that recruitment is constant, even at low stock size. It has subsequently been pointed out that this is the wrong null model, with the more appropriate one being a straight line through the origin implying no density–dependence (Sissenwine and Shepherd 1987; Sissenwine et al. 1988; Fogarty et al. 1991). In fact, this is an example where fisheries scientists made a risk-prone choice in their response to uncertainty (Holt 1990).

Due to the ease of application of yield per recruit analysis, its utility for indexing growth overfishing, and the complexity of stock–recruitment data, stock–recruitment relationships were largely ignored for several decades. Nowadays, recruitment overfishing is widely recognized as a reality based on studies that have developed the underlying theory (e.g., Sissenwine and Shepherd 1987), and from empirical demonstrations of the phenomenon such as the meta-analyses conducted by Myers et al. (1994). Consequently, stock–recruitment relationships are now a primary focus of uncertainty modeling. Advances in population dynamics theory and improved computer capabilities have also made it possible to model other sources of natural variation, including density–dependent growth, survival and maturity, multispecies interactions and environmental effects on demographic parameters.

(2) Observation errors in input data

The term, "observation error" incorporates both measurement error (e.g., errors in counts, length measurements or temperature readings) and sampling error (errors associated with sampling from unknown or unobserved distributions). In most fisheries applications, sampling errors are likely to be of greater magnitude than measurement errors. Of all the uncertainties in fisheries science and management, techniques for dealing with observation errors have the longest history of development both in the fisheries literature and in the statistical literature. Overall, fisheries scientists (who generally have at least basic training in mathematical statistics) have fairly routinely dealt with uncertainty in the form of observation errors in their assessments, at least for selected variables. Standard statistical techniques such as data fitting procedures and residual analysis are employed in the majority of assessments. There are many review papers and books that address observation errors in input data; for example, Doubleday and Rivard (1983) and FAO (1999) for sampling catches; and Alverson (1971), Mackett (1973), Saville (1977), Doubleday and Rivard (1981), Sissenwine et al. (1983) and Barnard et al. (1985) for resource abundance surveys.

The foregoing is not meant to imply that observation errors are always easy to deal with. To the contrary, the (common) case where both the independent and dependent variables are measured with error (the errors-in-variables problem) has historically not been paid due regard, even though it can be sufficiently serious that it masks the true relationship (Walters and Ludwig 1981). When the independent variables are measured with error, the estimates will be biased. Various methods for correcting the bias have been developed (e.g., Kendall and Stuart 1979; Ludwig and Walters 1981), but in general, these have not had widespread utility (Hilborn and Walters 1992). In some cases, sophisticated statistical methods such as Kalman filters can be used to separate the signal from the noise (Collie and Sissenwine 1983; Peterman et al. 2000). Kalman filters are particularly useful tools for parameter estimation in systems with time-varying (nonstationary) productivity parameters.

Despite the availability of techniques for representing and analyzing the consequences of observation error, until recently, fisheries scientists tended to present only "best estimates" of management reference points such as "the projected quota for next year based on an $F_{0.1}$ harvest strategy."

(3) Model misspecification

In essence, all models are misspecified in the sense that they are abstractions of reality. There are two basic ways in which this misspecification can occur. First, the assumed structural relationships between variables may be incorrect or oversimplified. An example is the common assumption that catch per unit effort is proportional to stock abundance. Second, potentially influential relationships and dynamics may have been totally excluded from the model. An example is single-species models that assume natural mortality (M) is constant over time when in fact it may vary substantially depending on the relative sizes of predator populations and alternative prey available to them. The distinction between the two forms of misspecification may or may not be important, but both represent trade-offs between simple (tractable) representations of reality and complex (potentially more realistic) representations. Although simple models may be biased, they tend to be more statistically stable than more complex models. And even in complex models, there is a possibility that some important aspect of the dynamics of the system will be excluded or misrepresented.

In terms of early representations of the uncertainty associated with model misspecification, the response was much the same as that to observation error. Although assessments might be conducted using alternative models and parameterizations of models in a sensitivity analysis framework, there was a general belief that fisheries managers should be presented with only a single "base case." In fact, even though uncertainties about model specifications were rarely presented in an operational format, they may often represent one of the most important sources of uncertainty overall in stock assessment models (Punt and Hilborn 1997). For example, Sissenwine et al. (1998) describe the implications of model misspecification (specifically for the spawner–recruit function) on rebuilding strategies for western

Atlantic bluefin tuna *Thunnus thynnus*. Growing evidence that ecological and/or climate regime shifts may fundamentally alter the production function of the species being assessed (Kawasaki 1992; Francis and Hare 1994) adds to the magnitude of the uncertainty associated with model misspecification.

(4) Translation of scientific advice into management actions designed to achieve stated objectives

Why did fisheries scientists model uncertainty but not present it? The primary reason is because of the difficulty of translating scientific advice into management action when the objectives of management are not stated explicitly, or unambiguously, or operationally. The first author of this paper recalls an instance in the mid-1980s when fisheries scientists presented 95% confidence intervals around the best estimate of next year's quota based on an $F_{0.1}$ harvest strategy. The managers asked the scientists to give a good reason as to why the upper limit of the quota range should not be used. Apparently, the scientists were unable to do so, because the quota was set near the upper end of the range. The form of the advice quickly reverted to "best estimates" without confidence limits.

In fact, until recently it might have been reasonable to generalize that in most situations, fisheries managers didn't ask for estimates of uncertainty, didn't want them, and probably wouldn't have known what to do with them. When stock assessments were presented, but managers chose to listen to fishing industry representatives who did not believe the stock assessments, risk prone decisions were often made. Such decisions generally led to overfishing. New England groundfish is one of the best known examples of protracted overfishing, for which the National Research Council (1997) concluded that "valid scientific warnings of imminent stock collapse were not acted on" due to political pressure. In this case, both of the authors witnessed how the misuse of information about uncertainty (as well as unsupported assertions) led to the collapse.

(5) Imperfect implementation of management strategies

Finally, one source of uncertainty that has largely been ignored until very recently is the errors or uncertainties in implementing agreed management actions. Following the collapse of the northern cod stock off Newfoundland, there were assertions that one of the causes of the collapse was illegal, unreported catch outside of the Canadian fishing zone by non-Canadian vessels. Conjecture about unreported catch resulted in considerable tension between Canada and other countries that were fishing off Newfoundland. This situation provided impetus for the negotiation of a new international treaty for the management of straddling and highly migratory fish stocks (United Nations 1995). Continuing concern about imperfect implementation of management strategies has led to FAO agreeing to prepare an International Plan of Action to address illegal, unreported, and unregulated (IUU) fishing.

Imperfect implementation of management strategies has received considerable international attention lately, but it is very much a national problem too. Concerns about misreporting of catches are frequently noted in stock assessment reports and can be extremely high. For example, it was estimated that the Bay of Fundy herring purse seine fishery exceeded its quota by 77% in 1984, a year of augmented enforcement (Mace 1985). Discarding is also problematic. Sometimes management measures are implemented without considering the effect they will have on the amount of mortality that will be caused by discarding. For example, the fishery management plan for cod in the Gulf of Maine set the limit on the amount that could be landed per fishing trip at a very low level in order to achieve the objective of the fishery management plan. However, the objective would only have been achieved if the trip limits did not cause an increase in the amount of mortality caused by discarding. The fishing industry itself, as well as scientific observers placed on a sample of vessels, reported levels of discarding so high that the objectives of the plan were almost certainly undermined.

The main way in which imperfect implementation of management strategies has (sometimes) been incorporated into stock assessments is to add estimates of misreporting rates and discards to the official landings for assessment purposes. However, the extent of misreporting and discarding is often unknown and may not be incorporated into the assessment process due to the lack of a scientific basis for estimation.

Implementation problems are even more acute in multispecies fisheries where switching between stocks can occur rapidly depending on relative stock abundance, management regulations, and market conditions. In such cases, it is important that management strategies take account of the likely behavior of fishing fleets and the likely response to new regulations (Rosenberg and Brault 1993).

Alternative Classifications of Uncertainty

Most classification schemes for uncertainty in the fisheries and statistics literature distinguish between observation error and process error in the estimation process (e.g., Walters 1986; Hilborn and Walters 1992; Meyer and Millar 1999; Polacheck et al. 1993; Quinn and Deriso 1999). The term observation error appears to have a relatively consistent and unambiguous interpretation in the literature, more or less equivalent to that described in item (2) above. The only discrepancy is that the term, measurement error is sometimes thought of as synonymous with observation error, and sometimes as a sub-category within the more inclusive observation error category that includes both measurement error and sampling error. Process error, on the other hand, is a term used to represent the variation in the dynamics of a system that is a property of the system itself as distinguished from errors incurred by sampling or measurement of the system dynamics. In this respect, process error is most closely related to item (1) in the above list. More broadly, process error has also been used to refer to any situation in which a process is not represented exactly in a model. Using this broad interpretation, process error could include both categories 1 (natural variability) and 3 (model misspecification). In this regard, the utility of the term "process error" may be limited unless it is clear which process is being referenced. In the current paper, we prefer to explicitly distinguish natural variation and model misspecification and therefore will not use the term further.

Another way of categorizing sources of uncertainty is in terms of the frequencies at which they occur. For example, Hilborn (1987) suggested the following three categories: frequently observed uncertainties, referred to as "noise"; rarely observed uncertainties, referred to as "uncertain states of nature"; and uncertainties beyond the range of prior experience, referred to as "surprise." Although it would result in a very complex system, a temporal classification of this nature could potentially be nested in each of the five categories used above.

Uncertainty and the Burden of Proof

To date, the existence of uncertainty has often been used to forestall implementation of restrictive management measures. Debates about whether it was overfishing, or environmental effects, or habitat destruction that caused stock declines exist throughout the world. Pollution and consumption by marine mammals are also commonly claimed to have larger effects than fishing. Even when fishing is acknowledged as at least a partial culprit, there may be arguments about the relative destructiveness of large vessels versus small, or one gear segment versus another, or recreational versus commercial vessels. To stock assessment scientists and fisheries managers, however, the most important question that carries the burden of proof goes something like: "are you sure enough of your results to destroy the economic base of several communities and the livelihoods of hundreds of individuals?"

The burden of proof has usually been directed towards scientists and regulators rather than users of fisheries resources. Thus, the common response to uncertainty has been management inaction unless it can be proven beyond reasonable doubt that fish stocks have become depleted. John Gulland, one of the pioneers of fisheries science, described the history of fishery management as "interminable debate about the condition of fish stocks until all doubt is removed." Such was the case for New England groundfish fisheries where effective management was largely delayed until the condition of the stocks was "beyond denial" (Collins 1994).

There are several reasons why society has allowed the burden of proof to be placed on scientists and regulators. The effects of overfishing in marine systems are usually reversible, even at the extreme, and overfishing is generally not hazardous to human health and safety. In addition, until recently, there has been a widely held belief that economic constraints would restrain fishing effort prior to recruitment overfishing. Thus, biological resilience and economic constraints have been widely used as reasons to delay restrictive management action (e.g., in Atlantic Canada; Angel et al. 1994). Subsequently, it has been discovered that some marine resources (e.g., northern cod) may not be as resilient as once believed and that "recruitment overfishing is possible prior to economic overfishing." (Angel et al. 1994).

The argument of biological resilience may be valid for most species in most situations, but dramatic stock collapses can result when overfishing is combined with other factors. For example, it may be possible to fish stocks at unsustainably high fishing mortality rates for long periods of favorable environmental conditions, but when the environment reverts to normal or unfavorable conditions, the combined effect of high fishing mortality and reduced survival, growth, or fecundity may result in sudden stock declines. If additional factors such as overly optimistic stock assessments, misreporting, and discarding are operative, the joint effect may be even greater. A combination of such factors is likely to have resulted in the demise of the northern cod fishery, one of the most dramatic and devastating stock collapses in recent years (Hutchings and Myers 1994; Hutchings 1996).

With respect to the belief that economic constraints will likely restrain fishing effort before overfishing becomes severe, this may have been generally true at one time when the technology to catch fish was limited and the value of fish was lower. Today, however, seafood is in high demand and the technology to produce it has advanced greatly. The diminished ability of economic constraints to restrain fishing effort is compounded by the reality that many of the stakeholders in fisheries tend to have very high discount rates (i.e., they value the more certain returns from the near future much more highly than the uncertain returns from the distant future). Aside from the uncertainties in stock assessments about current stock status, fishers have to contend with uncertainties about future abundance and distribution of the resource, future stock assessment findings, future allocation decisions and competition from other sectors of the fishery, potential changes in government law or policy, concern about continued physical ability to participate in fishing, and potential downturns in market demand or prices (Mace 1993, 1997). Even politicians, and to some extent managers, may have high discount rates due to the limited time they are likely to be in office.

One of the most disturbing consequences of allowing short-term gains to take precedence over long-term sustainability is that once begun, it becomes progressively more difficult to make amends. The greater the degree of overfishing, the greater the short-term sacrifices that need to be made to ensure long-term sustainability, and as a result, fishers may become progressively more entrenched in maintaining a progressively worsening status quo.

Recent Responses to Uncertainty: Shifting Paradigms

The burden of proof and the precautionary approach

One reason that the burden of proof has been placed on the science rather than fisheries activities per se is the continued belief in the inexhaustibility paradigm, as articulated by Thomas Huxley in 1884 (cited in Smith 1994) and repeated with many variations since:

> . . . the cod fishery, the herring fishery, the pilchard fishery, the mackerel fishery, and probably all the great sea-fisheries, are inexhaustible; that is to say nothing we do seriously affects the number of fish. And any attempt to regulate these fisheries seems consequently . . . to be useless.

There has been a gradual move away from viewing the bounty of the oceans as essentially infinite to managing "on the edge," where fishers, scientists and managers have been gaining a more realistic appreciation of the potentially disastrous consequences of uncertainty. As a result, there has been a rapid evolution of paradigms primarily in the last decade – from "it is not possible to overexploit natural marine resources" to "it is not acceptable to overexploit natural marine resources" to "it is both possible and acceptable to under-exploit natural marine ecosystems [to ensure that they] are preserved in perpetuity while still contributing substantially to the food, recreational and livelihood requirements of the world's human population" (Mace 1999). Here the term "under-exploit" is not meant to imply an ultraconservative harvest strategy; rather it is meant to imply that "optimal" is not necessarily equivalent to "maximum." In other words, in order to optimally exploit marine resources, it may not be a good strategy to aim for the maximum exploitation rate due to the pitfalls of attempting to "manage at the edge." Attempts to squeeze out the last possible ton often result in overruns due to a combination of assessment errors, lags in reporting landings, misreporting, discarding, and other factors. Such attempts may also ignore species interdependencies and other ecological relationships.

The problem of the burden of proof being placed on fishery scientists and managers has long been recognized. Thompson (1919) wrote, ". . proof that seeks to change the way of commerce and sport must be overwhelming." More recently, several authors have argued to change the burden of proof (Sissenwine 1987; Mangel et al. 1996; and Dayton 1998). In fact, in 1991 the intent to make the change was clearly stated in the strategic plan of the National Marine Fisheries Service (NMFS 1991), which explicitly called for risk-averse decisions to help maintain productive fisheries. Internationally, the burden of proof was being discussed in conjunction with concerns about the applicability of the "precautionary principle" to fisheries. The precautionary principle was originally conceptualized as a means of managing highly polluting activities by controlling pollution at source even in the absence of "proof" of a causal link between emissions and environmental effects. There was concern that application of the precautionary principle, which was designed to guard against the possibility of making irreversible mistakes, might be too extreme in fisheries where most mistakes have a high probability of being reversed. This debate led to the evolution of the "precautionary approach," a "gentler" version of the precautionary principle, which was adopted as a key element of the United Nations Implementing Agreement for Straddling Fish Stocks and Highly Migratory Fish Stocks (United Nations 1995) and the FAO Code of Conduct for Responsible Fisheries (FAO 1995). Subsequently, FAO (1996) published technical guidelines on the application of the precautionary approach to capture fisheries. Serchuk et al. (1997), Restrepo et al. (1998), and Mace and Gabriel (1999) have summarized the development and key features of the precautionary approach to fisheries management.

Although the precautionary approach is an adaptation of the "precautionary principle," the difference is more than subtle. An example might serve to make the distinction. It is well known that certain fishing gears affect marine habitats by disturbing them, but the long-term effects on productivity and species composition are generally not well understood. Application of the precautionary principle might result in a complete prohibition of such gears or technologies (as was the case with large-scale high seas drift nets), whereas application of the precautionary approach could lead to the conclusion that fishing should continue but with additional cautionary measures such as area closures and limitations on fleet size. In fact, the precautionary approach is multifaceted and broad in scope. According to the Code of Conduct (FAO 1995), precaution is required in development planning, management, research, technology development and transfer, legal and institutional frameworks, fish capture and processing, fisheries enhancement, and aquaculture. Even when applied to fisheries management specifically, there are many facets and it is difficult to define succinctly. The following is one such attempt (Restrepo et al. 1999):

> the precautionary approach is about applying judicious and responsible fisheries management practices, based on sound scientific research and analysis, proactively (to avoid or reverse overexploitation) rather than reactively (once all doubt has been removed and the resource is severely overexploited) to ensure the sustainability of fishery resources and associated ecosystems for the benefit of future as well as current generations.

This definition emphasizes the need to be proactive with respect to uncertainty rather than awaiting the certainty that comes when the stock can no longer support a viable fishery. The precautionary approach is particularly significant to the issues being dealt with herein because it explicitly addresses uncertainty and states that greater uncertainty should result in greater caution in fishing activities.

FAO's (1996) technical guidelines on the precautionary approach state that fisheries should be developed cautiously pending scientific evidence that fishing activities are sustainable. In addition to the maxim that the greater the uncertainty, the greater the degree of caution that must be exercised, FAO suggests that all fishing activities must have prior management authorization and be subject to periodic review, and that those who exploit public resources have the responsibility of accountability for their actions, and to report their activities (where, when, how, and how much caught) to the public. The guidelines do not explicitly call for a reversal in the burden of proof (thus distinguishing the precautionary approach from the precautionary principle); rather they conclude that if the precautionary approach is properly applied, then the burden of proof will be appropriately placed. In essence, the distinction lies with the standard of proof to be applied, analogous to the difference between criminal and civil court cases in terms of the level of certainty required. The precautionary principle implies that activities should not proceed unless it can be shown beyond a reasonable doubt that the activity is safe; the precautionary approach implies that the preponderance of evidence should indicate that the activity is sustainable. While rebuilding of depleted stocks and reduction of overexploitation is the main priority of precautionary management, if there are options for achieving these objectives that enable a viable fishery to continue, these should be favored provided, of course, that the current fishing capacity can be supported by the resource in the long term.

The MSY transition

These paradigm shifts, along with a better understanding of how uncertainty should be treated, are reflected in several transitions in the perception of the concept of Maximum Sustainable Yield (MSY) (Mace 2001). MSY has gone in and out of vogue since its inception about 70 years ago and has been periodically subject to harsh criticism (e.g., Larkin 1997; Sissenwine 1978). However, MSY is the foundation of numerous international and national fisheries laws, it is intuitively appealing and logical to nontechnical audiences, and no better operational management goal has been found to be as widely applicable (Punt and Smith 2001). Lately, it has experienced a resurgence in application within the United States and elsewhere, due in part to the most recent step in the evolution of its interpretation, as described below.

A literal or *static* interpretation of MSY is that MSY is the maximum *constant* yield (MCY) that can be taken year after year (Mace and Sissenwine 1989). However, the maximum constant yield that can be taken from a naturally fluctuating resource is low because any constant catch will represent a relatively large fraction of the stock at the low points of the natural fluctuations. Thus, a constant catch will often drive a low population even lower, and therefore carries a high risk of stock collapse (Sissenwine 1978). It is much safer for the resource and also produces higher average yields if catches are more or less proportional to stock size. For this reason, the more modern *dynamic* interpretation of MSY is that it is the maximum *average* yield (MAY) that can be achieved by varying annual yields in response to fluctuations in stock size. One common, straightforward way to approximate MAY is to apply a constant fishing mortality rate (F_{MSY}). However, this will not necessarily achieve the "global" (highest possible) MAY. Even higher average yields can sometimes be obtained by applying a variable fishing mortality strategy (e.g., constant or proportional escapement), although these strategies tend to also have the disadvantage of high interannual variability in catch. Fishing mortality reference points can also be made to vary with some relevant environmental factors to account for phenomena such as environmental regime shifts that may substantially impact productivity. For example, the NMFS Pacific coastal pelagics FMP treats F_{MSY} as a function of sea surface temperature (Jacobson and MacCall 1995; PFMC 1999).

The most recent transition regarding the perception of MSY is to think of it (or associated reference points) in terms of limits rather than targets. This was first recommended in an international context in Annex II of the United Nations Implementing Agreement on Straddling Fish Stocks and Highly Migratory Fish Stocks (United Nations 1995). Paragraph 7 of Annex II states that "the fishing mortality rate which generates maximum sustainable yield [F_{MSY}] shall be regarded as a minimum standard for limit reference points." Subsequently, in 1996, the U.S. Sustainable Fisheries Act changed the definition of optimum yield (OY) from "... MSY as *modified* by [relevant factors]" to "... MSY as *reduced* by [relevant factors]," implying that MSY should be thought of as an upper bound on OY.

There are several reasons why it makes sense to treat F_{MSY} as a limit rather than a target (Mace 2001). Among these is that it is a more risk-averse way of dealing with uncertainty. If the true F_{MSY} is repeatedly overshot (due to assessment errors, implementation errors, or other factors), the cumulative effect is that eventually yields will fall below the associated long-term maximum average yield and the stock will become depleted relative to B_{MSY}, ultimately requiring implementation of economically and socially disruptive restrictive management measures in order to rebuild the stock. On the other hand, the cumulative effect of habitually undershooting the true F_{MSY} (e.g., by treating it as a limit to be avoided), is that the risk of future economic and social disruption is minimized and there may even be future opportunities for temporary increases in catches. In addition, foregone yields may be small or negligible because reductions in fishing mortality are compensated to some extent by increases in biomass (at least for cases where the yield curve is relatively flat near F_{MSY}).

The transition from a static to a dynamic interpretation of MSY stems largely from the scientific community. However, the transition from MSY-based reference points as targets to limits is more a result of increased public awareness and concern. Fisheries environmentalists have noted that while MSY has been embodied in several important international and national agreements since the mid-1970s, overfishing has increased in terms of both the number of fisheries and the magnitude of overfishing. This has resulted in periodic calls to abandon the concept but since no one seems to have been able to come up with a better substitute (e.g., one that is more conceptually appealing, operational and widely applicable), one alternative is to treat it as an upper limit and also to focus on other aspects of precautionary approaches to fisheries. Thus, this aspect of the evolution of the relationship between science and management has largely been spurred by increased public outcry about overexploitation of natural resources. However, it should be emphasized that the cases of overfishing that have raised concern are not the result of shortcomings in MSY as a management strategy; rather, they are a result of risk-prone management strategies that failed to achieve MSY.

Targets, limits, and control rules

Attempts to implement the precautionary approach have resulted in the development of harvest control rules that incorporate fishing targets and limits, with the difference between the two reflecting the degree of uncertainty. A harvest control rule should have the following two elements: (1) a limiting fishing mortality rate that may or may not be related to biomass over some or all of its range, and (2) a target fishing mortality rate that may or may not be related to biomass over some or all of its range, but must always be less than the limit fishing mortality over all parts of the range. The distance between the target and the limit should be sufficiently great that attempts to achieve the target on average will have low probability of exceeding the limit. One of the simplest control rules satisfying these criteria is the example provided by ICCAT (2000), in which the limit control rule specifies fishing mortality as independent of biomass over the entire range of biomass, and the target control rule specifies fishing mortality as being independent of biomass at high biomass, but tending towards zero with declining biomass in order to provide an additional margin of safety for severely depleted stocks (Figure 2).

In general, fishing mortality rates above the limit control rule correspond to the act of overfishing, while biomass levels below some threshold or limit denote an overfished or depleted stock.

Some paradoxes

Even though the precautionary approach dictates that management actions should be more conservative in situations where there is a paucity of data or great uncertainty about stock status, in reality the reverse may occur. For example, in theory, if there was no information about a particular fish stock other than some rough measure of annual landings, the fishery should be curtailed until adequate data can be collected to determine the status of the stock; in reality, it would be extremely difficult for the fisheries management authorities of most countries to implement any quota that was less than recent landings. In theory, the more the data, the greater the accuracy and precision of estimates of biological reference points such as fishing mortality targets, and therefore the lower the probability of exceeding the target on average; in reality, more data allow use of more complex models with more parameters and therefore more potential sources of uncertainty, so that estimated (not actual) confidence intervals around targets and limits may increase.

In most instances, fisheries management is still about preserving the status quo, and the transitions that have taken place are more evident on paper than in practice. In addition, the legacy of the inexhaustibility paradigm remains with us. Although there are increasingly many examples of

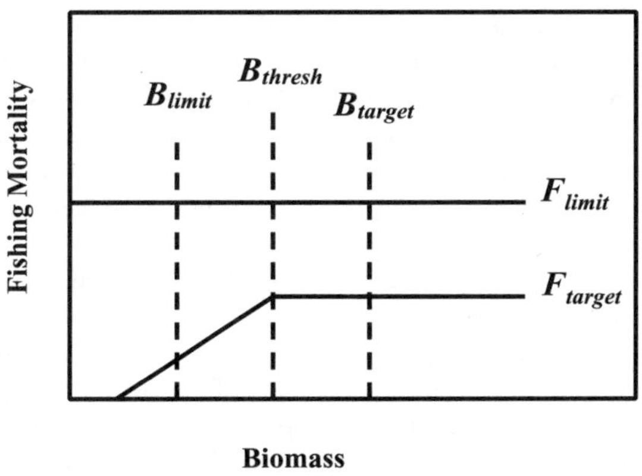

FIGURE 2. Simple example harvest control rule illustrating potential biomass targets, thresholds, and limits and fishing mortality targets and limits. After ICCAT (2000).

economically devastating stock collapses, there are no known examples where entire marine species have been fished to extinction (such is certainly not the case for freshwater species or for individual stocks of marine or anadromous species such as salmonids). For these reasons, risk prone decisions are still made today, especially in those cases where management objectives are still not fully and operationally defined. However, this practice is increasingly being judged to be unacceptable by the public and by the courts in the United States and elsewhere.

Modern Approaches to Portraying and Confronting Uncertainty

Even a cursory review of the fisheries science literature on risk and uncertainty is sufficient to reveal the wide diversity of ways of presenting and dealing with uncertainty (see, for example, the papers in Smith et al. (1993) on "Risk evaluation and biological reference points for fisheries management," and Volume 56, Issue 6 (1999) of the ICES Journal of Marine Science, which is devoted to papers from the symposium, "Confronting uncertainty in the evaluation and implementation of fisheries management systems"). In fact, techniques for presenting and dealing with uncertainty have existed for several decades and a number of papers from the 1970s and 1980s were already claiming that a revolution in the approach to uncertainty was taking place. This revolution has, however, been slow to catch fire. Even up until about 10 years ago, the most common ways that fisheries scientists would present uncertainty in stock assessment results to fisheries managers and stakeholders were to provide a range of quotas or other relevant metric, or a point estimate with confidence intervals. Many of the techniques being developed and advocated in the 1970s and 1980s (e.g., Monte Carlo techniques, Beddington and Cooke 1983; Multi-Attribute Utility Analysis, Healey 1984) have only begun to achieve widespread application within the last 5–10 years.

One of the more significant revolutions in the response to uncertainty and inadequate data has been to develop techniques for including data from as many sources as possible and involving stakeholders in the analysis of alternative hypotheses and management strategies. Rather than stakeholders using uncertainty as a lever to advance a particular course of action or inaction, it is preferable if all players work together to develop and evaluate alternative hypotheses and actions, so that they share a sense of involvement as well as gaining a better understanding of the complexities and trade-offs of the alternatives.

The potential pitfalls of attempting to produce a single best estimate of an assessment output have been pointed out by a number of authors including Hilborn et al. (1993) and Schnute and Hilborn (1993). Of particular concern are situations involving contradictory data (Richards 1991; Schnute and Hilborn 1993). The uncertainty in the divergent data will be masked, rather than being presented as two alternative hypotheses that decision makers and stakeholders should be cognizant of throughout the entire decision process. Hilborn et al. (1993) suggest that for the northern cod stock assessments conducted just prior to the collapse of that stock, it may have been better to present two alternative hypotheses about stock size rather than averaging the contradictory signals from the commercial CPUE (which suggested an increasing stock) and research survey data (which suggested a decreasing stock). Alternatively, contradictory results can be consolidated under a single analysis that allows varying degrees of credibility for each data set (Schnute and Hilborn 1993).

Rather than attempting to produce a "best" estimate of some management attribute such as a quota, a simple way to explicitly deal with uncertainty and risk is to construct a decision table (Walters and Hilborn 1976; Hilborn et al. 1993, 1994). Decision tables can take many forms, but one basic approach is to specify a number of alternative system states based on biological and/or economic and/or social criteria as columns, then a range of possible management actions (e.g., harvest levels) as rows, with entries in the cells of the table being expected consequences (in terms of, for example, catch, risk of stock collapse, or employment levels) of each management action if each of the alternative system states is true (Hilborn et al. 1993). Decision makers and stakeholders can then develop a process for weighting the risks and reaching agreement on the best management action(s) given all of the inherent uncertainties in the analysis.

Multi-Attribute Utility Theory (MAUT) analysis is a related but more formal approach. Healey (1984) outlined the six steps involved in the application of this theory to the resolution of multiobjective problems, using estimations of optimum yields based on biological, economic and social considerations as examples. The six steps include bounding the problem, determining the range of feasible yields, establishing the criteria against which the yields will be judged, weighting the criteria, scoring the yields, and applying the decision rule. Healey (1984) pointed out that the methodology was already well-developed by the mid-1970s and wondered why it was not more widely used in fisheries given its intuitive appeal, its ability to incorporate a wide range of types and qualities of data, and its adaptability to a variety of decision rules.

The decision analysis approach that seems to have gained the largest foothold in recent years is Bayesian analysis (Punt and Hilborn 1997; McAllister and Kirkwood 1998; Deriso, and Parma, both this volume). Bayesian methods have been widely promoted as approaches for representing and taking account of a wide range of uncertainties and incorporating existing knowledge and expert judgment. Bayesian decision analysis has proved particularly useful in data-poor situations where prior distributions based on incomplete data or data from ecologically similar species can be used to assist in differentiating between alternative hypotheses about stock status.

Overall, there has been widespread adoption of more informative approaches to portraying uncertainty. These often utilize some form of risk analysis format for portraying the consequences of decisions about current and future catches or fishing mortalities. Generally, these consequences are presented in terms of the risk that something bad may occur (e.g., $\Pr(B < B_{\text{limit}})$) or the probability of something good (e.g., $\Pr[B > B_{\text{MSY}}$ within 20 years]).

Three recent examples of graphic portrayals of the uncertainty associated with assessments of current and future stock status are provided below. These examples were chosen as "typical" of recent methods of depicting uncertainty, rather than as "ideal" methods that should be advocated.

In the northeastern United States, the NMFS Northeast Fisheries Science Center has developed a standard output for results from Virtual Population Analysis (VPA), which provides graphs of probability distributions around the most recent estimates of fishing mortality and biomass. Cumulative distributions enable determination of the probability that fishing mortality exceeds some specified rate or that spawning stock biomass is below a given level. For Gulf of Maine cod, it appears that there is about a 50% probability that the fishing mortality in 1997 exceeded 0.75 and about a 50% probability that spawning biomass was below 9,000 mt (Figure 3; Mayo et al. 1998). Projections of stock status under various scenarios are portrayed as the median and 80th percentile bootstrapped confidence limits. These methods of presenting uncertainty have been routine for about 5–6 years.

On the other side of the United States, at the NMFS Alaska Fisheries Science Center and elsewhere, there is much less use of VPA and increased recent emphasis on assessment methods based on Bayesian approaches. However, similar methods can be used to present uncertainty in current estimates and probabilities of future stock conditions, as illustrated for eastern Bering Sea walleye pollock *Theragra chalcogramma* (Figures 4 and 5; Ianelli et al. 1999). Figure 4 depicts the probability that yields in 1999 would have been less than some specified value if fishing occurred at one of three reference levels of fishing mortality. Figure 5 shows the probability that spawning biomass will be less than a given level in the year 2003 if fishing proceeds at a rate of $F_{40\%}$. It appears that the probability of spawning biomass being below $B_{40\%}$ is less than 50% and the probability of being below B_{MSY} is even lower.

The last example is from the Scientific Committee on Research and Statistics (SCRS) of the International Commission for the Conservation of Atlantic Tunas (ICCAT 1999), which uses a wide range of assessment techniques but almost always projects stock and fishery status in terms of constant quota scenarios rather than the constant fishing mortality scenarios that are more common elsewhere. At ICCAT, great emphasis is placed on stock projections using alternative quotas, as well as the uncertainty around those projections. The example provided is for the 1998 western Atlantic bluefin tuna assessment using bootstrapped VPA results (Figures 6 and 7; ICCAT 1999). The results indicate that the stock has high probability of rebuilding at a quota of 1500 mt, but depending on the stock–recruitment hypothesis used, it may or may not rebuild at the current quota level of 2500 mt.

There are still several unresolved questions related to presentations of uncertainty. For example, most calculations of uncertainty are known or believed to underestimate true levels of uncertainty because they do not consider the entire spectrum of possible models or model parameterizations. Another concern is the determination of appropriate time horizons for stock projections. Although it may be useful to know the long-term consequences of particular management strategies, projections become increasingly uncertain as the time horizon increases. Some stock assessment scientists feel that the credibility of the science may be adversely affected

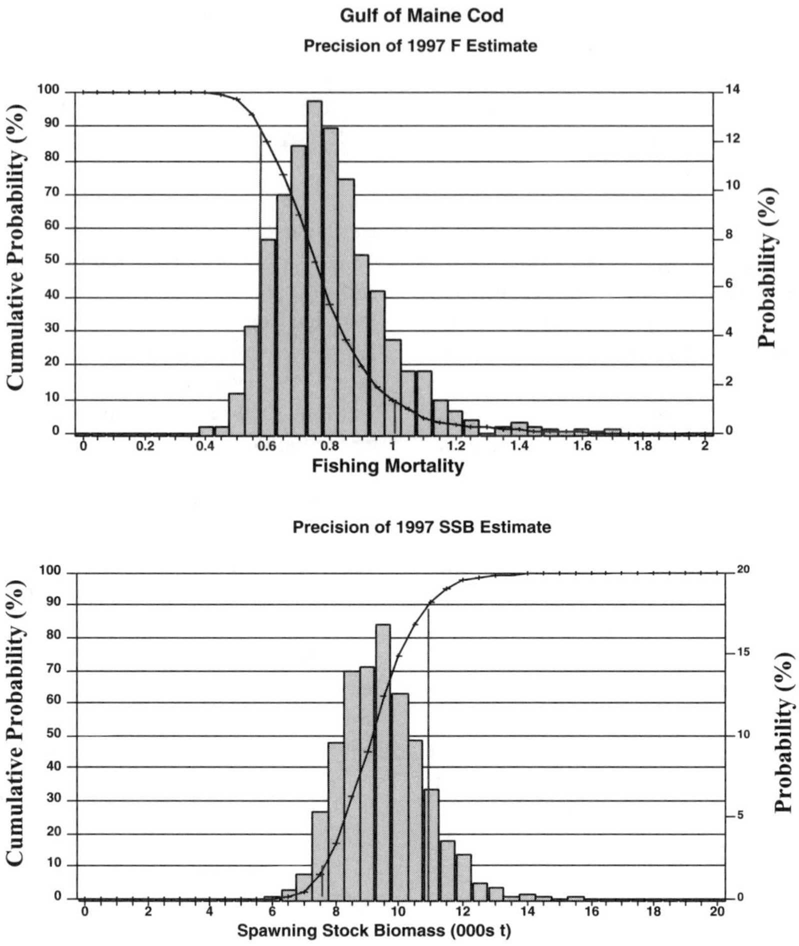

FIGURE 3. Presentations of uncertainty in estimates of current biomass and current fishing mortality. From Mayo et al. (1998).

by extrapolating more than a few years into the future. At the other extreme, projections that are too short may be of limited utility; for example, calculations of the probability of rebuilding to some target biomass within a single year.

As mentioned previously, the three examples presented above simply demonstrate current rather than "ideal" methods for portraying uncertainty and/or the results of risk analyses. None of the examples can be considered comprehensive examinations of the sources and implications of uncertainty and the most appropriate responses to them. One reason that such comprehensive analyses are not conducted on a routine basis is that they would require considerably more time and human resources, particularly if the full suite of biological, economic and social consequences of a wide range of different management actions were to be thoroughly examined in a forum involving representatives of all affected stakeholders. In addition, for key species, stock assessments are usually conducted every 1–2 years, thus allowing for frequent, if not comprehensive, evaluation of previous responses to uncertainty.

Beyond Simple Single Species Stock Assessments

So far this paper has focused on uncertainties arising from "simple" single species stock assessments, and how fisheries scientists and managers have dealt with uncertainty at this level. The situation becomes considerably more complex when one considers a broader spectrum of uncertainties. Currently, there are two distinct avenues of investigation that extend the realm of uncertainties. First is the

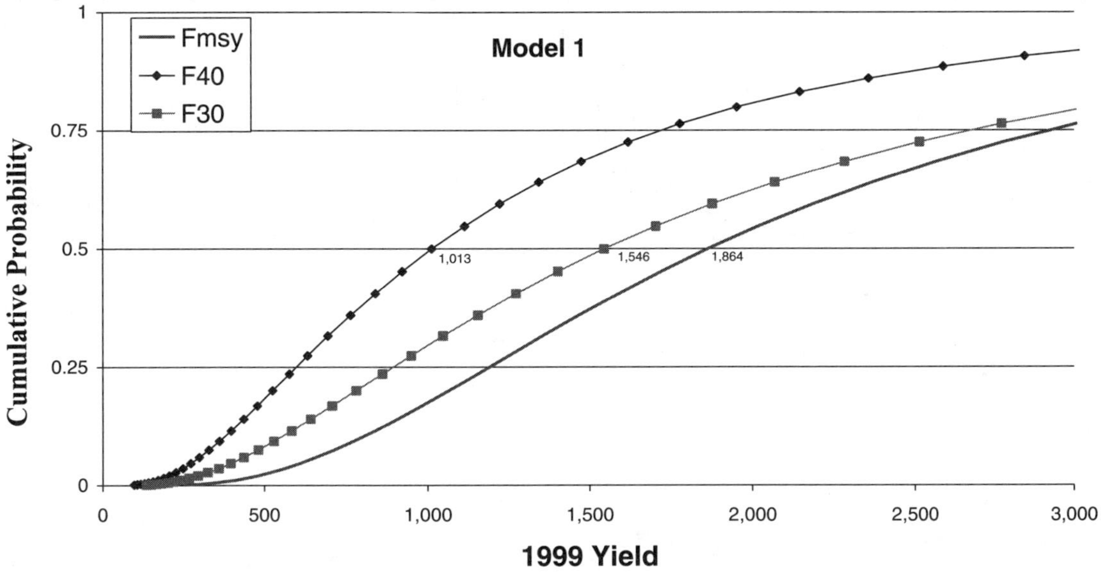

FIGURE 4. Measures of uncertainty in 1999 (unadjusted) yield for Eastern Bering Sea walleye pollock as a cumulative distribution. Values along the curve represent the estimated probability (vertical axis) that the 1999 yield will be lower than the corresponding value on the horizontal axis when fishing at three reference levels of fishing mortality: $F_{30\%}$, $F_{40\%}$ and F_{MSY}. From Ianelli et al. (1999).

continued development of feedback control policies over the past 2–3 decades. The name and acronym given to the procedure varies; for example it was called a fisheries control system by Hilborn (1979), a revised management procedure or RMP by the International Whaling Commission (IWC 1993 1994), a system for evaluating management strategies by ICES (1994), a management procedures simulation model approach by ICCAT (2000), a management strategy evaluation or MSE by Smith et al.

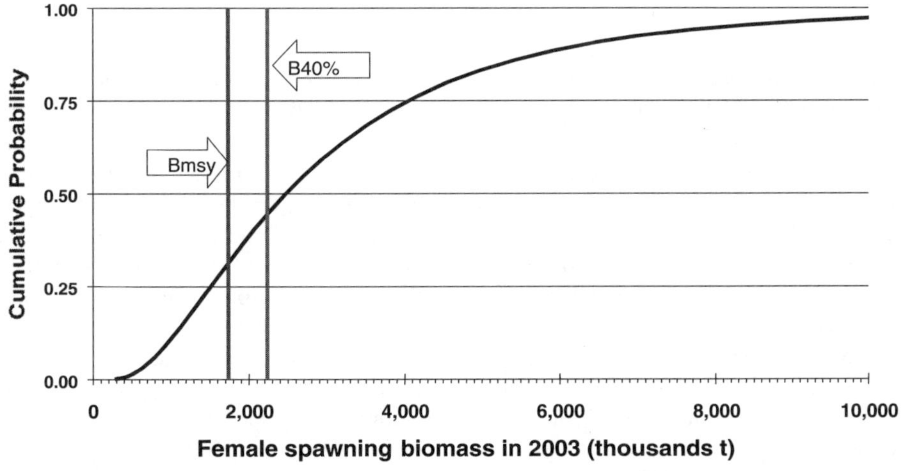

FIGURE 5. Approximate cumulative probability distribution of future female Eastern Bering Sea pollock spawning biomass relative to B_{MSY} and $B_{40\%}$ based on projections assuming future fishing mortality rates are set to adjusted $F_{40\%}$ values. Points along the curve represent the estimated probability (vertical axis) that the female spawning biomass in 2003 will be lower than the corresponding value on the horizontal axis. From Ianelli et al. (1999).

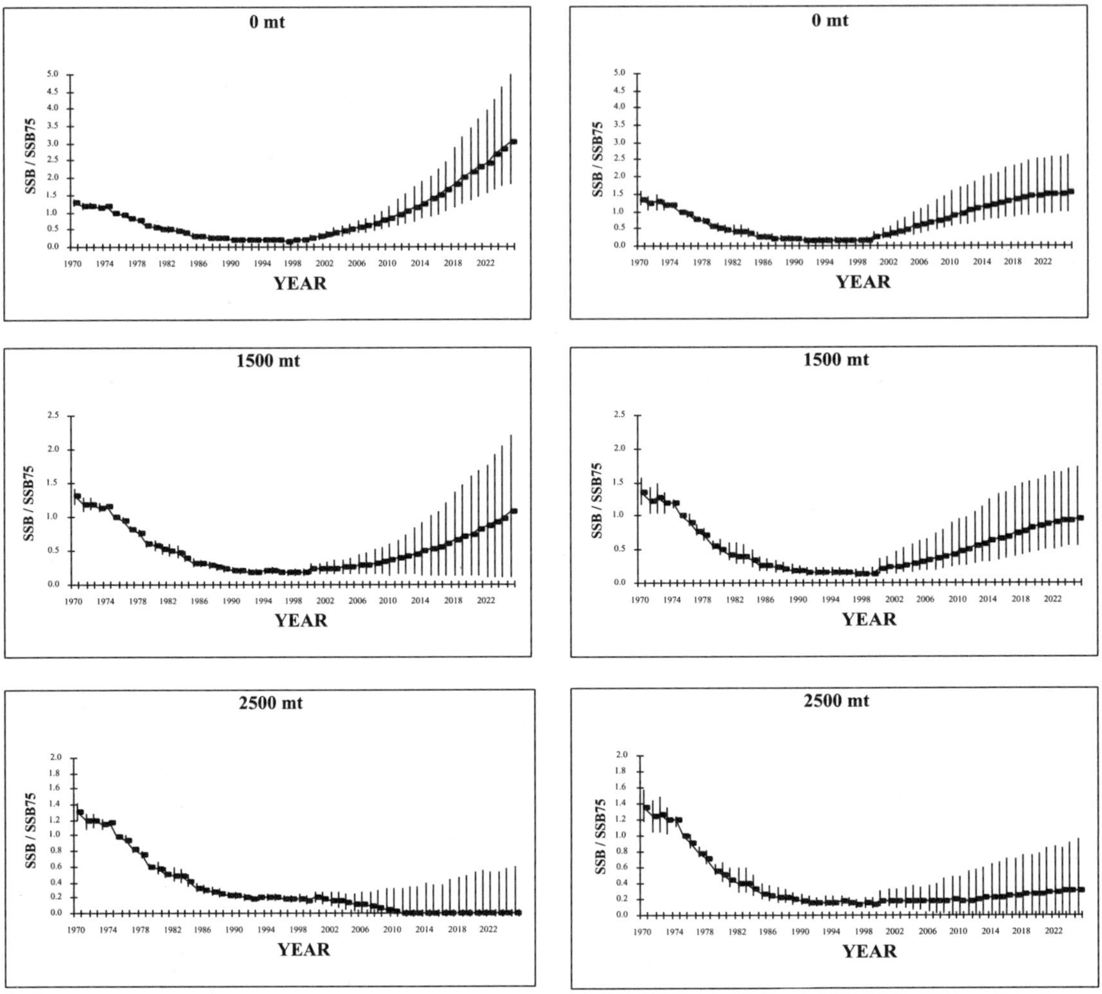

FIGURE 6. Stochastic projections of spawning stock biomass (SSB) relative to the 1975 level for western Atlantic bluefin tuna using two alternative stock recruitment relationships: a Beverton–Holt relationship (left panels) and a two line model (right panels). Error bars represent 80% bootstrapped confidence intervals. From ICCAT (1999).

(1996) and Polacheck et al. (1999), and an operational management procedure or OPM by Barnes (1999) and others in South Africa. However, the structure is generally similar, consisting of an operating model that provides a simulation of a "true" population, a procedure for sampling the true population, an assessment model that uses the sampled data to produce a "perceived" population, a management model that implements specific harvest rules, and performance statistics and feedback associated with each of these components (Figure 8). Although this type of model tends to be oriented to single species, it goes beyond uncertainties in stock assessments alone by also considering uncertainties in other aspects of the assessment and management system such as process errors and implementation errors, and feedback between the real system, the perceived (assessed) system, and the management system. It is a particularly useful framework for investigating the robustness of various types of biological reference points and management actions.

The second distinct avenue for adding complexity is to adopt an ecosystem approach by including multispecies interactions and environmental effects. Common examples of multispecies or ecosystem approaches include multispecies virtual population analysis (MSVPA; Sparre 1991) and the EcoPath, EcoSim, and EcoSpace models developed from the original EcoPath formulation (Polovina 1984) by scientists at the University of British

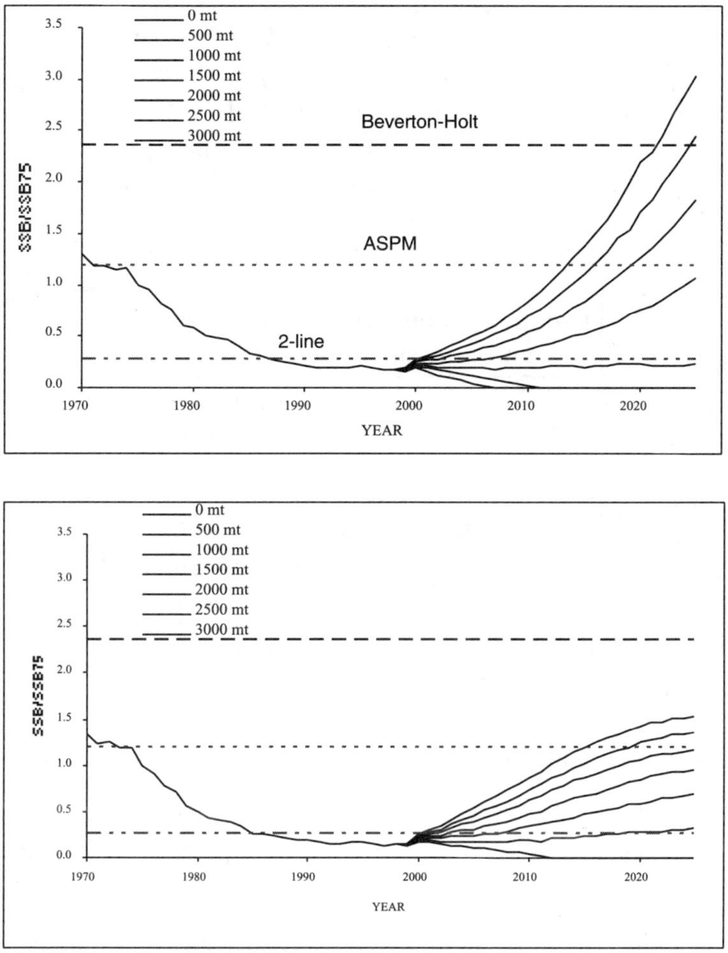

FIGURE 7. Median projections of spawning stock biomass (SSB) relative to the 1975 level for western Atlantic bluefin tuna assuming a Beverton–Holt stock recruitment relationship (upper diagram) and a two line stock recruitment relationship (lower diagram) for constant catches in the range 0–3000 mt. Horizontal lines represent three different estimates of the SSB at MSY relative to the 1975 SSB, one based on a Beverton–Holt relationship fit to stock and recruitment estimates from virtual population analysis, another on a piecewise linear (2 line) relationship fit to a subset of the same estimates, and a third one based on stock and recruitment estimates from an age-structured production model (ASPM). From ICCAT (1999).

Columbia Fisheries Center and elsewhere (Pauly et al. 2000). The complexity of ecosystems and the large number of alternative plausible formulations of ecosystem structure and function makes representation of uncertainty particularly difficult. As a result, the utility of multispecies and ecosystem models for providing operational objectives, performance measures, and management advice has, to date, been very limited. In recognition of this difficulty, a committee of the National Research Council and a congressionally mandated panel formed by the National Marine Fisheries Service both concluded that applying the precautionary approach to single species management was a necessary step toward an ecosystem approach (Ecosystem Principles Advisory Panel 1999; National Research Council 1999). An alternative, or perhaps a complementary approach, is to develop adaptive management strategies as mechanisms for dealing with uncertainty. The focus of adaptive management is to treat fisheries as experiments that are designed to provide information as they proceed (Walters 1986). For example, fisheries can be experimentally manipulated in ways

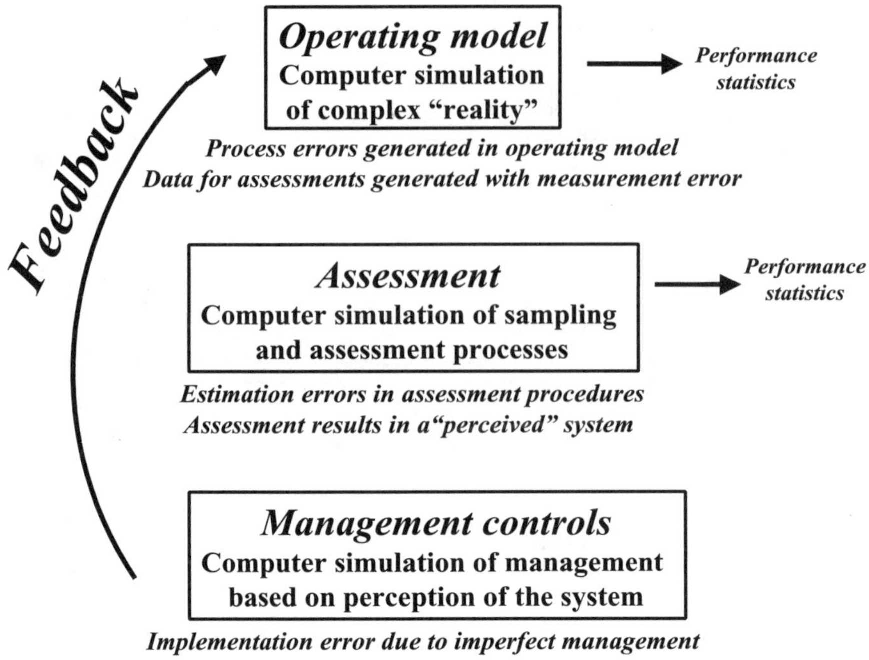

FIGURE 8. Schematic diagram of a management procedures simulation model. From ICCAT (2000).

that will enable evaluation of competing hypotheses about system function (Sainsbury et al. 1997).

Concluding Remarks

Scientists, managers, and fishers have always known that uncertainty is an integral part of nature. In addition, most fisheries scientists have strong quantitative backgrounds and have received training in methods of presenting uncertainty. Therefore, why isn't the presentation of uncertainty more routine in fisheries? One of the primary reasons suggested in this paper is that it has generally been used as a reason for continuing the status quo, rather than as a reason for exercising restraint. Recently, however, increased public awareness of, and concern about, the limits to the productivity of natural marine resources has resulted in a transition in paradigms from one of "it is not possible to overexploit" to one of "it is not acceptable to overexploit." Overall, in most cases, the practice has yet to catch up with the theory in terms of this paradigm shift.

One of the most important keys to the future elimination and/or prevention of overfishing is to shift from risk-prone to risk-averse decision making as called for by the precautionary approach. There are two general ways to accomplish this. One is to devise strict, unambiguous, enforceable laws (as exemplified by the U.S. Marine Mammal Protection Act). The other is to develop and foster a strong conservation ethic to which all or most stakeholders subscribe (as exemplified by the Pacific halibut fishery; McCaughran 1997). The latter option is to be preferred because it is likely to be less controversial, and therefore less costly to the nation, and more likely to succeed. However, there are considerable challenges inherent in developing a strong conservation ethic in fisheries where generations of fishers have operated on the premise that fisheries resources should be capable of providing livelihoods for all who wish to pursue this way of life.

Another difficult challenge is to determine whether it was overfishing or environment that caused a stock to become depleted. The actual cause of depletion may be largely irrelevant in terms of the subsequent management actions required, in that a depleted stock is likely less productive and harvest levels will need to be reduced to either maintain the stock at its depleted level or to rebuild it. However, the one respect in which the overfishing/environment dichotomy may matter regards the choice between an active and a passive rebuilding strategy (Sissenwine et al. 1998). Passive rebuilding strategies are potentially the most pragmatic approach to use in instances where environmental regime shifts are suspected. If a regime shift to lower productivity has occurred, and the target rebuilding biomass is based on the period of higher productivity, it will be impossible to reach. Alternatively, if the stock is in a phase of high productivity, and an estimate of target biomass from a low productivity phase is used, it will be very easy to achieve but may well be far from optimal. When, as is generally the case, it is not known whether the stock has experienced a regime shift or has actually been depleted by fishing and other factors, it may be particularly difficult to estimate an appropriate target biomass for rebuilding. In such cases, the most pragmatic management strategy may be one that ensures steady progress, including milestones along the way to an initial target (e.g., the level of biomass based on a low productivity phase). Once the initial target is achieved, the situation can be reevaluated to determine whether the target biomass should be even higher. A passive rebuilding strategy would require that fishing mortality be kept within reasonable bounds (e.g., those appropriate to reaching the low productivity target), while not requiring drastic reductions in fishing mortality (e.g., those appropriate to reaching the high productivity target), but would also operate by capitalizing on years of good recruitment and "banking" them for future stock rebuilding rather than cropping them off as soon as they become available to the fishery.

Now and into the future, it is essential that scientists, managers, the industry and the public be cognizant of the risks and the benefits associated with fisheries management decisions, and that such risks and benefits therefore be presented in a rigorous, risk-averse decision analysis framework that is readily interpretable.

References

Alverson, D. L. 1971. Manual of methods for fisheries resource survey and appraisal. Part I. Survey and charting of fisheries resources. FAO Fisheries Technical Paper 102. Food and Agriculture Organization of the United Nations. Rome, Italy.

Angel, J. R., D. L. Burke, R. N. O'Boyle, F. G. Peacock, M. Sinclair, and K. C. T. Zwanenburg. 1994. Report of the workshop on Scotia-Fundy groundfish management from 1977 to 1993. Canadian Technical Reports in Fisheries and Aquatic Sciences No. 1979.

Barnard, J., W. Myers, J. Pearce, F. Ramsey, M. Sissenwine, and W. Smith. 1985. Surveys for monitoring changes and trends in renewable resources. Forests and Marine Fisheries. American Statistician 39(4), Part 2:363–374.

Barnes, W. R. 1999. Viewpoint: an industry view of the application of operational management procedures to setting total allowable catches for the South African pelagic fishery. ICES Journal of Marine Science 56:1067–1069.

Beddington, J. R., and J. G. Cooke. 1983. The potential yield of fish stocks. FAO Fisheries Technical Paper 242.

Collie, J. S., and M. P. Sissenwine. 1983. Estimating population size from relative abundance measured with error. Canadian Journal of Fisheries and Aquatic Sciences 40(11):1871–1879.

Collins, C. H. 1994. Beyond denial: the northeastern fisheries crisis: causes, ramifications and choices for the future. National Fish and Wildlife Foundation, Washington, D.C.

Dayton, P. K. 1998. Reversal of the burden of proof in fisheries management. Science 279:821–822.

Doubleday, W. G., and D. Rivard. 1983. Sampling commercial catches of marine fish and invertebrates. Canadian Special Publications of Fisheries and Aquatic Sciences 66.

Doubleday, W. G., and D. Rivard. 1981. Bottom trawl surveys. Canadian Special Publications of Fisheries and Aquatic Sciences 58.

Ecosystems Principles Advisory Panel. 1999. Ecosystem-based fishery management: a report to Congress. United States Department of Commerce, National Oceanic and Atmospheric Administration, National Marine Fisheries Service. Silver Spring, Maryland.

FAO. 1999. Guidelines for the routine collection of capture fishery data. FAO Fisheries Technical Paper 382. Food and Agriculture Organization of the United Nations. Rome, Italy.

FAO. 1996. Precautionary approach to capture fisheries and species introductions. Elaborated by the Technical Consultation on the Precautionary Approach to Capture Fisheries (Including Species Introductions). Lysekil, Sweden, 6–13 June 1995. Published as FAO Technical Guidelines for Responsible Fisheries No. 2. Food and Agriculture Organization of the United Nations, Rome, Italy.

FAO. 1995. FAO Code of Conduct for Responsible Fisheries. Food and Agriculture Organization of the United Nations, Rome, Italy.

Fogarty, M. J., R. K. Mayo, L. O'Brien, F. M. Serchuk, and A. A. Rosenberg. 1996. Assessing uncertainty and risk in exploited marine populations. Reliability Engineering and System Safety 54:183–195.

Fogarty, M. J., M. P. Sissenwine, and E. B. Cohen. 1991. Recruitment variability and dynamics of exploited marine populations. Trends in Ecology and Evolution 6(8): 241–246.

Francis, R. C., and T. H. Hare. 1994. Decadal-scale regime shifts in the large marine ecosystem of the northeast Pacific. A case for historical science. Fisheries Oceanography 3:279–291.

Healey, M. C. 1984. Multiattribute analysis and the concept of optimum yield. Canadian Journal of Fisheries and Aquatic Sciences 41:1393–1406.

Hilborn, R., E. K. Pikitch, and M. K. McAllister. 1994. A Bayesian estimation and decision analysis for an age-structured model using biomass survey data. Fisheries Research 19:17–30.

Hilborn, R., E. K. Pikitch, and R. C. Francis. 1993. Current trends in including risk and uncertainty in stock assessment and harvest decisions. Canadian Journal of Fisheries and Aquatic Sciences 50:874–880.

Hilborn, R., and C. J. Walters. 1992. Quantitative fisheries stock assessment: choice, dynamics and uncertainty. Chapman and Hall, New York.

Hilborn, R. 1987. Living with uncertainty in resource management. North American Journal of Fisheries Management 7:1–5.

Hilborn, R. 1979. Comparison of fisheries control systems that utilize catch and effort data. Journal of the Fisheries Research Board of Canada 36:1477–1489.

Holt, S. J. 1990. Recruitment in marine populations. Trends in Ecology and Evolution 5:231.

Hutchings, J. A. 1996. Spatial and temporal variation in the density of northern cod and a review of hypotheses for the stock's collapse. Canadian Journal of Fisheries and Aquatic Sciences 53:943–962.

Hutchings, J. A., and R. A. Myers. 1994. What can be learned from the collapse of a renewable resource? Atlantic cod, Gadus morhua, of Newfoundland and Labrador. Canadian Journal of Fisheries and Aquatic Sciences 51:2126–2146.

Ianelli, J. N., L. Fritz, T. Honkalehto, N. Williamson, and G. Walters. 1999. Eastern Bering Sea walleye pollock stock assessment with yield considerations for 1999. Alaska Fisheries Science Center, National Marine Fisheries Service.

ICCAT 2000. Report of the meeting of the ICCAT ad hoc working group on the precautionary approach (Dublin, Ireland, 17–21 May.) 1999. ICCAT Collective Volume of Scientific Papers 51:1941–2057.

ICCAT. 1999. Executive summary of the Atlantic bluefin tuna assessment. Pages 54–68 *in* International Commission for the Conservation of Atlantic Tunas Report 1998/99. Part 1 (English), Volume 2 SCRS.

ICES. 1994. Report of the working group on long term management measures. ICES CM 1994/Assess:11.

IWC. 1994. The revised management procedure (RMP) for baleen whales. Report of the International Whaling Commission 44:142–152.

IWC. 1993. Report of the Scientific Committee. Report of the International Whaling Commission 43:57–64.

Jacobson, L. D., and A. D. MacCall. 1995. Stock-recruitment models for Pacific sardine (*Sardinops sagax*). Canadian Journal of Fisheries and Aquatic Sciences 52: 566–577.

Kawasaki, T. 1992. Mechanisms governing fluctuations in pelagic fish populations. South African Journal of Marine Science 12:873–879.

Kendall, M., and A. Stuart. 1979. Advanced theory of statistics, Volume II: Inference and relationship. Charles Griffin and Sons, Bucks, UK.

Larkin, P. A. 1997. An epitaph for the concept of maximum sustainable yield. Transactions of the American Fisheries Society 106(1):1–11.

Ludwig, D., and C. J. Walters. 1981. Measurement errors and uncertainty in parameter estimates for stock and recruitment. Canadian Journal of Fisheries and Aquatic Sciences 38:711–720.

Mace, P. M. 2001. A new role for MSY in single-species, and ecosystem approaches to fisheries stock assessment, and management. Fish and Fisheries 2:2–32.

Mace, P. M. 1999. Current status and prognosis for marine capture fisheries. Fisheries 24(3):30.

Mace, P. M., and W. L. Gabriel. 1999. Evolution, scope and current applications of the precautionary approach in fisheries. Pages 65–73 *in* V. R. Restrepo, editor. Proceedings of the Fifth National NMFS Stock Assessment Workshop: Providing Scientific Advice to Implement the Precautionary Approach under the Magnuson-Stevens Fishery Conservation and Management Act. NOAA Technical Memorandum NMFS-F/SPO-40.

Mace, P. M. 1997. Developing and sustaining world fisheries resources: the state of science and management. Pages 1–20 *in* D. A. Hancock, D. C. Smith, A. Grant, and J. P. Beumer, editors. Developing and sustaining world fisheries resources: the state of science and management. Proceedings of the Second World Fisheries Congress, Brisbane, Australia, 1996. CSIRO Publishing, Australia.

Mace, P. M. 1993. Will private owners practice prudent resource management? Fisheries 18(9):29–31.

Mace, P. M., and M. P. Sissenwine. 1989. Biological reference points for New Zealand fisheries assessments. New Zealand Fisheries Assessment Research Document No. 89/11.

Mace, P. M. 1985. Catch rates and total removals in the 4WX herring purse seine fisheries. Canadian Atlantic Fisheries Scientific Advisory Committee Research Document 85/74.

Mackett, D. J. 1973. Manual of methods for fisheries resource surveys and appraisal. Part 3. Standard methods and techniques for demersal fisheries resource surveys. FAO Fisheries Technical Paper 124. Food and Agriculture Organization of the United Nations, Rome, Italy.

Mangel, M., and 41 coauthors. 1996. Principles for conservation of wild living resources. Ecological Applications 6(2):338–362.

Mayo, R. K., L. O'Brien, and S. Wigley. 1998. Assessment of the Gulf of Maine Atlantic cod stock for 1998. Northeast Fisheries Science Center Reference Document 98-13.

McAllister, M. K., and G. P. Kirkwood. 1998. Using Bayesian decision analysis to help achieve a precautionary approach for managing developing fisheries. Canadian Journal of Fisheries and Aquatic Sciences 55:2642–2661.

McCaughran, D. A. 1997. Seventy-five years of halibut management success. Pages 680–686 *in* D. A. Hancock, D. C. Smith, A. Grant and J. P. Beumer, editors. Developing and sustaining world fisheries resources: the state of science and management. Proceedings of the Second World Fisheries Congress, Brisbane, Australia, 1996. CSIRO Publishing, Australia.

Meyer, R., and R. B. Millar. 1999. Bayesian stock assessment using a state-space implementation of a delay-difference model. Canadian Journal of Fisheries and Aquatic Sciences 56:37–52.

Myers, R. A., A. A. Rosenberg, P. M. Mace, N. Barrowman, and V. R. Restrepo. 1994. In search of thresholds for recruitment overfishing. ICES Journal of Marine Science 51:191–205.

NMFS. 1991. Strategic Plan of the National Marine Fisheries Service. Department of Commerce, National Oceanic and Atmospheric Administration, National Marine Fisheries Service. Silver Spring, Maryland.

National Research Council. 1999. Sustaining marine fisheries. National Research Council. National Academy Press. Washington, D.C.

National Research Council. 1997. Striking a balance: improving stewardship of marine areas. National Research Council. National Academy Press. Washington, D.C.

Pauly, D., V. Christensen, and C. Walters. 2000. Ecopath, Ecosim, and Ecospace as tools for evaluating ecosystem impacts of fisheries. ICES Journal of Marine Science 57:697–706.

Peterman, R. M., B. J. Pyper, and J. A. Grout. 2000. Comparison of parameter estimation methods for detecting climate-induced changes in productivity of Pacific salmon (*Oncorhynchus* spp.). Canadian Journal of Fisheries, and Aquatic Sciences 57(1):181–191.

PFMC. 1999. The coastal pelagic species fishery management plan, Amendment 8. Pacific Fishery Management Council, Portland, Oregon.

Polacheck, T., N. L. Klaer, C. Millar, and A. L. Preece. 1999. An initial evaluation of management strategies for the southern bluefin tuna fishery. ICES Journal of Marine Science 56:811–826.

Polacheck, T., R. Hilborn, and A. E. Punt. 1993. Fitting surplus production models: comparing methods and measuring uncertainty. Canadian Journal of Fisheries and Aquatic Sciences 50:2597–2607.

Polovina, J. J. 1984. Model of a coral reef ecosystem. I. The ECOPATH model and its application to French Frigate Shoals. Coral Reefs 3:1–11.

Pope, J. G. 1975. Estimation of unknown natural mortality. International Commission for the Northwest Atlantic Fisheries Dumm Document 75/2.

Punt, A. E., and A. D. M. Smith. In Press. The gospel of maximum sustainable yield in fisheries management: birth, crucifixion, and reincarnation. In J. D. Reynolds, G. M. Mace, K. R. Redford, and J. R. Robinson, editors. Conservation of exploited species. Cambridge University Press, Cambridge, UK.

Punt, A. E., and R. Hilborn. 1997. Fisheries stock assessment and decision analysis: the Bayesian approach. Reviews in Fish Biology and Fisheries 7:35–63.

Quinn, T. J. III, and R. B. Deriso. 1999. Quantitative fish dynamics. Oxford University Press, New York.

Restrepo, V. R., P. M. Mace, and F. M. Serchuk. 1999. The precautionary approach: a new paradigm or business as usual? Reviewed Feature Article for Our Living Oceans 1999. NOAA Technical Memorandum NMFS-F/SPO-41.

Restrepo, V. R., G. G. Thompson, P. M. Mace, W. L. Gabriel, L. L. Low, A. D. MacCall, R. D. Methot, J. E. Powers, B. L. Taylor, P. R. Wade, and J. F. Witzig. 1998. Technical Guidance on the use of precautionary approaches to implementing national standard 1 of the Magnuson-Stevens Fishery Conservation and Management Act. NOAA Technical Memorandum NMFS-F/SPO-31.

Richards, L. J. 1991. Use of contradictory data sources in stock assessments. Fisheries Research 11:225–238.

Rosenberg, A. A., and S. Brault. 1993. Choosing a management strategy for stock rebuilding when control is uncertain. Pages 243–249 *in* S. J. Smith, J. J. Hunt and D. Rivard, editors. Risk evaluation and biological reference points for fisheries management. Canadian Special Publication of Fisheries and Aquatic Sciences 120.

Sainsbury, K. R. A., R. Campbell, R. Lindholm, and A. W. Whitlaw. 1997. Experimental management of an Australian multispecies fisheries: examining the possibility of trawl-induced habitat modification. Pages 107–112 *in* E. K. Pikitch, D. D. Huppert, and M. P. Sissenwine, editors. Global trends in fisheries management. American Fisheries Society, Symposium 20, Bethesda, Maryland.

Saville, A. 1977. Survey methods of appraising fishery resources. FAO Fisheries Technical Paper 171. Food and Agriculture Organization of the United Nations. Rome, Italy.

Schnute, J. T., and R. Hilborn. 1993. Analysis of contradictory data sources in fish stock assessments. Canadian Journal of Fisheries and Aquatic Sciences 50:1916–1923.

Serchuk, F. M., D. Rivard, J. Casey, and R. Mayo. 1997. Report of the ad hoc working group of the NAFO Scientific Council on the precautionary approach. NAFO SCS Document 97/12.

Sissenwine, M. P., P. M. Mace, J. E. Powers, and G. P. Scott. 1998. A commentary on western Atlantic bluefin tuna assessments. Transactions of the American Fisheries Society 127:838–855.

Sissenwine, M. P., J. J. Fogarty, and W. J. Overholtz. 1988. Some fisheries management implications of recruitment variability. Pages 129–152 *in* J. Gulland, editor. Fish population dynamics. Wiley Limited, Sussex, England.

Sissenwine, M. P. 1987. Councils, NMFS and the Law. Pages 203–204 *in* R. Stroud, editor. Recreational Fisheries Number 11. Sport Fishing Institute, Washington, D.C.

Sissenwine, M. P., and J. G. Shepherd. 1987. An alternative perspective on recruitment overfishing and biological reference points. Canadian Journal of Fisheries and Aquatic Sciences 44:913–918.

Sissenwine, M. P. 1984. The uncertain environment of fishery scientists and fishery managers. Marine Resource Economics 1(1):1–30.

Sissenwine, M. P., T. Azarovitz, and J. B. Suomala. 1983. Determination of fish abundance. Pages 51–101 *in* Experimental biology at sea. Academic Press, London.

Sissenwine, M. P. 1978. Is MSY an adequate foundation for optimum yield? Fisheries 3(6):22–42.

Smith, A. D. M., A. E. Punt, S. E. Wayte, and N. L. Klaer. 1996. Evaluation of harvest strategies for eastern gemfish *(Rexea solandri)* using Monte Carlo simulation. Pages 120–164 *in* A. D.M. Smith, editor. Evaluation of harvesting strategies for Australian fisheries at different levels of risk from economic collapse. Fisheries Research and Development Corporation Report T93/238, Australia.

Smith, S. J., J. J. Hunt, and D. Rivard, editors. 1993. Risk evaluation and biological reference points for fisheries management. Canadian Special Publication of Fisheries and Aquatic Sciences 120.

Smith, T. 1994. Scaling fisheries. Cambridge University Press. Cambridge, UK.

Sparre, P. 1991. Introduction to multispecies virtual population analysis. ICES Marine Science Symposium 193:12–21.

Thompson, W. F. 1919. The scientific investigation of marine fisheries, as related to the work of the Fish and Game Commission in southern California. Fisheries Bulletin (California) 2:3–27.

United Nations. 1995. Agreement for the Implementation of the Provisions of the United Nations Convention on the Law of the Sea of 10 December 1982 Relating to the Conservation and Management of Straddling Fish Stocks and Highly Migratory Fish Stocks. Rome, Italy. September 1995.

Walters, C. J. 1986. Adaptive management of renewable resources. MacMillan, New York.

Walters, C. J., and D. Ludwig. 1981. Effects of measurement errors on the assessment of stock-recruitment relationships. Canadian Journal of Fisheries and Aquatic Sciences 38:704–710.

Walters, C. J., and R. Hilborn. 1976. Adaptive control of fishing systems. Journal of the Fisheries Research Board of Canada 33:145–159.

American Fisheries Society Symposium 27:29–35, 2002

Management in the Face of Uncertainty: A Case Study with Amberjack

GREGG T. WAUGH
South Atlantic Fishery Management Council, 1 Southpark Circle, Suite 306
Charleston, South Carolina 29407-4699, USA

SUSAN SHIPMAN
Georgia Department of Natural Resources, Coastal Resources Division, One Conservation Way
Suite 300, Brunswick, Georgia 31520-8687, USA

Abstract.—This paper focuses on greater amberjack management in the South Atlantic region, with an examination of the deficiencies in data collection and resulting impacts on the amberjack stock assessment and management decisions. This is a case study of uncertainty leading to more conservative, risk-averse management. Amberjack management in the South Atlantic is illustrative of the debate of whether you do what you can with what you have versus not doing anything until you have good information. Ultimately one has to ask what are the consequences to resource management. While National Standard #2 says you must base your decisions on the best scientific information, scientific information entails more than just the stock assessment results. It entails other biological information as well as social and economic information. The South Atlantic Fishery Management Council went as far as we could without opening ourselves up to a legal challenge. We attempted to balance concern for the resource, Florida's desire to be more conservative, skepticism about the data, and the fishermen's desire to fish.

Introduction

This paper focuses on greater amberjack management in the South Atlantic region, with an examination of the deficiencies in data collection and resulting impacts on the amberjack stock assessment and management decisions. This is a case study of uncertainty leading to more conservative, risk-averse management.

Greater amberjack *Seriola dumerili* are widely occurring in temperate and tropical waters from Massachusetts to the Caribbean and into the Gulf of Mexico. The largest of the amberjacks, it is popular as a game and food fish. These fast swimming jacks school around reefs, rock outcrops, and structure. Females mature at three to four years of age (785–861 mm FL; 31–34 inches FL) and males at age 4. Greater amberjacks are long lived. Specimens up to 86 kg (190 pounds) and two meters (six feet) in length have been recorded. The species is an aggressive, top order piscivore.

The greater amberjack fishery as described by several authors is summarized in Cummings (1998). Recreational harvest of greater amberjack from New York to Texas, mostly incidental to other species, occurred for decades; however, amberjack only became a target species in the 1970s. Increasing catches of amberjack for sport and food in the 1970s prompted the collection of catch data and biostatistics. Despite its increased popularity, problems plagued fishery dependent data collection for amberjack, not the least of which has been proper identification of the Seriola species (greater, lesser, and banded rudderfish). Historical landings are believed to have been under reported prior to the 1980s.

Traditionally, private anglers and charter boats accounted for the majority of the recreational catch, although since 1991, greater amberjack have been caught primarily by charter boat anglers. Most of the recreational harvest is attributed to Florida anglers.

The various reports show that the Atlantic greater amberjack commercial harvest is primarily landed in Florida, with significant catches in the Florida Keys (Monroe County). Commercial harvest increased substantially in the mid-1980s, especially off South Florida as amberjack became an important restaurant item. Landings peaked in 1991 at 1,043.3 mt (2.3 million pounds), and have declined since. To what degree this is reflective of regulation is not known. Commercial harvest is primarily hook and line, with peak harvest during March through May, the spawning season for this species.

Federal management of greater amberjack in the South Atlantic is under the South Atlantic Fishery Management Council's Fishery Management Plan (FMP) for the Snapper Grouper Fishery (SAFMC, 1983). This plan was implemented in 1983 and was designed to prevent growth overfishing of 13 reef fish species and to establish a strategy for preventing overfishing in the 56 other species included in the plan. The original FMP was based on yield-per-recruit (YPR) analysis. There were data for YPR analysis for 17 species; 13 of the 17 were considered likely to be growth-overfished. Amberjack was not one of these; thus, no management measures were proposed for this species. At that time, amberjacks were caught incidental to other species and were viewed largely as an underutilized species. However, greater amberjack was included in the management unit. As early as 1983, the Snapper-Grouper Plan recognized the data necessary to document growth or recruitment overfishing was very limited. The Plan authorized statistical reporting and data collection for all the species in the management unit.

The National Marine Fisheries Service (NMFS) Beaufort Laboratory, with input from the South Atlantic Council's Plan Development Team (PDT), prepared the first assessment of species in the snapper-grouper complex in 1990. The assessment (based on data from 1972 through 1988/1989) changed the biological reference point from YPR to SSR (Spawning Stock Ratio; [SSR] = Spawning Stock Biomass per Recruit [SSBR] = Spawning Potential Ratio [SPR]). The biological simulation analysis produced in 1990 used a standard age-structured population model that employs catch at age. The data necessary to document growth and/or recruitment overfishing and to calculate SSR was very limited. Nowhere was this more true than with amberjack. The amberjack analysis was calculated with a quasi-virtual population analysis using 1988 catch data for both recreational and commercial sectors.

Despite authorization for data collection in the original plan, in 1990 management continued to be hindered by a lack of fishery dependent and fishery independent data needed for an assessment. Nonetheless, the 1990 stock assessment estimated SPR by sector as follows: 7–18% for the recreational sector, depending on geographic area; 27% SPR for the commercial fishery.

The PDT recommended reductions in fishing mortality through minimum sizes and/or bag/catch limits for the various species including amberjack. The council undertook development of Amendment 4 to the Snapper-Grouper FMP (SAFMC 1991). The council set the overfishing definition for all the FMP species other than jewfish *Epinephelus itajara* (also known as Goliath grouper) at 30% SPR. Based on this definition, greater amberjack were considered overfished. The council set the rebuilding time frame for recovery of amberjack at a maximum of 10 years, with 1991 as year one.

The council added and modified minimum size limits based on the assessment. Amendment 4 established a minimum size of 709 mm (28 inch) fork length (FL) in the recreational fishery and 912 mm (36 inch FL) or 709 mm (28 inch) core length (length of fish after head and tail are removed) in the commercial fishery. This afforded additional protection given the high catches during spawning aggregations. The minimum sizes were projected to achieve SPRs of 27% and 43% for the recreational and commercial sectors, respectively.

Amendment 4 also established a bag limit of three greater amberjack for the recreational fishery. The bag limit, coupled with the 709 mm (28 inch) FL recreational size limit, was deemed sufficient to yield a SPR greater than 30% and thereby prevent overfishing. Although the bag limit was expected to impact a minimum number of trips, it would spread out the harvest in the recreational sector.

The amendment also imposed a greater amberjack spawning closure during April south of Cape Canaveral which was where spawning occurred. During April, both recreational and commercial harvests were limited to the three fish bag limit. Sale of the bag limit was allowed in conformance with state law and commercial size limit. Spawning closure coupled with size and bag limit was deemed sufficient to achieve the combined SSR greater than 30%. Because dense aggregations at specific sites and aggressiveness during the spawning period make amberjack especially vulnerable to fishing mortality, the spawning closure would help prevent excessive harvest. The spawning closure was estimated to reduce the instantaneous rate of fishing mortality (F) by 19%. Amendment 4 became effective in January 1992. Although the plan provided for mandatory landing of fish in the round for data collection, it is unknown whether that was ever required.

While Amendment 4 was under development and subsequent implementation, NMFS conducted a 1991 stock assessment (Anonymous 1991) based on 1988 data, and a 1992 assessment (Huntsman et al. 1992) based on 1990 data. The 1991 and 1992 assessments produced a combined SSR value for greater amberjack of 79% and 51% (62% with minimum size limit), respectively. Instantaneous F for fully recruited ages (about age five in these studies) was estimated at 0.07 and 0.69. Amberjack went from "overfished" to "not overfished" status. Using 1991 catch data, Huntsman et al. (1993) re-estimated SPR to be 0.45 (0.54 with minimum size) using a different age length key. The revised estimate of F for fully recruited age classes was less than 0.10. Greater amberjack remained not overfished. It should be noted the initial estimates were based on South Florida growth information, while Huntsman's second estimate of SPR was based on an age-length key derived from only 127 fish from the Atlantic.

In 1993, Parrack presented stock size estimates using a Simple Likelihood Method (SLM). This analytical method was applied as opposed to typical Cohort or Virtual Population Analyses (VPA) or stock production models due to the paucity of data. Adequate biostatistics were not available nor was adequate growth information or comprehensive size composition data available to reconstruct historical population and exploitation levels for the Atlantic group amberjack fishery. The resource was not believed at that time to be in an equilibrium state. Non-equilibrium production models were considered but not used because the data series of complete statistics was extremely short (six years).

Analysis of Atlantic greater amberjack stock condition from 1986 to 1991 indicated that, despite minimum size regulations implemented in 1990, sample average weight decreased in both the commercial and headboat fisheries in 1991 which could have been due to low sampling effort rather than a real decline in average weight. In addition, the trend in estimated stock sizes suggested large variation in stock sizes, on the order of 100%+ annually. Parrack (1993) noted the degree of uncertainty was too large for the estimates to be useful in measuring changes in abundance between years. The uncertainty resulted from insufficient monitoring of the basic data required for stock assessments: reported yields, estimated catches, fishing effort, size and weight samples.

Study results indicated that the Atlantic greater amberjack stock dynamics are recruitment driven, with recruitment highly erratic in magnitude. Because of lack of size data, estimates of recruitment magnitude were not possible. Recruitment was only indirectly estimated. From 1986 to 1991, overall trend in stock size appeared to be up from the initial year of study (1986). Despite these large changes, the council kept significant management measures in place to remain risk averse.

In 1993, Parrack noted all amberjack fisheries, and especially the commercial fishery, were severely under-sampled for both length and sample average weight throughout the time series, 1979–1992. The inadequacy of biostatistical sampling was pointed out to managers at that time. The SLM model did not include information on size or age in the model estimate; thus, the separation of recruitment and/or the level of adult stock was not possible. Parrack concluded the basic fisheries data, especially size and age data, would have to be dramatically improved in scope and quality before accurate assessment of stock condition and sound management advice would be possible for Atlantic greater amberjack stock.

The reductions anticipated from Amendment 4 were never realized due in part to changes in fishing pattern whereby the commercial fishery expanded beyond South Florida. In the 1990s, anecdotal information from the recreational sector suggested large declines in amberjack catches in the Atlantic. Although the commercial fisheries did not perceive a similar decline, available data did support a consistent decline for the Atlantic group overall, with a more pronounced decline in certain recreational sectors. Consequently, a reassessment was undertaken in 1996 and 1997. The 1996 assessment by Cummings and McClellan was the first comprehensive age based population analysis undertaken for Atlantic group greater amberjack. Applying the ADAPT Virtual Population Analysis to amberjack statistics from 1987 through 1995, Cummings and McClellan (1996) calculated SPR for the 1993 year at 0.84 (subsequently recalculated at 0.81%). Natural mortality was assumed to be 0.30, and the results were considered imprecise for ages three and four. Despite the decline for the Atlantic group overall, the estimated SPR increased which created more uncertainty in terms of understanding the stock status. Again the Council maintained risk averse management measures.

Despite the major weakness noted in the level of sampling of the fishery in the 1993 assessment and the recommendation for enhanced data collection, there were still problems in the data for the 1996 assessment. The Marine Recreational Fishery Statistics Survey (MRFSS) catch estimates were still noted to be of concern; some of the amberjack values for state waters were suspect. The recreational catches showed a decline of differing magnitude since 1987, although it was impossible to discern whether the cause was changes in abundance or regulations. Length sampling of greater amberjack actually declined in the years subsequent to the 1993 assessment (Table 1A and B). Overall recreational sampling rates of Atlantic greater amberjack remained at an extremely low level (described as "deplorably" low by the NMFS assessors) at less than 1% by number. Length sampling of the commercial fisheries declined 60% between the 1993 and 1996 assessments, at more than 1% of the estimated catch.

Of necessity, catches were matched to size samples; however, it was noted this likely masks part of the variation in catch at size. Nonetheless, the data available dictated this procedure. Fishing mortality was, in general, very low over the time frame of the

TABLE 1A. Commercial greater amberjack size samples. Source: NMFS.

Year	Monroe County	Florida East Coast	Georgia	South Carolina	North Carolina	All areas
1980	25					25
1981	73					73
1982						
1983						
1984					1	1
1985		1			23	24
1986		1	56		7	64
1987		4	32	2	23	61
1988		7	37	12	56	112
1989	1	46	18	29	20	114
1990	55	60		67	45	227
1991	483	65	194		73	815
1992	425	879	100	96	19	1,519
1993	199	736	192	90	31	1,248
1994	182	152	193	26	70	623
1995	25	263	41	116	58	503
1996	89	306	38	185	27	645
1997	367	159	136	197	26	885
1998	189	388	43		12	632
Total	2,113	3,067	1,080	820	491	7,571

1993 and 1996 analyses. Trends in estimated stock size indicated a general increase in adult stock between 1987 and 1992. Most of the increase in stock production was from fish recruiting into the fishery and becoming available to the gear as opposed to recruitment of age zero fish. The data available for this analysis suggested a significant decline in recruitment of the Atlantic greater amberjack and in abundance of age one to three fish. The 1996 assessment suggested that although SPR levels were reasonably high, future declines in SPR were imminent.

The 1996 assessment recommended using the 1993 figures for F (0.193) and SPR (84%) as the best estimates for management. Council members, council staff, and the council's Scientific and Statistical Committee (SSC) questioned whether a stock could sustain removals from a population at the levels observed (i.e., current F levels) and have these high SPR results. It was suggested that the delay in

TABLE 1B. Recreational greater amberjack size samples. Source: NMFS.

Year	Monroe County	Florida East Coast	Georgia	South Carolina	North Carolina	All areas
1980	6	4			1	11
1981	8	248		1	51	308
1982	3	101	2	1	16	123
1983	5	257		12	15	289
1984	22	223	4	40	14	303
1985	7	194	11	10	50	272
1986	19	214	3	10	41	287
1987	22	220	25	12	89	368
1988	11	72	5	30	88	206
1989	10	193	26	28	79	336
1990	5	127	4	25	60	221
1991	18	104	2	26	57	207
1992	15	142	21	63	47	288
1993	12	87		78	109	286
1994	6	102	8	46	103	265
1995	16	43	4	36	88	187
1996	8	28	1	33	118	188
1997	28	40	5	33	40	146
1998	39	24	6		40	109
Total	260	2,423	127	484	1,106	4,400

vulnerability of the fish to the fishery until a much older age than 100% maturity because of size regulations could be creating the very high SPR ratios. There was a lopping off of small fish in the age distributions. From 1992 forward, there are some year classes which are not vulnerable to the fishery due to size limits, for which little was known.

Normally in a fishery with similar levels of removals, declines in catch per unit effort (CPUE) would be expected. The lack of any decline in the greater amberjack CPUE was suggested by some to be the result of fishing on a spawning aggregation, given that a large portion of the commercial catch was from spawning areas. In such a situation, one would not expect a decline in CPUE, until the stock was severely overfished. In the face of such a high SPR, some members of the public requested the Council relax regulations. However, the Council once again maintained risk averse management measures to address the uncertainty about stock status.

In the 1996 assessment, major segments of the recreational catches for the Atlantic side of the Florida Keys were not included. It was suggested that the underestimate of catch data could have resulted in an underestimated F, and an artificially high SPR.

Some of the SSC members suggested the data available did not accommodate the appropriate use of a VPA. Some opined that the utilization of the data was overextended, that errors were extrapolated forward, and assumptions compounded. The assessors concluded that the exact level of stock size estimates was not necessarily dependable. The SSC emphasized their conclusion was not a negative comment on the assessment methodology but a recognition of the need for more data before an analysis of this type could be completed that would accurately portray stock status.

There was trepidation on the part of the council and council staff over fishermen's perception that we were throwing out an assessment showing such a high SPR, while readily accepting other assessments with low SPRs. The council had the SSC revisit the assessment in April 1997. The SSC qualified their endorsement of the 1996 assessment, accepting it as the best data "under the circumstances" and recommending that a reassessment using non-age based techniques be requested. In fact, NMFS had serious concerns about their stock assessment, and considered the stock status unknown in their 1997 Report to Congress (NMFS 1997).

In short, the level of uncertainty about the conclusions of the stock assessment clearly reflected the poor state of data collection for greater amberjack. Florida council member Ben Hartig, a commercial fisherman who had been a historical participant in the amberjack fishery and was seeing a substantial decline in the population, perhaps best captured the council's sentiments about the state of data collection: "Despite the weaknesses in the 1993 assessment and the recommendations for improved data, three years later we had even less data."

The Florida Marine Fisheries Commission had conducted workshops and public hearings in the summer of 1996 to assess public opinion on amberjacks. Simultaneously, the South Atlantic Fishery Management Council was contemplating additional protection for greater amberjack. Given the level of data uncertainty and the sentiment of many constituents, including the State of Florida, the council concluded that further management was necessary and appropriate to ensure the long-term sustainability of the greater amberjack resource. Development of Snapper-Grouper Amendment 9 was initiated.

Amendment 9 (SAFMC 1998) incorporated measures to reduce the recreational bag limit from 3 to 1; prohibit the harvest and possession in excess of the bag limit during April throughout the EEZ; establish a 453.6 kg (1,000 pound) daily commercial trip limit and a quota equal to 63% of the 1995 landings (530.68 mt; 1,169,931 pounds); and move the beginning of the fishing year from 1 January to 1 May. Available data indicated the quota would be met during March, and the fishery would be closed at the start of spawning season. The amendment prohibited all sale during the season closure and prohibited coring, thereby increasing the potential for biostatistics collection, especially aging. The projected reductions in catches were 41% from the commercial sector based on decreased catches under the quota as compared with the previous year's catches, 11% from the recreational sector (assumes constant number of anglers; as angler numbers increase, the reduction will be much less) based on a bag limit analysis using MRFSS data, and 29% in total F across the entire fishery assuming there is a one to one correlation between percentage reduction in catch and percentage reduction in F.

During development of Amendment 9, there was some dispute over the stock status of greater amberjack given the 1996 stock assessment's estimated SPR of 84%. While the unusually high SPR estimates were attributed to problems in sampling, nonetheless the commercial sector questioned the need for what they considered extreme management measures. The commercial sector agreed additional management was warranted but not to the degree being considered by the council. Testimony during the public hearings convinced the council that the amberjack stock had declined and needed further management. The numbers in the assessment didn't mesh with what the council was hearing from the fishermen and seeing in the fishery. The extent of that management was what was debated, largely by the commercial sector.

The initial preferred alternative was a three month March–May spawning season closure, which tracked Florida's and the Gulf of Mexico Fishery Management Council's regulations. However, public input made it clear it would have disproportionately affected the Florida Keys—both recreational and commercial sectors. The council ultimately elected to go with the Florida Keys commercial fishing industry's proposal of the one month closure and trip limit. The council added the quota to ensure the desired reduction in mortality would be reached and changed the fishing year to assure access to the fishery throughout the region. The council changed the fishing year to begin in May, which would provide additional biological protection since the quota was projected to be met prior to the start of the spawning season in March. This was acceptable to the commercial industry whereas a two-month spawning closure in March and April was not acceptable. This is a good example of how the council was able to implement precautionary management measures in the face of uncertainty in stock status.

Amendment 9 was adopted in August 1997 and submitted to NMFS in February 1998. However, seizing upon new mechanisms available through the SFA, the council had also voted in November 1997 to request NMFS implement the measures in Amendment 9 by emergency interim rule. The request for interim rule was submitted in mid-January 1998. In April 1998, the State of Florida filed a Coastal Zone Management (CZM) non-consistency objection to the amberjack provision of Amendment 9, citing that the council measure afforded less protection to amberjack than the rejected alternative of the three month closure (the State of Florida and the Gulf of Mexico Fishery Management Council had implemented this alternative). This controversy over the amberjack measures was a factor in the Secretary's delay in implementation of the interim rule.

In May 1998, the NMFS suspended action processing the interim rule request and elected instead to address the measures through implementation of Amendment 9. That implementation continued to languish, and in September 1998 the council voted to implement Amendment 9 by emergency action. The decision was by unanimous vote, in the absence of the NMFS and should have been binding on the Secretary. It wasn't. The Secretary elected not to act on the emergency request, and, instead, Amendment 9 was finally approved in December 1998, except for the greater amberjack trip limit. Environmental groups issued notice to the Secretary of their intent to sue if the final rule was not published by 27 January 1999. NMFS published the Final Rule on 25 January 1999, and Amendment 9 became effective February 1999, except for the amberjack trip limit. NMFS cited disproportionate economic impacts to the commercial fishermen north of the Florida Keys. However, a subsequent in-depth review by NMFS economists reversed this conclusion, and NMFS requested the council resubmit the amberjack trip limit, which was done in March 1999. This measure is currently being redrafted by NMFS as Amendment 12. The uncertainty concerning stock status was a factor in the delay in NMFS decision making. Fishermen raised concerns to NMFS about the impacts of management measures (i.e., spawning season closure and trip limit) in the face of extremely high SPR levels, which ultimately led to the disapproval of the greater amberjack trip limit.

In response to SFA mandates and amberjack management needs, assessment of the greater amberjack stock has continued. Potts et al. (1998) applied a VPA analysis to commercial and recreational catch data from 1986 through 1996. Using two values of natural mortality, estimates of SSR and F were calculated to be 0.58 and 0.20 for $M = 0.20$, and 0.71 and 0.16 for $M = 0.25$. Since the 1991 peak harvest, landings through 1997 declined by 39.9% by weight.

Amberjack management in the South Atlantic is illustrative of the debate of whether you do what you can with what you have versus not doing anything until you have good information. Ultimately one has to ask what are the consequences to resource management. While National Standard #2 says you must base your decisions on the best scientific information, scientific information entails more than just the stock assessment results. It entails other biological information as well as social and economic information. The SAFMC went as far as we could without opening ourselves up to a legal challenge. We attempted to balance concern for the resource, Florida's desire to be more conservative, skepticism about the data, and the fishermen's desire to fish. This approach was successful in preventing any legal challenges, however, it was not successful in preventing fishermen contacting NMFS and ultimately being successful in having NMFS disapprove the trip limit.

How much progress has been made toward addressing the assessment data needs thereby reducing uncertainty about stock status? Following the SSC's review of the 1996 assessment, the council sent a letter to NMFS requesting implementation of a program to collect adequate age and size data ASAP. However, little improvement has been made in sampling levels (Tables 1A and 1B). In its report to the council in March 1999, the Council's Snapper Grouper Assessment Group noted that size/age data should be collected, species identification should be improved, and fecundity and reproductive studies should be undertaken. The council can suggest target sampling levels but this is an area NMFS considers to be a part of their responsibility in implementing fishery management plans and amendments. NMFS has been unable to collect adequate age and size data in the Southeast Region due to the lack of adequate funding. While the council can propose reporting requirements, and these can ultimately lead to regulations, the council has no authority or direct involvement in enforcement. NMFS, Coast Guard, and State enforcement agents (in some States) provide enforcement of federal regulations. It is up to the NOAA attorneys to determine the level of fines and/or permit sanctions that result from not meeting reporting requirements.

Where do we go from here? What are the plans to address data needs? The Atlantic Coastal Cooperative Statistics Program (ACCSP) Biological Sampling Panel is addressing the data needs for stock assessments for all species under interstate and federal management. Their report to the partner agencies of the ACCSP, including the East Coast fishery management councils, came out in 2000. This plan includes a blueprint for sufficient trip sampling of catch to overcome existing data deficiencies. The remedy to uncertainty in amberjack management largely hinges on implementation of the ACCSP Panel's recommended sampling strategy. In the interim, the South Atlantic Council will hopefully stay the course of risk averse management.

References

Anonymous. 1991. South Atlantic snapper grouper assessment 1991. Unpublished manuscript presented to the South Atlantic Fishery Management Council, June 1991. U.S. Department of Commerce, NOAA, NMFS, SEFSC, Beaufort Laboratory, Beaufort, North Carolina.

Cummings, N. J. 1998. Updated landings, catches, and CPUE statistics for the Atlantic greater amberjack, *Seriola dumerili*, fishery through 1997. U.S. Department of Commerce, NOAA, NMFS, SEFSC, Miami Laboratory, unpublished manuscript.

Cummings, N. J., and D. B. McClellan. 1996. Stock Assessment of the Atlantic Ocean Greater Amberjack through 1995. U.S. Department of Commerce, NOAA, NMFS, SEFSC, Miami Laboratory Contribution No. MIA-96/97-04.

Huntsman, G. R., J. Potts, R. Mays, R. Dixon, P. Willis, M. Burton, and B. Harvey. 1992. A stock assessment of the snapper-grouper complex in the U.S. South Atlantic based on fish caught in 1990. U.S. Department of Commerce, NOAA, NMFS, SEFSC, Beaufort Laboratory, unpublished manuscript.

Huntsman, G. R., J. C. Potts, and R. W. Mays. 1993. Estimates of spawning stock biomass per recruit ratio based on catches and samples from 1991 for five species of reef fish from the U.S. South Atlantic. Report to the South Atlantic Fishery Management Council, June 1993. NMFS Beaufort Lab, 101 Pivers Island Road, Beaufort, North Carolina, 28516-9722.

NMFS. 1997. Report to Congress: Status of Fisheries of the United States. September 1997.

Parrack, N. C. 1993. The exploitation status of the Atlantic amberjack fisheries through 1991. DOC\NOAA\NMFS \SEFSC\Miami Laboratory Contribution MIA-92/93-77, unpublished manuscript.

Potts, J., M. L. Burton, and C. Manooch. 1998. Trends in catch and estimated static SPR values for fifteen species of reef fish landed along the southeastern United States. U.S. Department of Commerce, NOAA, NMFS, SEFSC, Beaufort Laboratory, unpublished manuscript.

SAFMC. 1983. South Atlantic Fishery Management Council. Fishery management plan, regulatory impact review, and final environmental impact statement for the snapper-grouper fishery of the South Atlantic region. South Atlantic Fishery Management Council, Charleston, South Carolina.

SAFMC. 1991. South Atlantic Fishery Management Council. Amendment Number 4, regulatory impact review, initial regulatory flexibility analysis and environmental assessment for the fishery management plan for the snapper grouper fishery of the South Atlantic region. South Atlantic Fishery Management Council, Charleston, South Carolina.

SAFMC. 1998. South Atlantic Fishery Management Council. Amendment 9 to the fishery management plan for the snapper-grouper fishery of the South Atlantic region, including a final supplemental environmental impact assessment, initial regulatory flexibility analysis, regulatory impact review, and social impact assessment/fishery impact statement. South Atlantic Fishery Management Council, Charleston, South Carolina.

American Fisheries Society Symposium 27:37–44, 2002

Mixed Monte Carlo/Bootstrap Approach to Assessing King and Spanish Mackerel in the Atlantic and Gulf of Mexico: Its Evolution and Impact

CHRISTOPHER M. LEGAULT AND JOSEPH E. POWERS
National Marine Fisheries Service, Southeast Fisheries Science Center, 75 Virginia Beach Drive Miami, Florida 33149, USA

VICTOR R. RESTREPO
University of Miami, Rosenstiel School of Marine and Atmospheric Science Division of Marine Biology and Fisheries, 4600 Rickenbacker Causeway, Miami, Florida 33149, USA

Abstract.—The stock assessment analyses of king and Spanish mackerel fisheries of the southeastern United States have a long history of incorporating uncertainty. The development of this philosophy resulted from a number of unique circumstances, both biological and historical, that encouraged the incorporation of stochastic approaches and risk evaluation to the assessment and management process. The progression from simple discrete decision tree analysis to delta methods to Monte Carlo/bootstrap methods was due not only to advances in assessment technology but also to changing requirements for management. The current method for mackerel stock assessment is a tuned virtual population analysis with uncertainty incorporated via a mixed Monte Carlo/bootstrap algorithm. Through this procedure, uncertainty in the tuning indices, catch-at-age and natural mortality rate are directly incorporated into the advice provided to management. The management advice is given in terms of probability statements, as opposed to point estimates, to reflect this uncertainty in the stock assessments. This approach is a result of the evolution of the assessment and management and provides a pragmatic alternative in the "frequentist versus Bayesian" debate.

The Evolution of Stochastic Approaches for Assessment and Management

King mackeral *Scomberomorus cavalla* and Spanish mackerel *Scomberomorus maculatus* are migratory coastal pelagic fish found from Maine to Texas in the Atlantic Ocean and Gulf of Mexico. They perform annual migrations from southern regions in the winter to northern regions in the summer to spawn. For management purposes, there are two groups for each species, Atlantic and Gulf of Mexico, which are separated by a time-dependent boundary in southeast Florida. These four management groups have been the focus of the Migratory Coastal Pelagic Fishery Management Plan developed jointly by the South Atlantic and Gulf of Mexico Fishery Management Councils.

The stock assessment analyses of king and Spanish mackerel fisheries of the southeast United States have a long history of incorporating uncertainty. The development of this philosophy resulted from a number of unique circumstances, both biological and historical, that encouraged the incorporation of stochastic approaches and risk evaluation to the assessment and management process. The king mackerel fishery has always been more valuable than the Spanish mackerel fishery, and thus most of the examples presented herein use king mackerel, but the similarities between the fisheries have allowed the same methodological approaches to be used for assessments and management advice. In this section we review the history of mackerel management and assessment history and the characteristics of the fishery that led to the present assessment approach.

Like many fisheries in the southeast United States, the mackerel fisheries are characterized by having a large recreational sector. The recreational fishery has shaped the assessments and historical databases. For example, about two thirds of the U.S. Gulf of Mexico catches of king mackerel (in weight) results from recreational fishers including private boat operators, charterboats, headboats or party boats and fishers operating from shore. This, in turn, poses some well-known statistical problems in determining the annual and seasonal catch levels by fishing sector. Recreational fisheries are characterized by having a very large number of participants with a wide variety of skill levels and numerous geographically dispersed landing points. Such a situation does not lend itself to logbook or dealer reporting systems as used in commercial fisheries. The option most utilized in marine recreational fisheries is a statistically designed survey. The Marine Recreational Fisheries Statistics Survey (MRFSS) was established in the southeast United States in 1979 (MRFSS 1991). Therefore, databases necessary for assessments were nonexistent prior to 1979. Catch histories prior to 1979 could only be surmised from commercial records; but, since subsequent estimates indicated that recreational catches were larger than commercial, predicting recreational catches from commercial effort was anecdotal at best.

A survey for recreational catches necessarily results in estimates with standard deviations associated with them. In the case of the existing survey vehicle in the southeast United States (MRFSS 1991) coefficients of variation are often around 30–40%. This fact alone limits the usefulness of the point estimates in a deterministic assessment. Methods that utilize the precision estimates are important.

Additionally, there were no systematic programs to collect effort data from the commercial sector in the early 1980s. Subsequently, trip-ticket programs and logbook programs were established for many fisheries, which allowed statistical analysis of catch-per-unit-effort (CPUE) indices of abundance. In some instances there were individual fishers who recorded their own catch and effort data in personal logbooks and shared these with assessment scientists. But, obviously, samples sizes were limited. Thus, initial assessments were extremely uncertain due to lack of basic catch and effort data and measures of precision associated with them.

Early king mackerel assessments reflected these data limitations. An initial assessment of southeastern U.S. king mackerel was conducted in 1983 using equilibrium production models (Powers and Eldridge 1983, 1984). The data available at the time were catch estimates for 1979 and 1980 (the accuracy and precision of which were severely questioned), long term logbook data from two individual charterboat fishers from the early 1970s, growth data, tagging data designed to address distributional issues and a scattering of size frequency data in various locations and times from various gears. Not surprisingly the status estimates that were derived from these data were extremely uncertain. This uncertainty was evaluated using *ad hoc* scenario tests; for example, alternative assumptions on recreational catch levels in the 1970s, the 1979–1980 catch levels, and so on (Powers and Eldridge 1984). Ranges of uncertainty based upon these scenarios were presented to the fisheries managers and formed the basis of an amendment to the management plan.

The amendment formalized the process of management objectives (based on an $F_{0.1}$ policy), the role of a scientific panel to determine a range of acceptable biological catch (ABC) based upon that policy, and the role of the managers in selecting a total acceptable catch (TAC) from the range of ABC. Amendment development was prolonged, requiring two years. The ensuing time was utilized to collate and integrate an assessment database for use in age-structured analyses. The integration required solicitation of age samples collected by various state and academic scientists and updating the MRFSS data through 1984. This culminated in an age-structured analysis (ad hoc tuned virtual population analysis; Nichols 1985). Ranges of ABC for each mackerel stock were specified using sensitivity analyses over alternative specifications of the natural mortality rate, alternative historical catch scenarios and alternative recruitment scenarios. Unfortunately, the delays in data acquisition coupled with delays in management plan implementation coincided with large declines in recruitment and in catch. Thus, in 1985 shortly after implementation of the management plan, king mackerel catches of approximately 6 million pounds were being taken from the U.S. Gulf of Mexico, but an ABC of about 2 million pounds was being recommended (Powers 1985).

The response to this age-based assessment was two-fold. First, the TAC was substantially reduced over the following two years to more closely coincide with the ABC recommendation. Secondly, an agency task team was charged to review the science and management processes that led to this situation. The review team reemphasized the need to characterize the uncertainty in assessments and to convey this to managers. Additionally, the team emphasized to managers that they should take account of these uncertainties (including lags in data and management and allocation processes) when making their decisions. Essentially, the review team called for risk analyses associated with ABC and TAC selection and for precautionary management procedures in the face of this uncertainty.

As indicated above, the initial ABC's of mackerels were conveyed to managers as a range of values associated with sensitivity analyses on natural mortality rate, catch levels and future recruitment (Powers 1985). This range was designed to include the relevant probability that a chosen TAC would achieve the target fishing mortality rate ($F_{0.1}$). However, the relevant probability was largely undefined. Also, although the managers were informed that outcomes near the borders of the range were less likely than those mid-range, these statements were not quantified and the managers tended to view the range as a uniform distribution with each value being equally likely. Thus, the sensitivity analysis range was an initial approximation to specifying risks and uncertainties. Nevertheless, there was a need to quantify these risks.

Therefore, an initial *quantification* of the uncertainty in ABC was done using discrete decision tree methods (Raiffa 1970). The problem was partitioned into the uncertainty in the estimate of stock size in the year for which ABC was being estimated and the uncertainty in the estimation of the management criteria (Powers 1986). The former was evaluated by assigning probabilities for three discrete levels of the current stock size (based on F estimated, 2*F and 0.5*F with probabilities approximated from scaled variance in 1985 yield), two recruitment predictions (high and low based upon scaled values and variance from a CPUE index of small fish, a proxy for recruitment strength), two values for the underlying natural mortality rate (based on assigned probabilities given the life history of the species) and two forms of the stock structure. The uncertainty in the management criteria were approximated from the variance of previous annual catches. Note that these probabilities were conditional on the branch of the decision tree relating to the estimate of current stock size. Twenty-four ABC outcomes were computed (weighted by the relevant branch probabilities). The results were smoothed with a gamma distribution and presented to the managers as both density and cumulative probability functions. Thus, probability approximations to risks of achieving the target fishing mortality rate were being presented as early as 1986.

Over the next few years, assessment technology improved considerably in terms of both the models being used and the computing ability to utilize these models. Assessments of the mackerel stocks then used formal statistical fitting procedures between a virtual population analysis model and indices of abundance, rather than *ad hoc* tuning (Parrack 1986; Gavaris 1988). Uncertainties in the ABC determination were approximated using delta method estimates of current stock size from the fits coupled with discrete alternatives for biological parameters (Powers 1989). Subsequently, technological innovation allowed uncertainty characterization using Monte Carlo and bootstrap methods (Restrepo, et al. 1991, Restrepo, et al. 1992, Powers and Restrepo 1993). Reasons that this approach was taken included 1) it was a natural extension of the discrete decision trees; 2) computing technology in the late 1980s and early 1990s allowed rapid evaluation of a substantial number of simulations within a 3–4 day scientific workshop framework; and 3) computing technology limited the use of Bayesian approaches (McAllister and Ianelli 1997; Punt and Butterworth 1997) for models with many estimated parameters. The Monte Carlo/bootstrap approach was a pragmatic method to integrate viewpoints from various scientists contributing to the process.

Management, also, evolved over this time with implementation of overfishing limits and spawning potential ratio (SPR) based F targets computed conditionally on selectivity and management goals for recreational, commercial and bycatch sectors. More recently, the management and assessments have been expanded to encompass control rules (Restrepo, et al. 1998; MSAP 1999). Thus, the system of ABC determination and uncertainty related to those ABC estimates involved various constraints and objectives that were implemented into the Monte Carlo/bootstrap framework (Powers 1995).

Where Are We Now?

The current assessment and projections procedures for mackerel fisheries in the southeast United States utilize a mixture of Monte Carlo simulation and bootstrap to incorporate uncertainty into the estimates of recent abundance and potential results from different management strategies. The tuned virtual population analysis (VPA) program Fadapt (Restrepo 1996), modified to allow Monte Carlo simulation and bootstrapping, is used to assess the current and historical status of the stock. Each Monte Carlo/bootstrap run produces estimates of the full age by year matrices for population numbers and fishing mortality rates, and the natural mortality values chosen. These sets of matrices are used as inputs to a projection program to reach the current year and forecast potential yields under given F strategies or future population changes under constant yield strategies. Full descriptions of the data sources for the mackerel assessments are given in Legault et al. (1998). Current conditions of the mackerel stocks are given in Legault (1999a). A detailed examination of the Gulf of Mexico king mackerel fishery is provided in Legault (1999b).

The modified Fadapt program allows for variability in the tuning indices, natural mortality rate, and catch-at-age. The tuning indices are bootstrapped over all index residuals. The indices are first scaled such that each has a mean of one in order to make the residuals comparable across all indices. This bootstrapping uses the predicted values from the deterministic Fadapt run as the initial values and then adds a randomly chosen residual to each predicted value to form the bootstrapped indices. Failure to standardize the indices first could result in high levels of bias in the results. An alternative would be to bootstrap only within an index, but this could artificially emphasize some indices over others simply based on the (usually few) number of observations within some indices relative to others.

The natural mortality rate (M) for each age, year combination is chosen randomly from a uniform distribution. This will result in most runs having overall natural mortality rates similar to the deterministic value, but allows for changes in the population dynamics. Using Monte Carlo simulation to select a single natural mortality rate to be applied to all age, year combinations results in a highly predictable relationship between M and the current population abundance, which also translates directly into the predicted yield under any constant F strategy. This single value Monte Carlo simulation approach requires good understanding of the probability density function (pdf) for M, while the multiple M selection used forms approximately a normal distribution of overall natural mortality rate through the central limit theorem. Thus, the multiple M selection approach is less demanding of the pdf for M. The uniform distribution was chosen for M because life history characteristics and maximum ages observed could be used to generate ranges of possible values for the natural mortality rate, but each value within the range was thought to be equally likely to be the true value.

Catch is treated as coming from two sources in the mackerel assessments using the modified Fadapt, the directed catch from the commercial and recreational fisheries and the nondirected catch of juvenile mackerels by the shrimp trawl fleet, referred to as bycatch. Both catch-at-age and bycatch-at-age are chosen randomly from lognormal distributions, with mean set equal to the deterministic values and coefficients of variation corresponding to overall sampling levels. This means that both the relative frequency of catch or bycatch-at-age within a year changes with each run and the total catch or bycatch in numbers of fish for the year changes with each run. This approach was chosen based on considerations of how the data are collected and sampled. A common alternative is the multinomial approach, whereby the relative frequency of catch-at-age changes but the total catch is held constant. The multinomial approach was deemed insufficient for the mackerels because of the relatively high degree of uncertainty associated with the total recreational catches.

The modified Fadapt program is used to generate 400 Monte Carlo/bootstrap results of population abundance, directed F, bycatch F and the natural mortality rates by age and year through a minimization routine to produce the best fits to the indices. The choice of 400 runs was originally made based on computation time as well as some experiments to determine how variable the 80% confidence intervals were under different numbers of runs. Larger confidence intervals, for example 95%, would require more runs, on the order of 1000, to generate the same level of stability in the results due to the difficulty in estimating the tails of empirical distributions.

The results from each run are used to initialize a projection of fixed catches, F strategies, or both, into the future. Variability in the projections is derived from the future recruitment values and the future natural mortality rates at age and year. The M values are chosen in the same manner as in the modified Fadapt program. The future recruitment values are not chosen from a stock recruitment relationship because the time series for the mackerels and the regression range with each time series are too small to allow for robust estimation of a stock recruitment curve. Instead, future recruitment is generated from a lognormal distribution with mean and variance set at the values from the historical recruitments estimated by Fadapt for that run. Thus, any trends in recruitment are not modeled, the future recruitment is assumed to be a constant value with noise. This assumption is not particularly important for short-term projections used to generate advice on acceptable biological catches (ABC) because most of the exploitable fish are already in the population, but becomes important in medium and long term projections.

The mackerel assessments are conducted annually such that the following year's total acceptable catch (TAC), or quota, can be chosen by the management councils from a distribution of acceptable biological catches provided by the stock assessment panel. However, the most recent catch-at-age data available corresponds to either one or two years ago. The total catch in weight for the year or years between the last year in the Fadapt results and the quota setting year can be estimated and used to project the population forward to the quota generating year. The directed F patterns are averaged over the most recent three to five years in the Fadapt results for each run and rescaled to a maximum of one to produce the overall selectivity pattern. This overall selectivity pattern is used in conjunction with a directed F

multiplier to produce F at age. The directed F multiplier is searched until the resulting yield at age from the F at age, given a population abundance at age and a bycatch F at age for that run, when summed matches the estimated total catch for that year. The population abundance is then projected forward under this directed F at age and bycatch F at age. The future bycatch F at age is generated as an average over the most recent three to five years for that run multiplied by any bycatch reduction due to the implementation of bycatch reduction devices in the shrimp trawl fishery. This process is repeated until the population abundance at age at the start of the quota setting year is generated for a run.

The projection methods differ for the quota setting year, and any further projection years, relative to the years between the end of the VPA and the current year just described. Instead of using a single total selectivity pattern for the directed fishery, two selectivity patterns are created, one for the commercial sector and the other for the recreational sector. These selectivity patterns are generated from the deterministic partial catches-at-age by sector and the total directed F at age over the most recent three to five years in the Fadapt run. These selectivity patterns have two associated F multipliers that are used to produce the commercial and recreational F at age. These F multipliers are searched until both the desired static spawning potential ratio (SPR) is achieved and the proportion of total catch in weight due to the recreational sector matches the allocation for the recreational sector (Legault 1998), which range from 50 to 64% for the mackerels depending on species and stock. The total F to generate a 40% SPR is used as a proxy for the F at maximum sustainable yield, but this total F can be created in many ways when there are two selectivity patterns and two F multipliers, thus the requirement to also match the recreational allocation. The bycatch-at-age is treated in the same manner as the years when total catch in weight is matched. When the F multipliers are found that both generate the desired static SPR and give the correct portion of total catch in weight to the recreational fishery for that run, the total catch in weight becomes the acceptable biological catch for that run. Repeating this procedure for all 400 runs produces the distribution of acceptable biological catch.

Obviously, the results from all 400 Fadapt runs and projections cannot be presented individually. Instead, the distributions of results are used to generate confidence intervals for relevant stock measures using the percentile method (Efron and Tibshirani 1993). The percentile method is the easiest way to produce confidence intervals from mixed Monte Carlo/bootstrap analyses. Results from the Fadapt analyses that are presented include the trend in recruitment, spawning stock, average F, and transitional SPR with 80% confidence intervals displayed with the median for each year. The acceptable biological catch is presented as a cumulative frequency distribution truncated at the upper 84th percentile. The 16th and 84th percentiles were chosen by the stock assessment panel historically as the range of acceptable biological catch that is presented to the management councils because it corresponds roughly to one standard deviation from the mean in a normal distribution. The distributions of ABC are rarely normally distributed, however, there is often a long tail of large values that are not shown due to the truncation. It was thought that these tails are not well estimated and thus should not be presented. These plots of confidence intervals and cumulative frequency distributions were well received by the management council and presented the uncertainty estimated in the results in an easy to understand manner.

Recent changes to the National Standard Guidelines for fishery management plans require the use of control rules. This change is well suited to the mixed Monte Carlo/bootstrap approach used for the mackerels. The technical guidelines default control rule (Restrepo et al. 1998) was selected for the mackerels. It consists of a constant F strategy that is designed to recover overfished stocks to levels corresponding to the biomass at maximum sustainable yield. Since stock recruitment curves cannot be estimated for the mackerels, a two line model is used. This two line model is determined by constant recruitment at the average of observed recruitment over some recent years for high stock sizes. When stock size falls below the average of the five lowest observed stock sizes, recruitment decreases linearly to the origin. The F at maximum sustainable yield is approximated by the F that generates a 30% static SPR while the F at optimum yield is approximated by the F that generates a 40% static SPR. The biomass at MSY (Bmsy) is approximated by the product of the constant recruitment from the two line model and the spawning stock per recruit that results from application of F30%SPR in equilibrium. Following the recommendations in the technical guidelines, the minimum stock size threshold (MSST) is defined as $(1\text{-}M)$*Bmsy proxy.

All of the values used to determine the control rules change over Fadapt runs, but can be computed

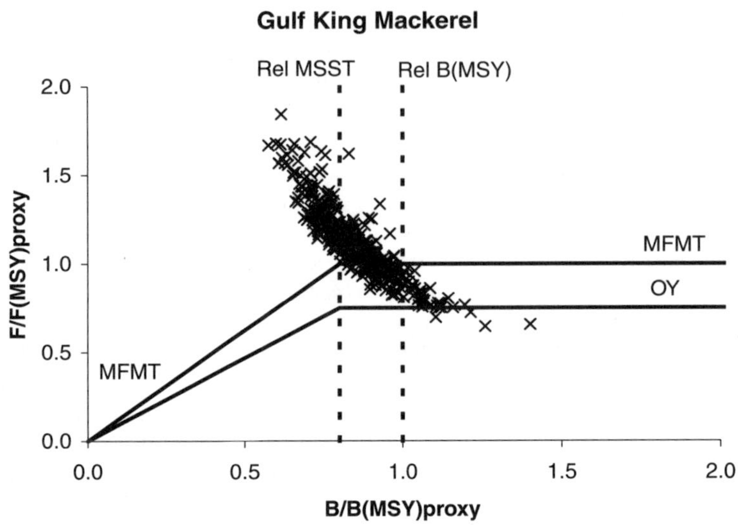

FIGURE 1. Example of uncertainty in current stock status relative to control rule for Gulf of Mexico king mackerel. Each of the 400 X's is determined by a Monte Carlo/bootstrap Fadapt run. MFMT = maximum fishing mortality threshold, OY = optimum yield, Rel MSST = relative minimum spawning stock threshold, Rel B(MSY) = relative biomass at maximum sustainable yield.

for each run. The status of the stock can be measured relative to the control rule by using ratios of current to MSY values for F and spawning stock. The 400 sets of ratios can then be plotted on a single graph with the control rule to demonstrate explicitly the uncertainty associated with current stock status (Figure 1, from Legault 1999a). The question that has not been answered to date is: How many of the 400 points have to be within an overfished or overfishing region to classify the stock as such? This important question of how to interpret the uncertainty is a management policy decision that will need to be answered soon in order for scientists to provide suitable advice.

Discussion

The Monte Carlo/bootstrap methods have been thrust into a frequentist versus Bayesian debate (Wade 1999). An important focus of the debate has been the way in which nuisance parameters are treated. Frequentist approaches usually perform poorly with nuisance parameters that are not updated by the data, which is a reason why nuisance parameters (e.g., natural mortality rate) are usually fixed. Bayesian approaches allow for more flexibility (and more elegance) in dealing with nuisance parameters, although it can be difficult to define noninformative priors for such analyses (Punt and Hilborn 1997). Poole et al. (1999) criticized the Monte Carlo approach as a bad method for dealing with the frequentist problem and gave examples of cases when the approach provides misleading results (e.g., when the data included in the objective function of the estimation procedure provide information on the parameters that are assigned prior distributions but are not updated by the procedure). Wade (1999) reviewed the uses of frequentist, Bayesian and likelihood methods in fitting population models to data. He referred to the Monte Carlo approach as an *ad hoc* method, which "may provide a pragmatic solution to this [frequentist] problem, but statistically this represents a method that is neither frequentist or Bayesian, but a hybrid between the two." Indeed, a prime motivation for applying Monte Carlo/bootstrap methods with mackerels was one of pragmatism (as noted above). We agree with Wade's (1999) comments: "numerical integration techniques and fast computers have made Bayesian methods practical for many applications. This was likely one reason that Bayesian methods were not used more often in the past." While we agree with Poole et al.'s (1999) criticism that the Monte Carlo/bootstrap methods are not statistically rigorous, we are convinced that these methods remain useful, especially when the linkages between likelihoods of data given parameters and priors are not strong. Perhaps Monte Carlo/bootstrap methods should be viewed just as what they are: one of many tools available to the stock assessment practitioner.

Clearly, historical and geographical circumstances led to the inclusion of risk and uncertainty into the fisheries management process at an early stage for the mackerels of the southeast United States. Short time series and incomplete databases precluded having much confidence in point estimates over other alternatives for parameter values. Additionally, the particular biological events emphasized to managers the importance of risks and costs of both negative and positive outcomes. The data, the scientific methods that utilize the data and management processes have all evolved to encompass these realities. This has resulted in the gradual recovery of U.S. Gulf king mackerel and relatively stable fisheries for the other mackerel stocks (MSAP 1999). The inclusion of uncertainty estimates has allowed managers to properly evaluate tradeoffs of risks to achieve a balanced approach to management.

The benefit of this mixed Monte Carlo/bootstrap simulation approach to assessment and projection has been demonstrated for other stocks as well, for example the swordfish and cod fisheries (Restrepo et al. 1992). These authors used simulation to quantify pragmatically how levels of uncertainty in the data related to uncertainty in the management advice and therefore which research areas provided the most benefit from increased funding. The mixed Monte Carlo/bootstrap simulation approach is well suited for such analyses due to the explicit control over the probability density functions for uncertain input parameters.

The ability to choose explicitly the pdfs for input parameters can also create unexpected results from this approach. One example is the use of a symmetric distribution for uncertainty in the natural mortality rate producing a nonsymmetric distribution of acceptable biological catches. This nonlinear relationship is common for many fisheries parameters and can lead to difficulty in using bias corrected confidence intervals (Legault 1999c). Bias corrected confidence intervals are recommended by Efron and Tibshirani (1993), but cannot be used directly with the mixed Monte Carlo/bootstrap simulation approach described here. Instead, the process would need to be nested, with bootstrapping conducted within a Monte Carlo simulation and the bias correction applied to just the bootstrapping. The bias corrected results from all Monte Carlo simulations would be combined at the end to produce the overall distribution of results. This nesting is required because the assumptions of known pdfs for input parameters in Monte Carlo simulation conflict with the idea of correcting results for differences between the mean and median in bootstrapping due to the nonlinear relationships between input parameters and results. This approach will be further developed in the future.

Another area that requires further work for the mackerels is the presentation of consequences for choosing a specific value from a distribution of possibilities. In the past, the management councils have treated each value within the 16–84% confidence interval for ABC as equally likely and have not necessarily considered the consequences to the fish population that arise from choosing the 16th versus the 84th percentile value. Easy to understand results demonstrating these consequences have not been routinely presented to the councils. This will have to change in the future. The new control rule requirements provide a means to make these consequences easily understood using the mixed Monte Carlo/bootstrap simulation approach. By projecting a given value of ABC for all runs and plotting the resulting F and spawning stock ratios on the control rule plots (as in Figure 1), the uncertainty in changes of stock status associated with any value within the ABC range can be immediately seen.

References

Efron, B., and R. J. Tibshirani. 1993. An introduction to the bootstrap. Chapman and Hall, New York.

Gavaris, S. 1988. An adaptive framework for the estimation of population size. CAFSAC (Canadian Atlantic Fisheries Scientific Advisory Committee) Research Document 88/29, Dartmouth, Nova Scotia.

Legault, C. M. 1998. Linking total catch quotas and allocation schemes. North American Journal of Fisheries Management 18:454–457.

Legault, C. M., N. Cummings, and P. Phares. 1998. Stock assessment analyses on Atlantic migratory group king mackerel, Gulf of Mexico migratory group king mackerel, Atlantic migratory group Spanish mackerel, and Gulf of Mexico migratory group Spanish mackerel. NMFS-SEFSC Sustainable Fisheries Division Contribution MIA-97/98-15. Mackerel Stock Assessment Panel Contribution MSAP/98/09.

Legault, C. M. 1999a. Updated Projections for King and Spanish Mackerel in the Gulf of Mexico and Atlantic Ocean. Sustainable Fisheries Division Contribution SFD-98/99-49. Mackerel Stock Assessment Panel Contribution MSAP/99/01.

Legault, C. M. 1999b. Status review of king mackerel in the Gulf of Mexico (Feature Article 3). Pages 81–86 *in* Our Living Oceans. Report on the status of U.S. living marine resources. NOAA Tech. Mem. NMFS-F/SPO-41.

Legault, C. M. 1999c. Simulation Study of Percentile and Bias Corrected Percentile Confidence Intervals for Allowable Biological Catch. Sustainable Fisheries Division Contribution SFD-98/99-48. Mackerel Stock Assessment Panel Contribution MSAP/99/02.

MRFSS (Marine Recreational Fishery Statistics Survey). 1991. Marine Recreational Fishery Statistics Survey, Atlantic and Gulf Coasts, 1987–89. National Oceanographic and Atmospheric Administration, National Marine Fisheries Service. Current Fisheries Statistics Number 8904.

MSAP (Mackerel Stock Assessment Panel). 1999. 1999 Report of the Mackerel Stock Assessment Panel. March 29–April 4, 1999. Gulf of Mexico and South Atlantic Fishery Management Councils.

McAllister, M. K., and J. N. Ianelli. 1997. Bayesian stock assessment using catch-age data and the sampling-importance resampling algorithm. Canadian Journal of Fisheries and Aquatic Science 54:284–300.

Nichols, S. 1985. Assessment of king mackerel via virtual population analysis. National Marine Fisheries Service, Southeast Fisheries Center, ML1-85-19, Miami, Florida.

Parrack, M. L. 1986. A method for analyzing catches and abundance indices from a fishery. Int. Comm. Cons. Atl. Tunas (ICCAT) Coll. Volume Sci. Pap. 24:209–221.

Poole, D., G. H. Givens, and A. E. Raftery. 1999. On a proposed stock assessment method and its application at the International Whaling Commission. Fish. Bull. U.S. 97: 144–152.

Powers, J. E., editor. 1985. Report of the Mackerel Stock Assessment Panel Meeting, August 14–15, 1985. Gulf of Mexico and South Atlantic Fishery Management Councils. National Marine Fisheries Service, Southeast Fisheries Center, ML1-85-23, Miami, Florida.

Powers, J. E. 1986. Discussion notes on king mackerel ABC risks. Report to Gulf and South Atlantic Mackerel Stock Assessment Panel. Southeast Fisheries Science Center, National Marine Fisheries Service, Miami, Florida.

Powers, J. E., editor. 1989. Report of the Mackerel Stock Assessment Panel Meeting, April 4–7, 1989. Gulf of Mexico and South Atlantic Fishery Management Councils. National Marine Fisheries Service, Southeast Fisheries Center, CRD-89/90-12, Miami, Florida.

Powers, J. E. 1995. Benchmark requirements for recovering fish stocks. N. American J. Fish. Man. 16:495–504.

Powers, J. E., and P. Eldridge. 1983. A preliminary assessment of king mackerel resources of the Southeast United States. National Marine Fisheries Service, Southeast Fisheries Center. SAW/84/GCP/1.

Powers, J. E., and P. Eldridge. 1984. Assessment of Gulf of Mexico and South Atlantic king mackerel. National Marine Fisheries Service, Southeast Fisheries Center. SAW/84/GCP/2.

Powers, J. E., and V. R. Restrepo. 1993. Evaluation of stock assessment research for Gulf of Mexico king mackerel: benefits and costs to management. N. American J. Fish. Man. 13:15–26.

Punt, A., and D. B. Butterworth. 1997. Assessments of the Bering-Chukchi-Beaufort Seas stock of bowhead whales (*Balaena mysticus*) using maximum likelihood and Bayesian methods. Rep. Int. Whal. Commn. 47:603–618.

Punt, A., and R. Hilborn. 1997. Fisheries stock assessment and decision analysis: the Bayesian approach. Rev. Fish Biol. Fish. 7:35–63.

Raiffa, H. 1970. Decision analysis. Addison-Wesley, Reading, Massachusetts.

Restrepo, V. R. 1996. Fadapt Version 3.0: a users guide. University of Miami, Florida.

Restrepo, V. R., J. E. Powers, and S. C. Turner. 1991. Incorporating uncertainty in VPA results via simulation. Int. Com. Cons. Atl. Tunas Coll. Volume Sci. Pap. 39:540–553.

Restrepo, V. R., G. G. Thompson, P. M. Mace, W. L. Gabriel, L. L. Low, A. D. MacCall, R. D. Methot, J. E. Powers, B. L. Taylor, P. R. Wade, and J. F. Witzig. 1998. Technical Guidance on the use of Precautionary Approaches to implementing National Standard 1 of the Magnuson-Stevens Fishery Conservation and Management Act. NOAA Tech. Mem. NMFS-F/SPO-31.

Restrepo, V. R., J. M. Hoenig, J. E. Powers, J. W. Baird, and S. C. Turner. 1992. A simple simulation approach to risk and cost analysis, with applications to swordfish and cod fisheries. Fish. Bull. U.S. 90:736–748.

Wade, P. R. In Press. A comparison of statistical methods for fitting population models to data. Pages 249–270 *in* L. Mcdonald et al., editors. Marine mammal survey and assessment methods. A. A. Balkema, Rotterdam, The Netherlands.

American Fisheries Society Symposium 27:45–62, 2002

Incorporating Uncertainty into Stock Assessments: A Case Study of Atlantic Striped Bass

GARY R. SHEPHERD

National Marine Fisheries Service, Northeast Fisheries Science Center, 166 Water Street Woods Hole, Massachusetts 02543, USA

Abstract.—A stock assessment of Atlantic striped bass *Morone saxatilis* was presented to illustrate potential sources of uncertainty in application of an age-based population model. Erroneous conclusions in stock assessment can result from incorrect model selection, input data that are not representative of the target population, and improper configuration of the selected model. Influence of incorrect input data and model configuration was investigated using striped bass catch-at-age data analyzed with a tuned virtual population analysis model (ADAPT VPA). Variations in model configurations were explored in addition to sensitivity to input parameters such as natural mortality. Violation of the assumption of constant natural mortality-at-age had a significant influence on the resulting estimates of F and stock size. Discard losses, particularly from the commercial fishery, were the largest source of uncertainty in the catch-at-age. Uncertainty due to process error in the VPA model was characterized by bootstrap realizations of the nonlinear least-squares estimates of fishing mortality. The implications associated with fishing at various Fs were also examined using a stochastic projection model. A comparison of fishing mortality estimates derived from two independent models, an age-structured population model and a tag-recovery model, indicated that both methods produced equivalent results. Evaluation of the striped bass stock assessment demonstrates that uncertainty could result from a variety of sources but this variability was only partially captured within the model framework. Understanding the possible sources of uncertainty and implications in interpreting model results should benefit the analyst in providing assessment advice to managers.

Introduction

The decline in marine fish stocks in the United States over the last decade has been well documented (Clark 1998; NMFS 1999). Despite increasing efforts to adopt more stringent management measures, relatively few examples exist where marine fish populations have recovered to predecline levels (Mace 1997). The Atlantic striped bass *Morone saxatilis* is one exception. Through aggressive management of commercial and recreational fisheries, improved habitat and good fortune, striped bass abundance has returned to levels comparable to the predecline period of the 1960s. By 1995, the Atlantic States Marine Fisheries Commission (ASMFC) Striped Bass Management Board declared, based on the recommendation of the ASMFC Striped Bass Technical Committee, that spawning stock biomass was at high enough levels that striped bass should be considered recovered. Restoration of striped bass, particularly the Chesapeake Bay stock, was an important accomplishment for fishery managers. Maintaining stock biomass at a self-sustaining level in the face of increasing fishing pressure on the resource may be even more challenging. This formidable task is dependent on adequate stock assessment information and an understanding by managers of uncertainty in those results as an element of decision making.

Coastal migratory Atlantic striped bass originate in the Chesapeake Bay and its tributaries, the Hudson River and the Delaware River. The stock from the Roanoke River/Albermarle Sound is believed to contribute insignificant numbers to the coastal migratory group at the present time. Despite many attempts to develop methods for stock delineation within the mixed coastal stock, stock identification techniques have not provided a high degree of correct categorization (Waldman and Fabrizio 1994; Waldman et al. 1997; Wirgin et al. 1997). Consequently, the fishery management plan and stock assessment considered the entire mixed stock migratory group as a single unit stock.

Striped bass have been subjected to management regulations since the time of the early settlers in New England. Concerns about overfishing along the New England coast were expressed as early as the 1700s (Bigelow and Schroeder 1953). More recently, declining striped bass abundance in the 1930s was the impetus for development of the ASMFC in 1942. The commission was organized to provide a compact among Atlantic coastal states to cooperatively manage coastal migratory stocks. The collapse of striped bass stocks in Chesapeake Bay during the late 1970s marked the beginning of state and federal efforts to restore the stocks to sustainable levels. These efforts culminated in the closure of striped bass fisheries in Maryland in 1985 and most other jurisdictions by 1989 (Richards and Deuel 1987). However, the fishery management plan (FMP) was structured with a regulatory mechanism that allowed a reopening of the fishery if above average recruitment occurred. The time-series average in the Maryland juvenile index (the mean number per tow of age-0 striped bass from a seine survey of Maryland estuaries begun in 1954) was exceeded in 1989, which led to a reopening of the fishery in 1990. However, strict management regulations were implemented that maintained large minimum sizes and a target instantaneous fishing mortality rate (F) at one half F_{msy} (0.38) (Field 1998). Populations of striped bass continued to rebuild and in 1993 and 1996 produced the two largest year classes in the time-series of juvenile indices.

A virtual (sequential) population analysis (VPA) of Atlantic striped bass was developed in 1997 as part of the ongoing work of the ASMFC Striped Bass Technical Committee (Shepherd and Lazar 1998). Previously, conclusions about the status of the stock were based on tagging conducted by state agencies and coordinated through the U.S. Fish and Wildlife Service (Young-Dubovsky et al. 1996). Although the tagging results provided useful information on survival rates, the type of information resulting from tagging studies was not sufficient to support various management requirements. Analysis of catch-at-age data with a VPA integrated catch, length and age data into year specific mortality and abundance estimates, beginning in 1982. These results provided the necessary information for managers to begin quota-based regulations of the entire coastal stock, as recommended in the FMP (ASMFC 1995).

Development of the VPA began with a compilation of available aging data, catch data and indices of abundance for the area between Maine and North Carolina. The assembled catch-at-age matrix was evaluated using the ADAPT method for virtual population analysis (Conser and Powers 1990; Gavaris 1988). The results of the analysis were reviewed and accepted during the 26th Northeast Fisheries Science Center Stock Assessment Workshop (NEFSC 1998). More detailed results are contained in the ASMFC Striped Bass Source document and the proceedings of the 26th Stock Assessment Workshop (Shepherd and Lazar 1998; NEFSC 1998).

The level of uncertainty in stock assessments depends on model selection and the quality of input data (Hilborn 1997). There are a variety of age-structured population models appropriate for assessing fish stocks, each with its own inherent suite of biases and assumptions (Megrey 1989). The intent of this paper is to explore the sources of uncertainty in the striped bass assessment associated with development of input data and model configuration within the ADAPT VPA framework.

Data Input

Commercial catch

Accuracy of the catch data were the first consideration in analysis of the catch-at-age matrix. Striped bass commercial landings data were least reliable during the period 1982–1990 when landings were collected by the National Marine Fisheries Service (NMFS) from records of fish dealers. Commercial striped bass fisheries were primarily small day-boat operations and were more likely to be missed or under-reported in the NMFS reporting system compared with landings in major ports. Since 1990, state agencies were responsible for monitoring striped bass landings and the methods more effectively sampled small dealers. In most states, licensed commercial fishermen were issued a limited number of locking jaw tags and these tags were required on each fish sold. This system provided essentially a census of the total number of striped bass landed. In states without tagging systems, reporting by licensed fishermen was required and strictly enforced. Landings were reported as numbers of fish and eliminated the error associated with converting landed weight to number using length frequency data and length-weight conversions. The strict regulations, limited quotas, and census-type reporting methods for landings increased the likelihood that commercial striped bass landings were relatively precise components of the catch matrix in recent years.

Commercial discards were the area of greatest uncertainty in the catch-at-age matrix. Limited quotas and increasing striped bass abundance raised the likelihood that a variety of fisheries would catch but discard striped bass. Sea sampling data were limited and expansion of sea sample data to total discard estimates required information on effort or some other measure across all fleets that may intercept striped bass. The problem was compounded by the large and diverse set of fleets that potentially catch striped bass owing to the dynamics of striped bass migrations. An alternative source of information available since 1987 was tag return data that contained information on the disposition of the fish. Total commercial bycatch discards were estimated using the ratio of commercial to recreational discards determined from tag returns. The ratio was expanded using total recreational discards estimated by the Marine Recreational Fisheries Statistics Survey (MRFSS) program. Tag reporting rate was assumed equivalent, although it is likely commercial reporting was lower. Total commercial discards were extrapolated for the period prior to the tagging programs using the trend in ratios from adjacent years. The proportion of discards by gear type was derived from tag recoveries, and gear specific survival rates were applied to discarded fish. Total estimated commercial discards followed expected trends over the time series. In the years with small minimum sizes and no quotas, discards were low but as the regulations were tightened and stock abundance increased, total discards also increased. Since 1994, higher discards followed years with large recruiting year classes. Several state marine fishery agencies had previously attempted to estimate discards in local fisheries using a combination of sea sample data and survey data. The sum of these estimates was comparable to the expanded tagging estimates. There were no expectations that the discard estimates would have a high degree of accuracy. However, since commercial bycatch was known to exist, an effort was made to incorporate that source of mortality to reduce a potentially significant source of bias.

Recreational catch

The MRFSS produced estimates of striped bass recreational catch and proportional standard errors since1982. Estimates were derived from expansion of survey results rather than the census approach used for commercial landing estimates. Total recreational landings declined steadily from 1982 to 1989. Despite the consistent trend in the total landing estimates, state specific landings were variable among years. For example, Virginia showed no recreational striped bass landings in 1982, 1983, and 1985 while the neighboring state of Maryland had significant landings. Landings in other year/state combinations, such as New Jersey in 1983, were three to four times greater than adjacent years. Associated with the reopening of fisheries in 1990 was a responsibility by states to collect additional recreational data. States with substantial recreational landings were required under the FMP to achieve a proportional standard error in their MRFSS estimate of 20% or less, which was accomplished by increasing the sample size of interviews. Consequently, total catch estimates became more precise and consistent since 1990. Recreational fisheries in the Hudson River were not included in the MRFSS program and were estimated by surveys conducted by the state of New York. The use of state specific landings rather than regional estimates in the development of a landings-at-age matrix prior to 1990 likely increased the variance in the analysis.

The MRFSS program was also the source for estimates of recreational discards. An assumption in the total discard estimate was that each release represented a unique fish. It was evident from examining tagging data that multiple recaptures of some fish occur during the year. The extent of double counting in the discard estimate is currently being investigated. Field experiments concluded that an average of 8% of released striped bass die due to hooking mortality (Diodati and Richards 1996). Although it has been shown that discard mortality varies by water temperature, terminal gear type, handling time, and a variety of other factors, the average rate (8%) was applied to total annual discard estimates.

Length data

Expansion of catch into length categories required representative samples by time, area and fishing gear. Commercial landings were divided into semiannual periods and length frequencies matched to fisheries by time and area cells whenever possible. Generally, the length data were adequate, with an average of 5% of landed fish measured. Smaller fisheries, such as fyke nets, tended to be poorly represented but since they contributed little to the total catch the consequences were minimal. The opportunity for collecting length frequency data from landings improved after 1995 as commercial quotas increased.

Length data for commercial discards were generally inadequate. Lengths of discarded striped bass in directed and nondirected commercial fisheries were expanded based on limited sea sampling or length frequencies of sub-legal size fish collected in fisheries independent surveys using comparable gear.

The number of length samples from recreational fisheries increased in recent years as states added larger numbers of samples to the intercept portion of the MRFSS survey. Since 1990, the MRFSS data have also been supplemented by information collected by volunteer fishermen. The length distributions of fish collected by volunteer anglers were assumed to be representative of recreational catches in that time and area. Length data to characterize discards were provided by anglers participating in a tag and release program managed by the American Littoral Society (Boreman and Lewis 1987). Since these fish were tagged and released by recreational anglers, they were by definition equivalent to the discarded (B2) fish in the MRFSS. Additional length data of discards were collected by volunteer fishermen from several states. The overall number of lengths available for recreational fisheries far exceeded the commercial data. In 1996, the recreational catch was expanded based on lengths of 11,240 landed and 19,303 discarded striped bass.

Catch at age

Uncertainty in the catch-at-age data are not incorporated into the ADAPT VPA model, however sensitivity to error in the catch data can be evaluated. Bias in commercial landings data are almost always a result of underestimating rather than overestimating the catch. Recreational catch estimates from survey results could be either under- or overestimated. The error in the striped bass catch-at-age matrix was likely most severe in the 1982–1989 period when sampling was not dictated by the FMP. Annual catch estimates for this period were varied ±50% to examine the influence on terminal year estimates of abundance and fishing mortality (Table 1). If catch were underestimated by 50%, the terminal year estimate of fishing mortality would be underestimated by 12% (0.03) and abundance by 12% (4.8 million). In contrast, if catch were overestimated by 50%, the terminal year F would be overestimated by 15% (0.05) and abundance by 13% (5.2 million). Numerous combinations of catch error could be explored but it is clear that catch for the first half of the time series would have to be severely misestimated to significantly impact F and N in the terminal year.

Both the catch data and tuning indices were categorized into ages using age-length keys. Semiannual age keys were compiled and divided into five geographic areas. All ages were determined from scales, which may have introduced error into the analysis. Aging error for scales increase beyond age 12, with scales biased toward underestimation of true age (Secor et al. 1995). However, an evaluation of scale ages using known age fish captured in New York (hatchery fish recovered with coded wire tags) showed a 90% agreement up to age 12 (Vecchio and O'Riordan 1999). A catch-at-age matrix with a 15+ category reduced the impact of aging error in older fish but undoubtedly some element of age error contributed to uncertainty in the assessment results.

The catch-at-age matrix from 1982 to 1997 is provided in Table 2.

Model Development

The ADAPT model framework was chosen for the striped bass assessment because of the flexibility it provided in evaluating a large set of abundance indices. This model assumes error in the indices is greater than in the catch-at-age matrix. Since the tuning indices were from multiple stocks and a variety of sampling designs, the assumption seemed reasonable. Initial configuration of the model included a matrix of 16 years × 15 ages and 108 potential age specific tuning indices. The catch-at-age matrix was restricted to 14 age groups with a 15+ category. During the late 1980s and early 1990s the minimum size in some coastal states was 91 cm (36″). Therefore any truncation of the matrix below age-14 would have incorporated a large percentage of the catch from the coastal recreational fisheries into the plus category, the age group that tends to be most poorly estimated in the model.

Tuning indices

The ADAPT model requires auxiliary abundance data for estimating population size. The striped bass assessment was unusual in the amount of information available, particularly juvenile indices. Each of the three stocks had long time series of surveys designed to measure juvenile abundance. The indices in the Hudson River were strongly correlated to abundance-at-age one measured in surveys of Long Island nursery areas (McKown 1992). Juvenile indices for the Maryland portion of the Chesapeake Bay were validated using relative abundance-at-age in subsequent commercial landings (Goodyear 1985). Juvenile surveys were also available from the Delaware River (Rago et al. 1995) and the Virginia portion of Chesapeake Bay (Rago et al. 1995). Although annual variance estimates were available for all these indices they were not incorporated into the tuning procedure.

Time series of fishery independent surveys designed to measure relative abundance of adult striped bass were available for the Maryland portion of Chesapeake Bay and coastal New York. The Maryland survey used a multiple meshsize gill net and the CPUE was adjusted for selectivity in different panels of the net (Helser et al. 1998; Helser and Waller 1999). The New York survey used a commercial haul seine deployed from beaches in

TABLE 1. Relative effect of changes in striped bass catch for 1982–1989 on estimates of age 1+ abundance (>000) and fishing mortality (age 4–13) from virtual population analysis.

Relative change in catch	Year																
	1982	1983	1984	1985	1986	1987	1988	1989	1990	1991	1992	1993	1994	1995	1996	1997	1998
Percent change in age 1+ abundance (000s)																	
−50%	−37%	−32%	−25%	−16%	−12%	−8%	−6%	−4%	−3%	−4%	−5%	−6%	−8%	−10%	−11%	−12%	−13%
−25%	−18%	−16%	−12%	−8%	−6%	−4%	−3%	−2%	−2%	−2%	−3%	−3%	−4%	−5%	−6%	−6%	−6%
−10%	−7%	−6%	−5%	−3%	−2%	−2%	−1%	−1%	−1%	−1%	−1%	−1%	−2%	−2%	−2%	−2%	−3%
0	5231	6513	7345	8427	9661	11459	13984	17091	22226	24289	25526	28047	37738	38692	37528	42578	40237
+10%	7%	6%	5%	3%	2%	2%	1%	1%	1%	1%	1%	1%	2%	2%	2%	2%	2%
+25%	18%	16%	12%	8%	6%	4%	3%	2%	2%	2%	3%	3%	4%	5%	6%	6%	6%
+50%	37%	32%	25%	16%	12%	9%	7%	5%	4%	5%	6%	7%	8%	9%	11%	11%	12%
Percent change in age 4–13 fishing mortality																	
−50%	−28%	−27%	−31%	−33%	−38%	−39%	−44%	−49%	1%	3%	3%	4%	6%	7%	10%	15%	
−25%	−12%	−11%	−13%	−13%	−18%	−18%	−22%	−22%	0%	1%	1%	2%	3%	3%	5%	8%	
−10%	−4%	−4%	−6%	−4%	−6%	−7%	−9%	−9%	0%	0%	1%	1%	2%	1%	1%	3%	
0	0.24	0.21	0.12	0.12	0.12	0.07	0.13	0.09	0.17	0.20	0.17	0.25	0.22	0.24	0.28	0.32	
+10%	3%	3%	3%	5%	6%	7%	9%	8%	−1%	−2%	−1%	0%	0%	−2%	−2%	−3%	
+25%	7%	9%	8%	14%	14%	18%	18%	21%	−1%	−2%	−2%	−2%	−2%	−4%	−5%	−6%	
+50%	14%	15%	18%	25%	26%	29%	36%	43%	−2%	−3%	−4%	−4%	−5%	−7%	−10%	−12%	

TABLE 2. Striped bass catch-at-age (000s); 1982–1997.

	Age															
Year	1	2	3	4	5	6	7	8	9	10	11	12	13	14	15+	Total
1982	1.8	105.6	256.8	221.0	58.5	19.2	24.3	16.8	11.7	10.6	11.0	13.7	3.4	4.1	8.1	2300.4
1983	3.6	110.4	178.4	193.3	150.3	39.4	18.7	4.1	2.9	3.7	4.6	5.7	4.9	4.1	4.6	2185.8
1984	5.6	543.1	303.0	82.7	60.6	51.9	18.4	4.7	2.1	2.1	0.7	0.3	2.2	4.3	4.7	3259.2
1985	1.3	72.6	102.1	40.6	58.8	43.2	43.6	17.3	6.4	3.4	1.0	0.8	0.5	0.9	8.9	1204.4
1986	11.4	21.1	64.0	133.0	50.0	32.1	20.4	24.1	9.2	5.3	3.4	1.6	0.9	2.5	6.7	1157.1
1987	1.4	11.1	38.0	51.8	67.6	25.2	13.3	6.6	6.5	3.0	1.5	2.0	3.4	2.1	7.7	723.2
1988	2.6	30.8	42.5	63.9	106.3	97.4	40.6	24.6	14.0	5.8	3.7	3.2	2.4	3.0	4.1	1334.6
1989	0.8	36.8	80.7	69.0	105.7	95.9	46.1	21.2	10.5	3.8	3.3	2.0	1.9	1.6	5.4	1454.4
1990	2.7	54.4	135.9	203.4	181.6	174.3	111.9	75.6	24.3	8.6	6.0	4.0	4.9	4.2	7.6	2998.2
1991	1.9	77.9	152.7	217.4	165.7	103.4	95.3	87.6	62.9	25.7	14.8	3.0	2.8	3.5	17.7	3097.3
1992	3.3	51.1	216.7	201.3	188.2	113.8	65.7	73.6	64.3	49.4	10.2	4.5	1.9	5.5	10.4	3180.0
1993	0.3	76.1	197.6	344.0	299.9	194.0	90.1	71.5	89.6	83.2	45.0	10.2	5.0	1.3	13.0	4562.1
1994	5.7	146.1	350.5	291.7	368.8	233.0	135.8	87.1	100.4	81.4	36.2	22.4	3.4	1.5	9.8	5621.3
1995	3.7	414.7	446.9	438.1	385.7	461.9	201.2	186.1	147.9	86.4	50.8	16.1	9.3	1.9	3.3	8561.9
1996	0.5	98.8	659.5	666.4	552.0	476.9	457.2	217.7	143.3	71.7	44.2	48.3	13.2	4.7	2.6	10371.3
1997	2.5	287.3	487.3	849.5	618.7	598.6	415.6	382.6	207.9	125.7	61.8	30.9	13.0	7.9	5.1	12283.6

southeastern Long Island. A source of uncertainty in this and all surveys is whether the survey gear captures fish proportional to stock size. It has been suggested that haul seine catches were biased toward small fish, but there were no data available to evaluate this potential bias. Other tuning indices in the model were derived from fishery dependent data. These included Hudson River shad gill-net CPUE, hook-and-line CPUE from Massachusetts commercial fisheries, and CPUE of Connecticut volunteer recreational fishermen.

Selection of the final tuning indices was based on the diagnostics from the VPA including CVs of N, mean square residuals, and partial variance estimates of indices. Indices with relatively high CVs were rejected. Time trended patterns in the residuals from stock specific indices were not considered in the diagnostics because of the mixed stock nature of the catch data. If a residual pattern existed, it may have been the result of changes in the contribution of that stock component over time relative to combined stock abundance. A total of 57 age-specific indices were chosen for inclusion into the model (Table 3; Appendix I).

Model configuration

The ADAPT software used in the striped bass assessment allowed evaluation of various model configurations that had important implications for the results. Some important choices included the use of constant or age-specific natural mortality (M); the assumption of a flat-topped or dome-shaped partial recruitment (PR) vector; use of reweighting algorithms for tuning indices; whether indices are tuned to population size at 1 January or 1 July; the

TABLE 3. Indices used in tuning the striped bass virtual population analysis. Age-0 indices advanced 1 year and 1 age in tuning procedure.

	Age								
Data Source	0	1	2	3	4	5	6	7	8 to 15+
MA commercial CPUE									X
CT recreational CPUE			X	X	X	X	X	X	X
Hudson River seine survey	X								
Hudson shad fishery CPUE							X	X	X
Long Island seine survey			X						
NY ocean haul seine survey						X	X	X	X
New Jersey seine survey		X							
Delaware seine survey	X								
MD spawning stock survey			X	X	X	X	X	X	X
Maryland seine survey	X	X							
Virginia seine survey	X								

TABLE 4. Effect of variable age-1 natural mortality (M) on striped bass virtual population analysis abundance at age estimates (000s) for 1998.

	M			
Age	0.2	0.3	0.45	0.6
1	7767	8583	9972	11585
2	11605	11605	11605	11605
3	4833	4833	4833	4833
4	4567	4567	4567	4567
5	6633	6633	6633	6633
6	1843	1843	1843	1843
7	735	735	735	735
8	773	773	773	773
9	847	847	847	847
10	453	453	453	453
11	228	228	228	228
12	140	140	140	140
13	78	78	78	78
14	81	81	81	81
15+	31	31	31	31

TABLE 5. Effect of variable age-1 natural mortality (M) on striped bass virtual population analysis estimate of fishing mortality at age for 1997.

	M			
Age	0.2	0.3	0.45	0.6
1	0.0	0.0	0.0	0.0
2	0.05	0.05	0.05	0.05
3	0.09	0.09	0.09	0.09
4	0.11	0.11	0.11	0.11
5	0.27	0.27	0.27	0.27
6	0.56	0.56	0.56	0.56
7	0.40	0.40	0.40	0.40
8	0.35	0.35	0.35	0.35
9	0.35	0.35	0.35	0.35
11	0.41	0.41	0.41	0.41
11	0.34	0.34	0.34	0.34
12	0.31	0.31	0.31	0.31
13	0.14	0.14	0.14	0.14
14	0.33	0.33	0.33	0.33
15+	0.33	0.33	0.33	0.33

criteria used to identify age at full recruitment to the fishery; and the calculation of F using a weighted or unweighted average across ages groups.

As with all assessments, natural mortality was one of the more elusive parameters in the model. The null hypothesis was an M that resulted in approximately 5% of a cohort surviving to maximum age. Maximum age was defined as 20 years, although some evidence suggested a higher maximum age was possible (Secor et al. 1995). Striped bass were assessed using a constant M of 0.15. An alternative hypothesis examined was the use of age specific values of M. Some evidence from field studies suggested natural mortality on juvenile and age one fish could be higher due to predation (Buckel et al. 1999). Beyond that age, the size of striped bass likely reduced their vulnerability to predation. Increasing the value of M on age-1 increased the abundance estimates of recruits but had no effect on estimates of fully recruited fishing mortality (ages 4–13) (Tables 4–6).

The consequences of misspecifying M in a sequential population analysis have been well documented (Mertz and Meyers 1997; Hilden 1988; Lapointe et al. 1989; Sims 1984). The impact of alternative Ms of 0.1, 0.2, 0.25, or 0.3 on the striped bass VPA results were examined. An M of 0.1 resulted in a decreased abundance estimate and an increased F in the terminal year while higher Ms resulted in higher estimates of N and lower Fs (Figures 1 and 2). The bias created by misspecification of M also increases with greater variability in F (Mertz and Meyers 1997). A change in M caused a similar directional change in the biological reference point (i.e., higher M resulted in higher value for F_{msy}). Therefore, if the constant M used in the assessment is greater than the "true" value there may be an increased risk of exceeding the biological reference point.

Identification of the age at full recruitment to the fishery has important implications in defining fully recruited F and development of biological reference points. One approach to calculation of partial recruitment to the fishery is the relationship between the highest F among ages and age specific Fs. For

TABLE 6. Effect of variable age-1 natural mortality (M) on striped bass virtual population analysis abundance at age estimates (000s) for 1982–1998.

	M			
Year	0.2	0.3	0.45	0.6
1982	1452	1605	1864	2166
1983	2888	3192	3708	4307
1984	2534	2800	3253	3779
1985	3267	3611	4195	4874
1986	2913	3219	3739	4343
1987	3626	4007	4656	5409
1988	4646	5134	5965	6930
1989	5727	6330	7354	8544
1990	8388	9270	10770	12513
1991	6311	6974	8103	9414
1992	5894	6513	7567	8791
1993	7440	8223	9553	11099
1994	15777	17436	20257	23534
1995	8409	9293	10797	12543
1996	7237	7998	9292	10795
1997	14178	15668	18203	21148
1998	7767	8583	9972	11585

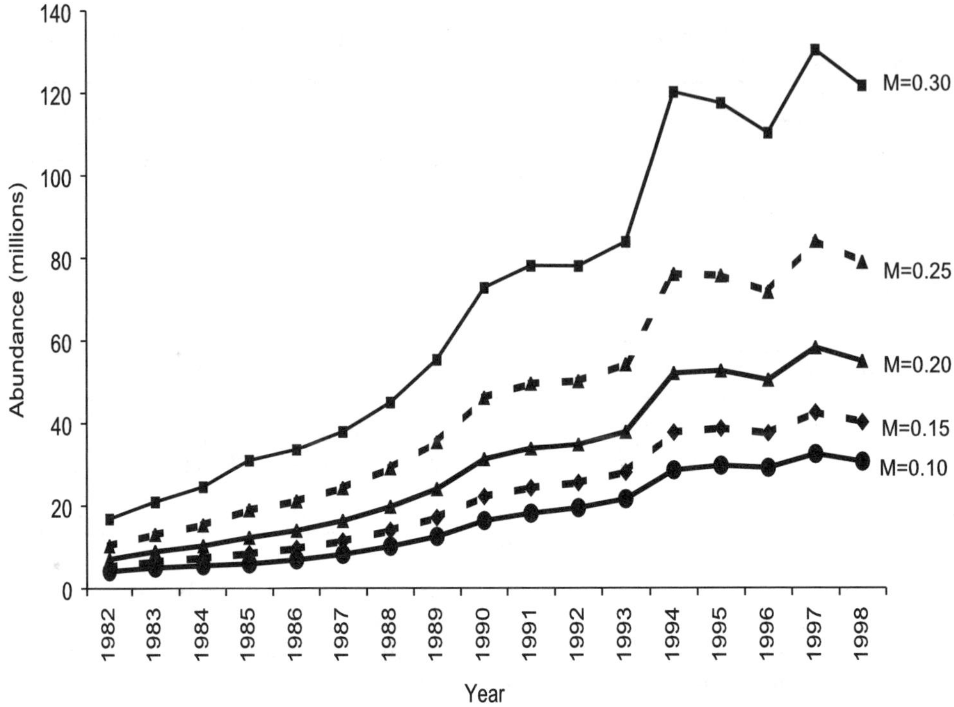

FIGURE 1. The impact of varying levels of instantaneous natural mortality (M) on virtual population analysis estimates of striped bass abundance (millions, age 1+).

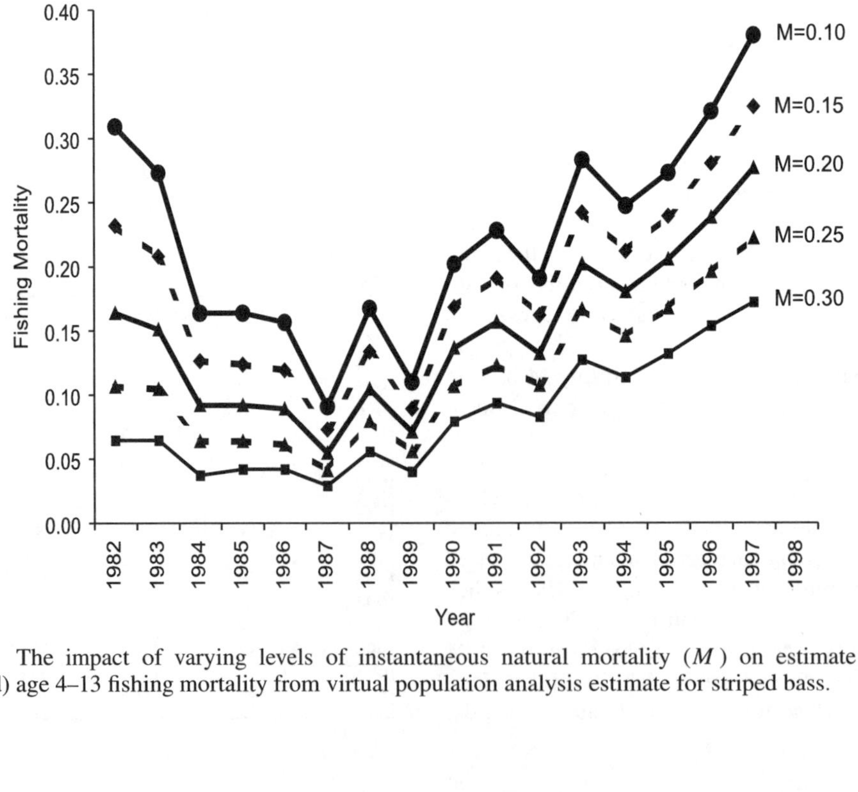

FIGURE 2. The impact of varying levels of instantaneous natural mortality (M) on estimate of average (unweighted) age 4–13 fishing mortality from virtual population analysis estimate for striped bass.

instance, if F at age 10 were the highest at 0.3 and F at age 4 was 0.15, then the partial recruitment at age 10 would be 1.0 and age 4 would be 0.5. An alternative approach was to use the age of maximum catch in the annual catch vectors. Depending on year class strength the maximum catch could occur at a significantly different age than the model calculation. For striped bass, the age at maximum catch was chosen to characterize age at full recruitment. The model calculations of partial recruitment reflected changes in the fishery regulations throughout the time series but were not used in calculation. The selection of age at full recruitment particularly affected calculation of fully recruited F in those years where the maximum catch occurred at ages 3 or 4 while the model PR equaled 1.0 at ages 8–10.

The shape of the partial recruitment curve also has important implications in the estimate of biological reference points. In the striped bass assessment a flat-topped PR at age vector was used in calculation of F_{msy}. The ADAPT VPA model does not use the partial recruitment values if the entire suite of Fs at age in the terminal year are estimated, as was the case in striped bass. However, if the true selectivity pattern was best defined by a dome shaped partial recruitment vector, the estimate of F_{msy} would be incorrect. The consequence could be allowable levels of fishing mortality that would not maintain an optimal level of spawning stock biomass.

Fully recruited F can also be calculated in several ways. Average fishing mortality could be an average of Fs on fully recruited fish or an average weighted by abundance or biomass. With a flat topped PR, all age classes were assumed to be fully recruited, therefore the associated Fs contributed equally to the total fishing mortality. The unweighted average F in 1997 is 0.33. The alternative was an average weighted by abundance-at-age; the fully recruited F in 1997 weighted by N was 0.24. The two approaches gave different results due to the presence of some large year classes that experienced low fishing mortality. The influence of the 1993 year class resulted in an overall weighted F lower than if the assumption were made that all fully recruited Fs were equivalent. The two estimates of F have different implications in comparison to the target fishing mortality of 0.31. A weighted average would downweight several older cohorts that were at or above the target F. An increase in mortality to attain the target F would subject these cohorts to an increased risk of overfishing. An unweighted average F was chosen to represent the fully recruited fishing mortality in striped bass. The use of this estimate may reduce the risk of increased mortality on cohorts already at or above the target.

Tuning indices in the model can be defined to reflect population abundance at either the beginning or middle of the year. Since the indices were not measured on 1 January or 1 July, the assumption was made that mortality was insignificant between the time the indices were collected and the tuning date. For instance, if indices were collected in October and a major fishery occurred in November, different levels of mortality could create an annual index that may not reflect population abundance as of 1 January. In the striped bass model all indices were tuned to population abundance on 1 January.

Iterative reweighting of the tuning indices is an option within the ADAPT framework and was used for the striped bass model. Re-weighting is done by weighting the relative contribution of an index by the inverse of its partial variance in minimizing the models objective function. Ideally, stock identification methods would distinguish among stocks in the coastal fisheries and the relative contribution of each stock could be used to weight stock specific indices. Since stock information was not available and the stock contributions were assumed to be unequal, the model was allowed to weight the indices based on relative fit to the total catch-at-age data. If annual production in all stocks was equivalent and reflected by the indices, then reweighting would have little impact. In striped bass, the relative contributions of the Chesapeake stock were generally highest. Therefore, the reweighting served to emphasize the stronger year classes produced by this stock. Although there was uncertainty associated with the reweighting scheme, it likely produced more accurate results in the mixed stock model than an assumption of equal contributions among stocks.

Characterizing the Uncertainty

A feature of the ADAPT software is the ability to characterize the uncertainty in model fit using a bootstrap resampling of the residuals from the observed-predicted tuning indices (Mohn 1993). In the nonlinear least-squares (nlls) estimate of a solution, estimates of population abundance were chosen that provided the best fit to the tuning indices. The residuals of that fit were bootstrapped 500 times and new values of N produced. The distribution of the associated Fs provided an indication of variation and the bias (bootstrap mean-nlls mean). Results of the bootstrap indicated an 80% probability that terminal values (1997) of F in the model ranged from

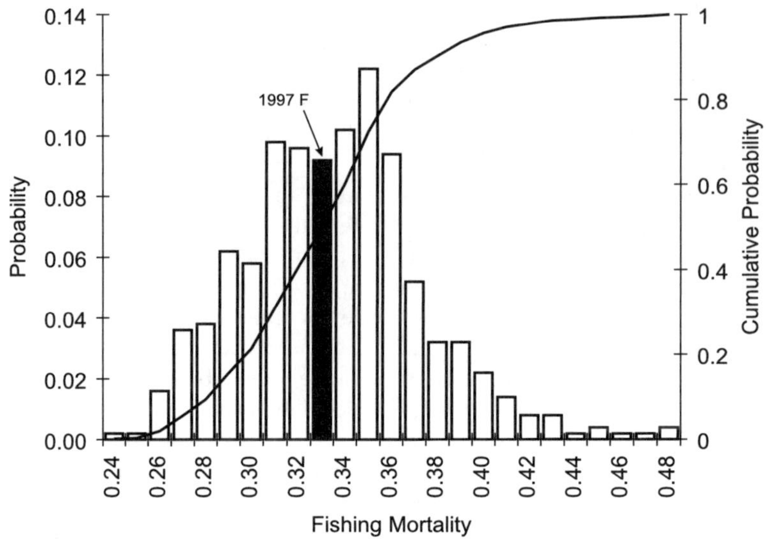

FIGURE 3. Bootstrap results of 1997 fishing mortality estimate for striped bass from virtual population analysis; $N = 500$.

0.30 to 0.36 (Figure 3), and bias in the F and N estimates was 2%. The model output also provided an estimate of the CV at age for N that ranged from 0.19 to 0.34.

The characterization of the uncertainty in the results based on the bootstrap estimates or CVs reflect the variability associated with the model fitting procedure. The bootstrapping did not account for most of the issues previously mentioned that could contribute to uncertainty in the assessment results. The model assumes that the catch-at-age matrix was measured without error. Therefore, any error associated with landings data, length data, and age data was not included. The error in the model configuration was assumed to occur in the time-series of the survey indices. Variances in the point estimates of the tuning indices were also disregarded although they were reduced by using a $\log_e$-$\log_e$ model in the fitting process. Another assumption in the bootstrapping was the correct configuration of the final model. If the model criteria, such as age at full recruitment and M, were incorrect it would not be reflected in the bootstrap results.

A common problem in virtual population analysis is bias associated with the terminal year estimates of fishing mortality and abundance (Mohn 1993). The striped bass VPA results were analyzed to determine if any retrospective patterns existed in the parameter estimates. The tendency of the model was to underestimate fishing mortality in the terminal year and over estimate recruitment of age-1 fish (Table 7). However, the retrospective pattern was not consistent among all years as there were also instances where the model over estimated the terminal year F. Estimates from the striped bass VPA model were not corrected for any retrospective patterns.

Implications

A series of issues pertaining to input data and model selection have been identified that contributed to uncertainty in the results. An important concern was the impact this had on the conclusion about the stock status. One method to assess the implication of this uncertainty is to compare estimates of F with an independently derived estimate. Such a comparison was possible with striped bass because there has been an extensive coast wide tagging program designed to estimate survival. The U.S. Fish and Wildlife Service (FWS) has coordinated the release of nearly 250,000 tagged striped bass by state fisheries agencies between North Carolina and Massachusetts since 1987 (Smith and He 1998). Annual estimates of survival were produced from analysis of tag returns from two programs sampling the mixed coastal stock; the New York ocean haul seine survey and a FWS tagging program off the coast of North Carolina. The results represented survival of fish 28 in and greater, fish approximately age 6 and older. Survival estimates from tag return data were adjusted to account for bias resulting from tag recoveries of fish that were released alive (Smith et al. 2000). The number of tag releases and recoveries were assumed to be proportional to abundance; therefore, results

TABLE 7. Retrospective analysis of striped bass virtual population analysis.

Terminal year	Year																
	1982	1983	1984	1985	1986	1987	1988	1989	1990	1991	1992	1993	1994	1995	1996	1997	1998
Estimates of abundance at age 1 (000s)																	
1992	1851	4856	4152	3729	2396	3756	5290	5589	7514	5603	4380	8270					
1993	1691	4321	3801	3562	2402	3223	4513	4952	6600	4983	4168	6039	16434				
1994	1591	3969	3524	3490	2814	3465	4597	5212	6472	4921	4737	5855	15021	8800			
1995	1382	3583	3176	3286	2804	3628	4785	5440	6686	4952	4714	6647	13804	7591	6578		
1996	1372	2746	2812	3109	2688	3346	4321	5279	7111	4953	4810	6747	13720	7055	6299	17885	
1997	1381	2747	2411	3108	2772	3449	4419	5448	7979	6003	5606	7078	15008	7999	6884	13486	7388
Estimates of fully recruited fishing mortality																	
1992	0.18	0.16	0.09	0.08	0.07	0.04	0.07	0.05	0.10	0.09	0.08						
1993	0.20	0.18	0.11	0.10	0.09	0.05	0.10	0.06	0.12	0.12	0.10	0.13					
1994	0.21	0.19	0.11	0.11	0.10	0.06	0.11	0.07	0.13	0.15	0.11	0.16	0.14				
1995	0.22	0.20	0.12	0.11	0.11	0.06	0.12	0.08	0.10	0.17	0.13	0.19	0.16	0.20			
1996	0.23	0.21	0.12	0.12	0.12	0.07	0.13	0.09	0.17	0.19	0.16	0.24	0.21	0.25	0.29		
1997	0.23	0.21	0.12	0.12	0.12	0.07	0.13	0.09	0.17	0.20	0.17	0.25	0.22	0.24	0.28	0.33	

were compared with the VPA average F weighted by abundance. The results showed a close comparison in estimates of total fishing mortality (Figure 4). The comparison was particularly encouraging because the estimates were made using different data input and models. There is extensive literature documenting the sources of error in tag survival estimates that include incomplete mixing of tagged fish into the population, tag loss, tag induced mortality, biased sampling and nonreporting of tag recoveries. Nevertheless, this type of independent verification of model results provided additional support that the VPA was correctly identifying the trend and relative magnitude in fishing mortality despite the different assumptions of both models.

Comparison to biological reference points

An overfishing definition for striped bass was defined as F_{msy} (0.38), calculated using a modified Thompson-Bell yield per recruit model and a Shepherd stock-recruitment model (Shepherd and Lazar 1998). A fishing mortality of 0.31 was chosen as the target fishing mortality in the FMP. Consequences of fishing at the target F were evaluated using a stochastic projection model incorporating the

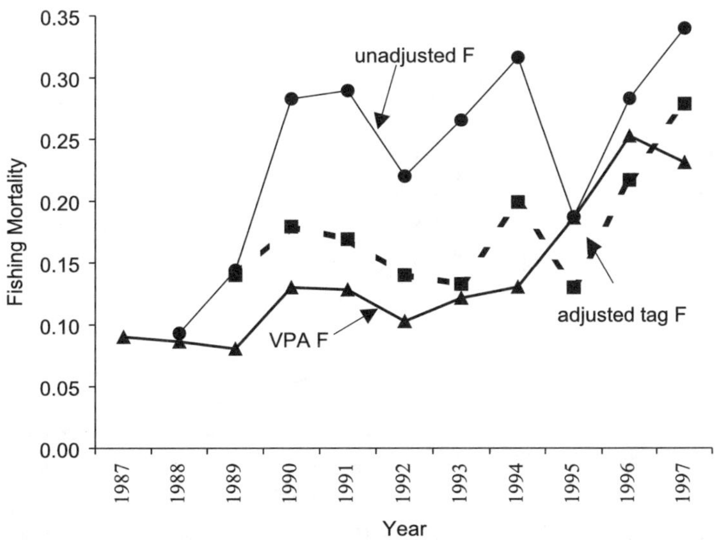

FIGURE 4. Comparison of tagging estimates of striped bass fishing mortality and VPA estimates of fishing mortality weighted by abundance.

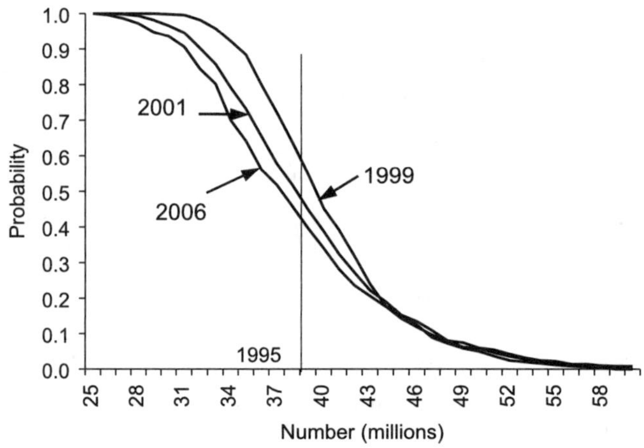

FIGURE 5. Probability of remaining at or above the 1995 level of striped bass population abundance if fishing mortality were held equal to 0.31.

bootstrapped terminal year VPA stock size and variability. Incoming annual recruitment was randomly chosen from log-normal recruitment distributions bounded by the levels occurring under moderate to high levels of spawning stock biomass. Starting stock size vectors were randomly chosen from a matrix of bootstrap realizations of population abundance. A Monte Carlo routine decremented stock size-at-age by the designated F and M and the 1997 exploitation pattern. Mean annual abundance and standard error were calculated from 500 abundance-at-age estimates for a ten-year projection period. The frequency distributions from the simulation were calculated to determine the probability of exceeding the benchmark level of 1995, the year the stock was declared restored (Figure 5).

The results provide insight into the short-term sustainability of the stocks at various levels of fishing mortality. Stock abundance would be expected to remain near current levels if fished at the target F of 0.31. The probability of remaining at or above the 1995 point estimate of abundance level was 55% in 1999, 47% in 2001, and 43% by 2006 (Figure 5). The projections at Fmsy followed a similar trend with a 53% probability of remaining at or above the benchmark level by 1999, 42% by 2001, and 34% by 2006 (Figure 6). Since large fish are targeted in several fisheries, abundance of fish greater than

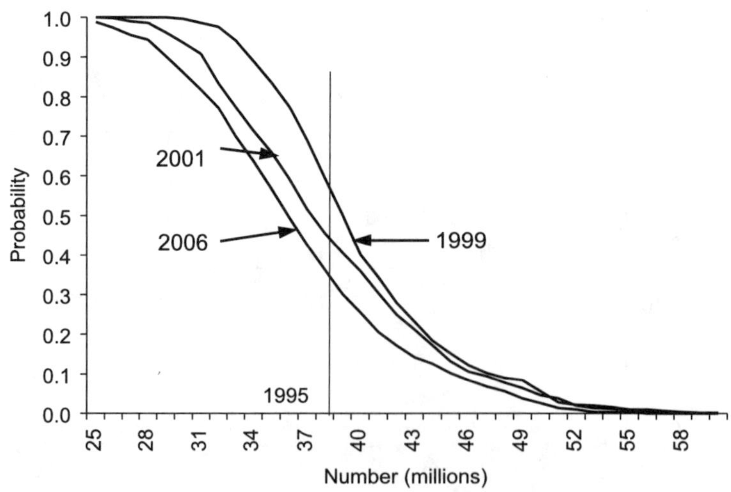

FIGURE 6. Probability of remaining at or above the 1995 level of striped bass population abundance if fishing mortality were held equal to F_{msy} (0.38).

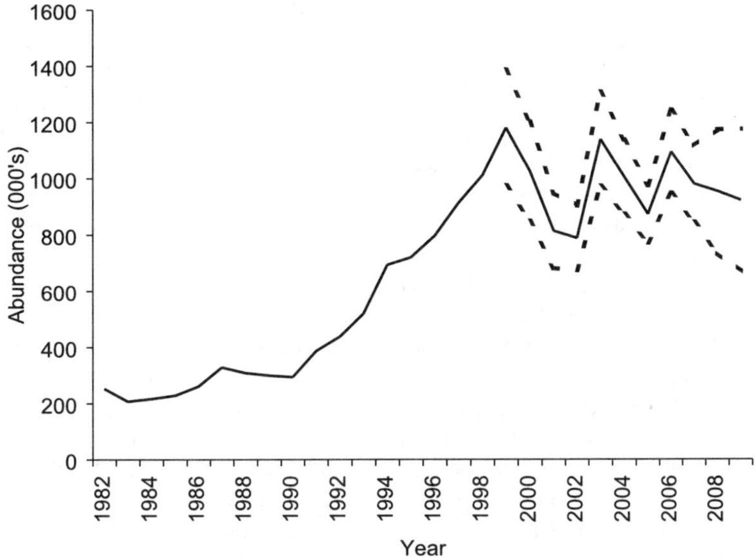

FIGURE 7. Projected abundance of age 10+ striped bass with fishing mortality equal to 0.31. Mean abundance and 80% prediction intervals with virtual population analysis estimates for 1982–1998.

age 10 was also examined. The presence of large year classes already in the population helped maintain the abundance of older fish, at least over the short term (Figure 7). However, increasing fishing mortality to F_{msy} resulted in a decline in the abundance of large older fish despite the influence of strong year classes (Figure 8). For the striped bass fishery, the level of acceptable risk and the long-term expectations of management are still being debated.

Conclusions

The striped bass assessment was presented as an example of an analysis with a variety of fishery dependent and independent data, reasonable catch

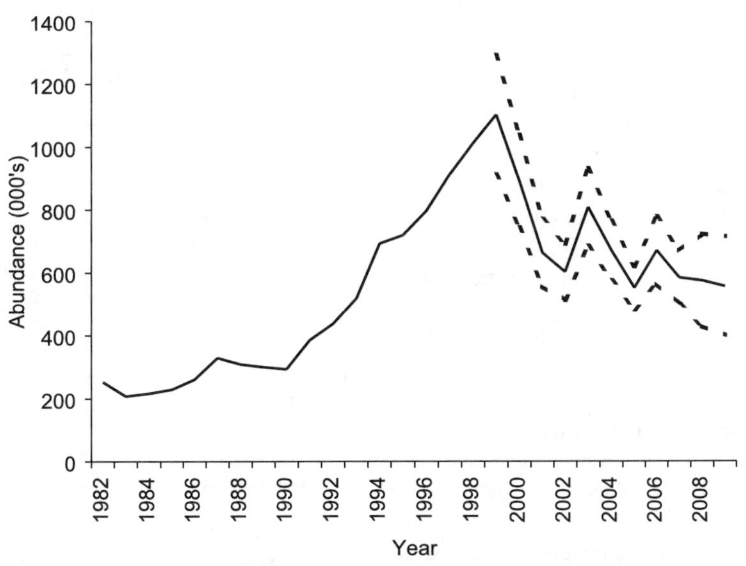

FIGURE 8. Projected abundance of age 10+ striped bass with fishing mortality equal to 0.38. Mean abundance and 80% prediction intervals with virtual population analysis estimates for 1982–1998.

data, and a clear contrast in population abundance during the time series. Beginning in the mid-1980s, stocks of striped bass increased dramatically and produced a consistent upward trend in abundance indices and landings. With such obvious signals from input data, most population models would reach similar conclusions. The bootstrap estimates of the VPA run provided some indication of the uncertainty in the model fit and for striped bass the results suggested the conclusions of the model were relatively precise. Yet, the uncertainty was evaluated relative to the model and did not fully incorporate variance surrounding the input data. Despite the uncertainty in the VPA, tagging models not dependent on catch matrices, produced comparable results. This type of redundancy shows that either both methods produced reasonably accurate results or were similarly biased.

As the population growth of striped bass begins to stabilize or decline, the implications of uncertainty in the results will become more critical. The difference between the target fishing mortality and the overfishing definition is only 0.07. The data and population models available for stock assessment are not adequate to estimate fishing mortality with that level of resolution. Although point estimates of *F* may be on target, the probability distribution from the bootstrap procedure may also exceed Fmsy. Consequently there will be some probability that overfishing is occurring in addition to risk associated with whatever variance was unaccounted for in the bootstrap.

The uncertainty in stock assessment results, as illustrated with the striped bass example using an age-structured model, can occur at several levels throughout the analysis; with the input data, the model selection and the configuration of the chosen model. The basis for any assessment model is representative sampling of the removals from a population. As that basic catch data is divided into sub-components of length and age, there becomes increased risk of introducing error into the model. Some model structures can incorporate information associated with error in the catch and age data, however, the analyst must have some knowledge of that uncertainty. Model selection can also contribute to the uncertainty in the result but this source of error is often not evaluated. Corroborating the assessment results using alternative models can provide some helpful insight regarding the conclusions. Another often overlooked aspect of the uncertainty in assessments is the effects of density changes on population parameters such as growth, maturity and natural mortality (Rosenberg and Restrepo 1994). Despite recent work demonstrating some strong population responses to density (Rochet 1998; Trippel et al. 1997), few assessment models directly account for these responses as a function of changes in fishing mortality (Trippel 1999). Knowledge of the plasticity of life history traits should be considered in evaluating possible sources of uncertainty.

When choosing the target mortality, fisheries managers should define the acceptable level of risk that they will exceed the overfishing level. It is the responsibility of the assessment scientists to characterize as much uncertainty in the results as possible and the risk of overfishing associated with various management approaches. Throughout the development of an assessment, scientists make both subjective and objective decisions that will affect the results in some fashion. A better understanding of the implications of those decisions should improve the chances that the correct choices are made.

Acknowledgments

The striped bass stock assessment was the result of cooperative work by members of the ASMFC striped bass stock assessment sub-committee, technical committee and tagging working group. I acknowledge John Field from ASMFC for his efforts in coordinating the striped bass program and Victor Vecchio of NYDEP for supplying tagging results. The manuscript benefited from comments from several reviewers, particularly Steven Murawski. The ADAPT software used in the analysis was developed by Laura Shulman of the NEFSC.

References

ASMFC. 1995. Amendment #5 to the Interstate Fishery Management Plan for Atlantic Striped Bass. ASMFC Fisheries Management Report No. 24, Washington, D.C.

Bigelow, H. B., and W. C. Schroeder. 1953. Fishes of the Gulf of Maine. U.S. Fish Wildlife Service Fish Bulletin 74(53): 1–577.

Boreman, J., and R. R. Lewis. 1987. Atlantic coastal migration of striped bass. Pages 331–339 *in* M. J. Dadswell, R. J. Klauda, C. M. Moffitt, R. L. Saunders, R. A. Rulifson, and J. E. Cooper, editors. Common strategies of anadromous and catadromous fishes. American Fisheries Society, Symposium 1, Bethesda, Maryland.

Buckel, J. A., D. O. Conover, N. D. Steinberg, and K. A. McKown. 1999. Impact of age-0 bluefish (*Pomatomous saltatrix*) predation on age-0 fishes in the Hudson River estuary: evidence for density-dependent loss of juvenile striped bass (*Morone saxatilis*). Canadian Journal of Fisheries and Aquatic Sciences 56:275–287.

Clark, S. H., editor. 1998. Status of the Fishery Resources off the Northeastern United States. NOAA Technical Memorandum NMFS-NE-115.

Conser, R. J., and J. E. Powers. 1990. Extension of the ADAPT VPA tuning method designed to facilitate assessment work on tuna and swordfish stocks. International Commission for the Conservation of Atlantic Tunas, Collected Volumes of Scientific Papers 32:461–447.

Diodati, P. J., and R. A. Richards. 1996. Mortality of striped bass hooked and released in salt water. Transactions of the American Fisheries Society 125(2):300–307.

Field, J. D. 1998. Atlantic striped bass management: Where did we go right? Fisheries 22(7):6–8.

Gavaris, S. 1988. An adaptive framework for the estimation of population size. Canadian Atlantic Fisheries Scientific Advisory Committee (CAFSAC) Research Document 88/29. Dartmouth, Nova Scotia.

Goodyear, C. P. 1985. Relationship between reported commercial landings and abundance of young striped bass in Chesapeake Bay, Maryland. Transactions of the American Fisheries Society 114(1):92–96

Helser, T. E., J .P. Geaghan, and R. E. Condrey. 1998. Estimating gillnet selectivity using non-linear response surface regression. Canadian Journal of Fisheries and Aquatic Sciences 55(6):1328–1337.

Helser, T., and L. Waller. 1999. Selectivity of experimental gill nets for Chesapeake Bay Striped Bass (*Morone saxatilis*): direct and indirect estimates. Report to the ASMFC Striped Bass Technical Committee, Baltimore, Maryland.

Hilborn, R. 1997. Uncertainty, risk and the precautionary principle. Pages 100–105 *in* E. K. Pikitch, D. D. Huppert, and M. P. Sissenwine, editors. Global trends: fisheries management. American Fisheries Society, Symposium 20, Bethesda, Maryland.

Hilden, M. 1988. Errors in perception in stock and recruitment studies due to wrong choices of natural mortality rate in virtual population analysis. Journal du Conseil, Conseil International pour l'Exploration de la Mer. 44:123–134.

Lapointe, M. F., R. M. Peterman, and A. D. MacCall. 1989. Trends in fishing mortality rate along with errors in natural mortality rate can cause spurious time trends in fish stock abundances estimated by virtual population analysis (VPA). Canadian Journal of Fisheries and Aquatic Sciences 46:2129–2139.

Mace, P. M. 1997. Developing and sustaining world fisheries resources: the state of the science and management. Commonwealth Scientific and Industrial Research Organization, Collingswood, Australia.

Megrey, B. A. 1989. Review and comparison of age-structured stock assessment models from theoretical and applied points of view. Pages 8–48 *in* E. F. Edwards and B. A. Megrey, editors. Mathematical analysis of fish stock dynamics. American Fisheries Society, Symposium 6, Bethesda, Maryland.

Mertz, G., and R. Meyers. 1997. Influence of errors in natural mortality estimates in cohort analysis. Canadian Journal of Fisheries and Aquatic Sciences 54:1608–1612.

McKown, K. 1992. Validation of the Hudson River young-of-the-year striped bass indices. New York Department of Environmental Conservation, Division of Marine Resources, Stony Brook, New York.

Mohn, R. K. 1993. Bootstrap estimates of ADAPT parameters, their projection in risk analysis and their retrospective patterns. Pages 173–184 *in* S. J. Smith, J. J. Hunt, and D. Rivard, editors. Risk evaluation and biological reference points for fisheries management. Canadian Special Publication of Fisheries and Aquatic Sciences 120.

NMFS. 1999. Our Living Oceans. Report on the status of U.S. living marine resources, 1999. U.S. Department of Commerce, NOAA Technical Memorandum NMFS-F/SPO-41 Northeast Fisheries Science Center. 1998. 26th Northeast Regional Stock Assessment Workshop: stock assessment review committee (SARC) consensus summary of assessments. NEFSC Reference Document 98–03.

Rago, P. J., C. D. Stephan, and H. M. Austin. 1995. Report of the juvenile abundance indices workshop, ASMFC Special Report 48.

Richards, R. A., and D. G. Deuel. 1987. Atlantic striped bass: stock status and the recreational fishery. Marine Fisheries Review 49(2):58–66.

Rochet, M. J. 1998. Short-term effects of fishing on life history traits of fishes. ICES Journal of Marine Science 55: 371–391.

Rosenberg. A. A, and V. R. Restrepo. 1994. Uncertainty and risk evaluation in stock assessment advice of U.S. marine fisheries. Canadian Journal of Fisheries and Aquatic Sciences 51:2715–2720.

Secor, D. H., T. M. Trice, and H. T. Hornick. 1995. Validation of otolith-based ageing and a comparison of otolith and scale-ageing in mark-recaptured Chesapeake Bay striped bass, *Morone saxatilis*. Fisheries Bulletin 93: 186–190.

Shepherd, G. R., and N. Lazar, editors. 1998. Source Document to Amendment 5 to the Interstate Fishery Management Plan for Atlantic Striped Bass. ASMFC Fisheries Management Report 34.

Sims, S. E. 1984. An analysis of the effect of errors in the natural mortality rate on stock-size estimates using virtual population analysis (cohort analysis). Journal du Conseil, Conseil International pour l'Exploration de la Mer 41:149–153.

Smith, D. R., and X. He. 1998. An overview of methods to estimate annual mortality of Atlantic Striped Bass from tag-recovery data. In G. R. Shepherd and N. Lazar, editors. Source Document to Amendment 5 to the Interstate Fishery Management Plan for Atlantic Striped Bass. ASMFC Fisheries Management Report 34.

Smith, D. R., K. P. Burnham, D. M. Kahn, X. He, C. J. Goshorn, K. A. Hattala, and A. W. Kahnle. 2000. Bias in survival estimates from tag-recovery models where catch-and-release is common, with an example from Atlantic striped bass (*Morone saxatilis*). Canadian Journal of Fisheries and Aquatic Sciences 57:886–897.

Trippel, E. A., M. J. Morgan, A. Frechet, C. Rollet, A. Sinclair, C. Annand, D. Beanlands, and L. Brown.

1997. Changes in age and length at sexual maturity of Northwest Atlantic cod, haddock and pollock stocks, 1972–1995. Canadian Technical Report of Fisheries and Aquatic Sciences 2157.

Trippel, E. A. 1999. Estimation of stock reproductive potential: history and challenges for Canadian Atlantic gadoid stock assessments. Journal of Northwest Atlantic Fisheries Science 25:61–81.

Vecchio, V. J., and H. O'Riordon. 1999. A study of striped bass in the Marine District of New York state-ocean haul seine study. NMFS Completion Report, 1997–1998. New York Department of Environmental Conservation, Division of Marine Resources, Stony Brook, New York.

Waldman, J. R., and Fabrizio, M. C. 1994. Problems of stock definition in estimating relative contributions of Atlantic striped bass to the coastal fishery. Transactions of the American Fisheries Society 123(5):766–778.

Waldman. J. R., R. A. Richards, W. B. Schill, I. Wirgin, M. C. Fabrizio. 1997. An empirical comparison of stock identification techniques applied to striped bass. Transactions of the American Fisheries Society 126(3):369–385.

Wirgin, I. J. R. Waldman, L. Maceda, J. Stabile, and V. J. Vecchio. 1997. Mixed-stock analysis of Atlantic coast striped bass *(Morone saxatilis)* using nuclear DNA and mitochondrial DNA markers. Canadian Journal of Fisheries and Aquatic Sciences 54(12):2814–2826.

Young-Dubovsky, C., G. R. Shepherd, D. R. Smith, and J. Field. 1996. Striped Bass Research Study. Report for 1994 U.S. Department of Commerce Report to Congress. Washington, D.C.

Appendix I. Indices of abundance used in tuning the striped bass VPA

Table A1. Young of year indices (lagged to January 1).

Year	Hudson River	Delaware	New Jersey	Maryland	Virginia
1982	8.86	–	0.00	0.59	1.57
1983	14.17	–	0.12	3.54	2.71
1984	16.25	–	0.03	0.61	3.40
1985	15.00	–	0.29	1.64	4.47
1986	1.92	–	0.02	0.91	2.41
1987	2.92	–	0.28	1.34	4.74
1988	15.90	–	0.41	1.46	15.74
1989	33.46	–	0.35	0.73	7.64
1990	21.35	0.42	1.03	4.87	11.23
1991	19.05	0.11	1.00	1.03	7.34
1992	3.60	0.18	0.47	1.52	3.76
1993	11.43	1.13	1.19	2.34	7.32
1994	12.59	1.14	1.78	13.97	18.12
1995	17.64	0.19	0.96	6.40	10.48
1996	16.23	0.42	1.98	4.41	5.45
1997	8.90	1.36	1.70	17.46	23.05
1998	22.30	0.14	1.01	3.91	9.35

Table A2. Long Island ocean haul seine abundance indices (number per haul).

	Age										
Year	5	6	7	8	9	10	11	12	13	14	15+
1988	8.71	7.71	2.89	1.13	0.28	0.15	0.00	0.01	0.00	0.01	0.00
1989	4.86	4.49	2.65	0.90	0.45	0.13	0.07	0.02	0.00	0.07	0.10
1990	1.27	2.03	1.42	1.18	0.32	0.09	0.11	0.02	0.01	0.01	0.05
1991	4.38	1.93	2.12	1.60	1.28	0.47	0.19	0.04	0.00	0.01	0.01
1992	5.12	1.64	0.77	1.05	1.46	0.80	0.36	0.09	0.08	0.02	0.10
1993	3.58	1.93	0.62	0.41	0.70	0.63	0.41	0.16	0.09	0.01	0.08
1994	6.65	2.75	1.80	0.74	0.46	0.57	0.45	0.31	0.12	0.04	0.05
1995	3.22	3.15	1.91	1.29	0.56	0.56	0.58	0.28	0.32	0.06	0.08
1996	2.34	0.70	0.76	0.39	0.21	0.14	0.16	0.14	0.05	0.06	0.02
1997	7.70	2.97	0.96	0.83	0.37	0.16	0.12	0.01	0.11	0.01	0.01
1998	34.96	5.40	1.57	0.34	0.36	0.37	0.17	0.06	0.13	0.00	0.05

Table A3. Maryland spawning stock gillnet survey CPUE (number per set).

	Age													
Year	2	3	4	5	6	7	8	9	10	11	12	13	14	15+
1985	72.83	243.05	41.79	19.02	8.88	8.25	1.44	1.83	2.19	0.39	1.74	1.31	0.31	7.01
1986	62.72	164.74	467.30	7.10	4.44	3.16	2.63	0.94	0.73	0.00	0.00	0.94	0.65	2.22
1987	60.93	204.10	128.14	335.33	3.72	2.95	3.48	0.12	0.00	0.00	0.00	0.00	7.25	4.94
1988	32.21	67.75	73.47	72.33	107.36	2.16	0.00	0.00	0.73	0.00	0.02	0.00	0.08	1.86
1989	15.52	121.59	100.51	71.51	91.10	59.62	0.38	0.00	0.37	0.00	0.19	0.00	0.00	0.34
1990	25.63	182.33	204.18	88.67	68.95	67.02	52.92	0.46	0.18	0.02	0.24	0.26	0.05	0.39
1991	40.31	186.40	72.20	68.43	40.60	38.94	35.44	14.97	0.43	0.30	0.00	0.11	0.10	0.45
1992	17.40	240.64	199.49	63.24	84.36	59.62	41.81	19.13	8.79	0.15	0.00	0.03	1.09	0.72
1993	33.40	130.16	222.42	98.53	60.37	57.34	46.52	22.28	7.92	3.27	0.33	0.31	0.46	0.35
1994	11.12	37.90	67.64	98.17	37.34	20.90	30.06	12.22	3.34	0.63	0.36	0.09	0.00	0.05
1995	42.90	110.10	71.81	72.13	56.29	49.63	33.55	41.38	17.83	24.86	8.21	2.14	0.00	0.34
1996	8.38	510.92	140.80	47.65	93.05	109.70	85.01	66.80	34.79	16.59	5.05	1.66	0.00	0.00
1997	33.46	27.73	181.71	63.84	29.59	32.17	41.79	32.57	22.26	8.80	6.69	3.38	0.44	0.00
1998	19.98	31.89	209.97	112.30	36.61	20.82	25.17	21.87	15.90	16.45	4.79	2.49	0.47	0.00

Table A4. Massachusetts CPUE (number per trip) from commercial hook and line fishery.

	Age							
Year	8	9	10	11	12	13	14	15+
1991	0.56	2.30	0.92	0.43	0.28	0.12	0.44	1.95
1992	0.86	3.51	4.99	0.92	0.31	0.31	0.07	2.04
1993	0.38	3.17	5.89	4.78	0.51	0.22	0.06	0.99
1994	0.19	1.97	6.41	8.59	5.33	0.86	0.17	0.50
1995	0.43	3.74	9.74	6.26	2.18	1.03	0.10	0.53
1996	1.13	5.62	9.13	6.75	2.84	1.08	0.27	0.11
1997	0.90	4.81	6.12	5.58	4.68	2.47	0.75	0.41
1998	1.28	9.48	10.36	7.68	5.40	3.00	1.80	1.00

Table A5. Connecticut volunteer angler CPUE (number per trip).

	Age													
Year	2	3	4	5	6	7	8	9	10	11	12	13	14	15+
1982	0.22	0.32	0.16	0.14	0.11	0.06	0.03	0.02	0.02	0.01	0.00	0.00	0.00	0.00
1983	0.33	0.21	0.11	0.09	0.08	0.04	0.02	0.01	0.01	0.00	0.00	0.00	0.00	0.00
1984	0.40	0.19	0.08	0.04	0.03	0.01	0.00	0.00	0.00	0.00	0.00	0.00	0.00	0.00
1985	0.12	0.33	0.23	0.14	0.05	0.04	0.01	0.00	0.00	0.00	0.00	0.00	0.00	0.00
1986	0.06	0.31	0.22	0.12	0.09	0.04	0.03	0.01	0.00	0.00	0.00	0.00	0.00	0.00
1987	0.08	0.20	0.47	0.45	0.18	0.05	0.01	0.05	0.02	0.00	0.00	0.00	0.00	0.01
1988	0.03	0.24	0.34	0.20	0.14	0.06	0.04	0.03	0.03	0.01	0.00	0.00	0.00	0.00
1989	0.02	0.52	0.28	0.18	0.15	0.12	0.05	0.03	0.01	0.00	0.00	0.00	0.00	0.00
1990	0.27	0.48	0.47	0.16	0.18	0.13	0.09	0.03	0.01	0.01	0.00	0.00	0.00	0.00
1991	0.17	0.58	0.55	0.27	0.12	0.13	0.15	0.13	0.05	0.02	0.01	0.01	0.00	0.01
1992	0.15	0.67	0.43	0.35	0.14	0.07	0.09	0.13	0.09	0.03	0.01	0.00	0.00	0.00
1993	0.17	0.48	0.57	0.29	0.23	0.11	0.09	0.16	0.15	0.09	0.02	0.00	0.00	0.00
1994	0.07	0.70	0.62	0.49	0.28	0.22	0.09	0.08	0.11	0.10	0.05	0.01	0.00	0.01
1995	0.21	0.61	0.88	0.46	0.57	0.35	0.23	0.16	0.19	0.14	0.07	0.05	0.01	0.00
1996	0.60	1.20	1.34	0.59	0.59	0.32	0.18	0.19	0.19	0.12	0.05	0.02	0.01	0.00
1997	0.47	1.09	2.39	0.90	0.84	0.37	0.59	0.37	0.23	0.10	0.08	0.10	0.02	0.02
1998	0.18	1.11	1.28	1.64	0.58	0.31	0.23	0.21	0.12	0.06	0.07	0.13	0.03	0.04

Table A6. Hudson River commercial shad gillnet CPUE (number per haul).

	Age		
Year	6	7	8
1982	0.012	0.023	0.010
1983	0.147	0.273	0.116
1984	0.443	0.110	0.005
1985	0.085	0.069	0.026
1986	0.248	0.080	0.031
1987	0.413	0.234	0.055
1988	0.536	0.327	0.164
1989	1.278	0.562	0.309
1990	0.856	0.682	0.365
1991	0.477	0.382	0.382
1992	0.707	0.200	0.354
1993	2.172	0.790	0.559
1994	1.755	1.047	0.277
1995	0.869	0.448	0.263
1996	0.000	0.000	0.000
1997	0.485	0.164	0.154

Table A7. Age-1 indices (data lagged to January 1).

Year	W. Long Island	Maryland
1982	–	0.02
1983	–	0.02
1984	–	0.32
1985	–	0.0
1986	0.61	0.15
1987	0.30	0.03
1988	0.21	0.05
1989	0.77	0.06
1990	1.73	0.15
1991	0.37	0.33
1992	1.24	0.19
1993	1.34	0.11
1994	0.72	0.19
1995	1.37	0.76
1996	1.26	0.12
1997	1.52	0.07
1998	0.99	0.26

American Fisheries Society Symposium 27:63–82, 2002

A Stochastic Decision-Based Approach to Assessing the Status of the Delaware Bay Blue Crab Stock

THOMAS E. HELSER[1]
Delaware Division of Fish and Wildlife, Post Office Box 330, Little Creek, Delaware 19961, USA

ALEXEI SHAROV
Maryland Department of Natural Resources, Tawes State Office Building, B-2, Annapolis, Maryland 21401, USA

DESMOND M. KAHN
Delaware Division of Fish and Wildlife, Post Office Box 330, Little Creek, Delaware 19961, USA

Abstract.—Stock assessment methodology has increasingly employed statistical procedures as a means to incorporate uncertainty into assessment advice. Deterministic values of fishing mortality rates (F_t) estimated from assessment models have been replaced by empirical distributions that can be compared with an appropriate biological reference point (F_{BRP}) to generate statements of probability (e.g., $\Pr[F_t \geq F_{BRP}]$) regarding the status of the resource. It must be recognized, however, that terminal year fishing mortality rates and the biological reference points to which they are compared are both estimated with error, which will impact the quality of decisions regarding the status of the stock. We propose a two-tier stochastic decision-based framework for a recently conducted stock assessment of the Delaware Bay blue crab stock that specifies not only the probability for the condition $\Pr(F_t \geq F_{BRP})$, but also the statistical level of confidence (i.e., 90%) in that decision. The approach uses a mixed Monte Carlo-bootstrap procedure to estimate probability distributions for both the terminal year fishing mortality rate (F_t) and the replacement fishing mortality rate, approximated by F_{MED} as an overfishing definition. Probability density functions (PDFs) for F_t and F_{MED}, generated using the mixed Monte Carlo-bootstrap procedure, show that recent fishing mortality rates (80% CI from 0.6 to 1.2) are generally below the F_{MED} overfishing definition (80% CI from 0.9 to 1.6), with significant overlap in the PDFs. Using the PDFs, the stochastic decision-based approach then generates a probability profile by integrating the area under the F_t PDF for different decision confidence levels (e.g. 90%, 80%, 70%, etc.), which can be thought of as one-tailed α-probability from standard statistical hypothesis testing. For example, at the 80% decision confidence level (value of F corresponding to the upper 20% of the F_{MED} PDF), $\Pr(F_t > F_{MED})$ is about 0.03. Thus, with high confidence (80%), we can state that the blue crab stock is not currently being overfished. This approach can be extended to decisions regarding control laws that specify both maximum fishing rate and minimum biomass thresholds.

Introduction

In recent years significant progress has been made in providing scientific advice to fishery managers in terms of estimating risk and uncertainty (Smith et al. 1993; Rosenberg and Restrepo 1994). Many of these advances have been in the area of incorporating uncertainty into assessment advice in the form of risk assessment (Francis 1992; Hilborn et al. 1993) and sensitivity analysis of population models. Rosenberg and Restrepo (1994) have identified five major types of uncertainty in stock assessment models. Among them, measurement and process errors have received the most attention by stock assessment scientists. Measurement error represent the uncertainty in assessment model inputs such as growth and abundance indices, which are used to tune Virtual Population Analysis (VPA), while process errors define the underlying stochastic nature of the stock's dynamics such as variability in recruitment. Uncertainty in fisheries science has largely focused on statistical methods to quantify measurement and process error, and to date analyses have included the jackknife (Claytor et al. 1993), Monte Carlo (Restrepo et al. 1992; Francis 1993), bootstrap (Mohn 1993; Conser 1994), and Bayesian (Hilborn et al. 1993; Walters and Ludwig 1994) approaches.

For many stock assessments, particularly in the Northeast United States and Canada, stock status is evaluated by comparing the empirical distribution of the current or terminal fishing mortality rate (F_t), generated by bootstrapping (Efron and Tibshirani 1993) residuals from the final assessment model, to the deterministic value of the fishing mortality associated with the biological reference point (F_{BRP}). Determining stock status is achieved by

[1]Present Address: NOAA/NMFS, Northwest Fisheries Science Center, 2725 Montlake Blvd, East Seattle, Washington 98122, USA.

calculating the cumulative probability distribution of realizations of F_t and evaluating the probability that F in the most recent year is equal to or greater than F_{BRP} [i.e., 1–Pr($F_t > F_{\text{BRP}}$)]. Advisory statements of this sort are then usually presented to managers in the form "there is an X% probability that the fishing mortality rates exceeded F_{BRP}" (Mace and Sissenwine, this volume). This type of inference regarding the status of the resource only accounts for uncertainty in the estimates of terminal fishing mortality, through bootstrapping errors from the assessment model, while treating the F associated with the biological reference point as deterministic.

Other work has recently focused on risk evaluation and uncertainty in biological reference points for fisheries management (Smith et al. 1993). This second tier of uncertainty analysis is important because models that generate biological reference points are also subject to considerable uncertainty. For instance, the calculation of F_{med} as a proxy for the replacement fishing mortality rate (Sissenwine and Shepherd 1987) contains uncertainty from both the survival ratio generated from the stock and recruitment estimates and the SSB-per-recruit analysis (Gabriel et al. 1989). In this regard, it must be recognized that terminal year fishing mortality rates and the biological reference points to which they are compared are both estimated with error, which will then impact the quality of decisions regarding the status of the stock. Thus, a more formal evaluation of risk must consider the uncertainty associated with both the calculation of the biological reference point and the estimation of current fishing mortality rates.

We propose a two-tier stochastic decision-based framework for a recently conducted stock assessment of the Delaware Bay blue crab stock (Helser and Kahn 1999). This statistical framework specifies not only the probability for the condition Pr($F_t >$ F_{BRP}) but also the statistical level of confidence (i.e. $\alpha = 0.05$) in that decision. Specifically, the approach we have applied to the Delaware Bay blue crab stock assessment compares the empirical distributions of the terminal year fishing mortality rate (F_t) and the replacement fishing mortality rate (F_{MED}) as an overfishing definition to assess the status of the stock. The terminal year fishing mortality rate is estimated using the catch-survey model (Collie and Sissenwine 1983), while F_{MED} is estimated using stock and recruitment plots coupled with SSB-per-recruit analysis (Gabriel et al. 1989). Variance estimators for both quantities are characterized numerically using the conditioned nonparametric bootstrap and Monte Carlo techniques.

Blue Crab Biology and Fishery

The blue crab *Callinectes sapidus* inhabits estuaries from Massachusetts to Uraguay (Williams 1974). The center of its distribution is tropical and severe winters can kill large proportions of overwintering blue crabs from Chesapeake Bay north (Van Engle 1982; Kennish et al. 1984; Kahn et al. 1998). Juveniles and adults primarily remain within one estuary for life, although adults and spawning females in particular may move into nearby coastal waters (Cargo 1954; Fischler and Walburg 1962). After hatching, larvae are carried onto the continental shelf along the North American Atlantic coast and into the Gulf of Mexico on the Gulf coast, where larval development occurs (Epifanio 1995; Perry et al. 1998). Stock mixing occurs at this stage; the current hypothesis for Delaware and Chesapeake Bays is that larvae are circulated on the shelf and reenter their estuary of origin (Epifanio 1995). In general, blue crabs grow to maturity approximately one year from appearance as juveniles (Tagatz 1968a; Perry 1975, Rothschild and Sharov 1996; Kahn et al. 1998). Peak settlement occurs from late summer to early fall from Delaware Bay through Mississippi (Kahn et al. 1998; Perry et al. 1998). If females mature in spring or early summer, they may spawn later that summer, while if they mature in midsummer to early fall, they overwinter before spawning the following year (Perry et al. 1998). Lifespan is usually reported as two to three years (Churchill 1919; Van Engle 1958), although a very few tag returns indicated four years in Florida (Tagatz 1968b) and five years in North Carolina (Fischler 1965).

The Delaware Bay fishery comprises crabbers from Delaware and New Jersey. Despite the fact that Delaware Bay is near the northern extremity of the range, blue crab catches currently produce the largest dockside value of any fisheries resources in Delaware. Commercial landings from 1988 to 1997 have averaged 4.9 million pounds (Figure 1), with an annual estimated ex-vessel value of \$3.3 million (Cole 1998). In 1997, hard crab landings were 4.8 million pounds of crab meat with a value of \$3.9 million. A two-year survey of the recreational blue crab fishery found recreational landings to be less than 1% of the commercial landings (Cole et al. 1997). These recent levels of blue crab harvest are the highest long-term yield ever documented for the

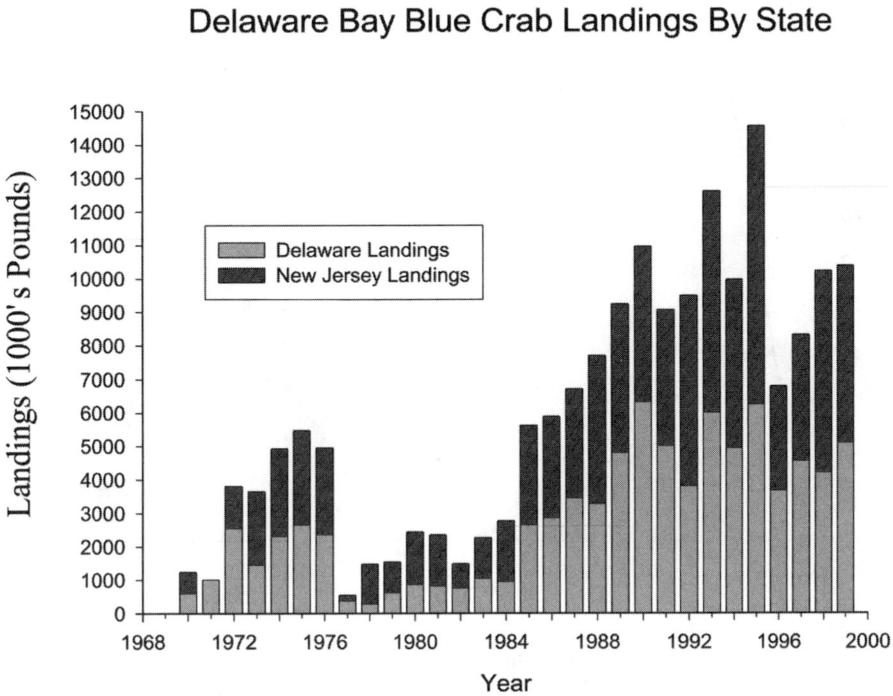

FIGURE 1. Delaware Bay blue crab landings (000's pounds) from Delaware and New Jersey, 1970–1999.

Delaware Bay resource based on historical landings dating back to 1941. New Jersey landings from Delaware Bay during the same period averaged 3.3 million pounds. The combined landings of the two states averaged 8.2 million pounds from 1988 through 1997.

Two commercial gear types are used to harvest blue crab in Delaware Bay. The crab pot fishery, which occurs between April and November, and dredges from November to December into March-April. Pots account for 84% of the annual harvest. The pot fishery harvests both hard and peeler crabs and tends to follow their seasonal migration pattern up bay in the spring and down bay in the fall. The pot fishery usually begins in April or May with a sook run where the fishery concentrates on mature, overwintering females inhabiting the higher salinity waters of lower Delaware Bay. These crabs are feeding intensively after inactivity during the winter months, and as these female crabs produce eggs in the spring, visible as a sponge mass under their bodies, they become protected from harvest. The fishery then moves further up in the bay between July and September and concentrates on the mature males or jimmies. Mature male crabs less than 127 mm in Delaware and less than 120 mm in carapace width in New Jersey are protected from harvest while mature females of any size can be harvested, except if they are sponge crabs. The crab dredge fishery, which is limited to 15 December–30 March, concentrates almost exclusively on adult female crabs that are overwintering in the lower Delaware Bay.

With an increasing demand for crab meat, coupled with declines of harvests in the Chesapeake Bay stock, effort on the Delaware Bay stock has increased markedly since the mid-1980s (Cole 1998). Very recent declines in landings and catch per unit effort with steady increases in fishing effort (Figure 2) has raised concerns that these declines may foreshadow future stock depletion. This concern prompted efforts to initiate the development of a bi-state (New Jersey-Delaware) fishery management plan and to conduct a comprehensive population assessment of the blue crab resource in the Delaware Bay (Helser and Kahn, 1999). Prior to this assessment there was no published analytical population dynamics model of any blue crab stock in the U.S. While numerous authors have employed stock assessment models of various types, including depletion models (Fischler 1965), stock-recruit models (Lipcius and Van Engle 1990; Rugolo et al. 1998; Uphoff 1998; Kahn et al. 1998),

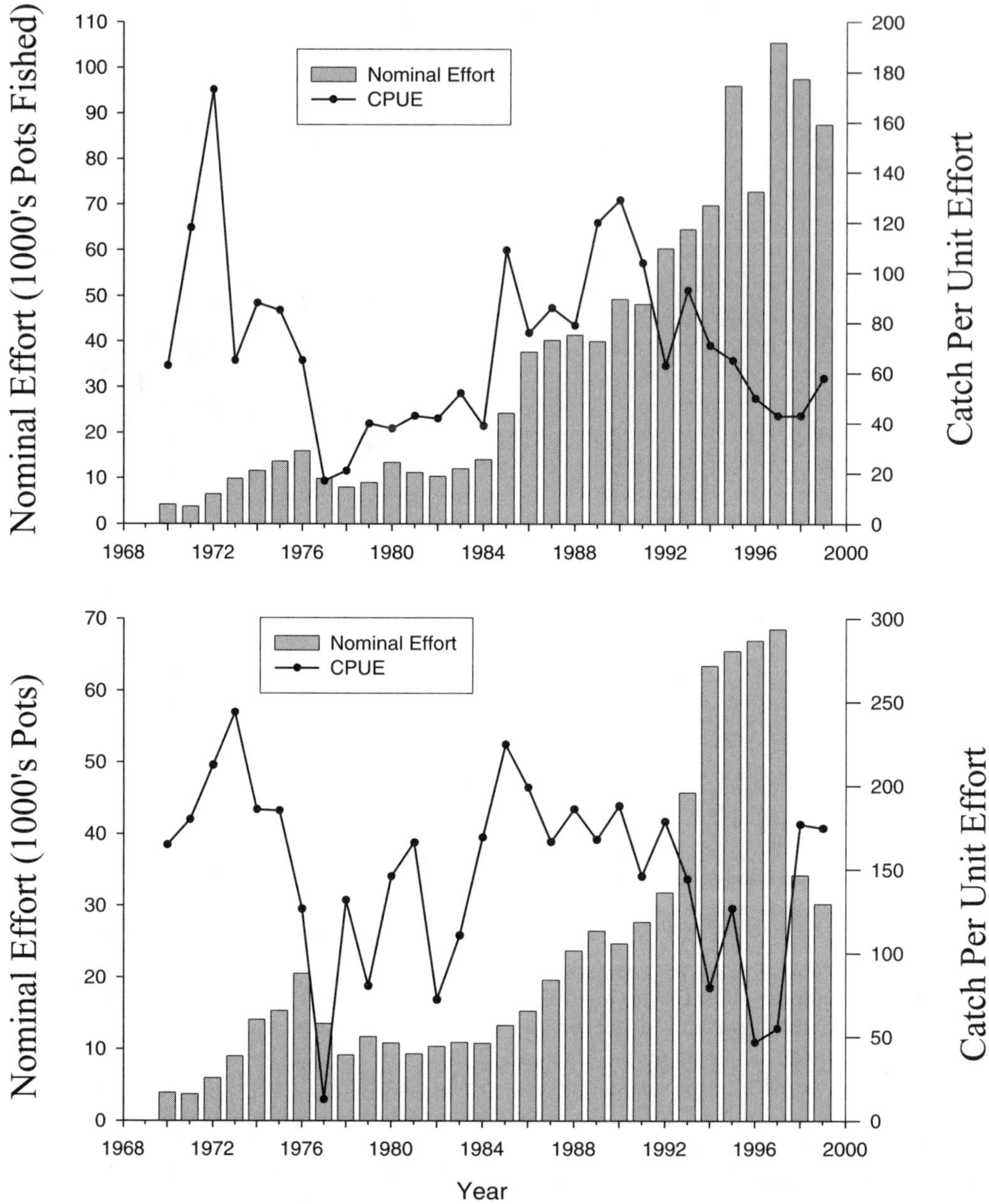

FIGURE 2. Nominal fishing effort (number of pots fished) and catch per unit effort of Delaware Bay blue crab caught in Delaware and New Jersey from 1970 to 1999.

and yield and spawning stock biomass-per-recruit models (Rugolo et al. 1998; Uphoff 1998), none have employed newer techniques using an observation-process error approach to modeling the population dynamics of blue crab. Helser and Kahn (1999) recently applied this approach using the catch-survey model of Collie and Sissenwine (1983), which models the population dynamics using only two age groups (recruit versus fully recruited) since there is currently no method of aging mature blue crabs. This basic assessment model for Delaware Bay blue crab served as the basis for the subsequent development of the uncertainty analysis presented in this paper.

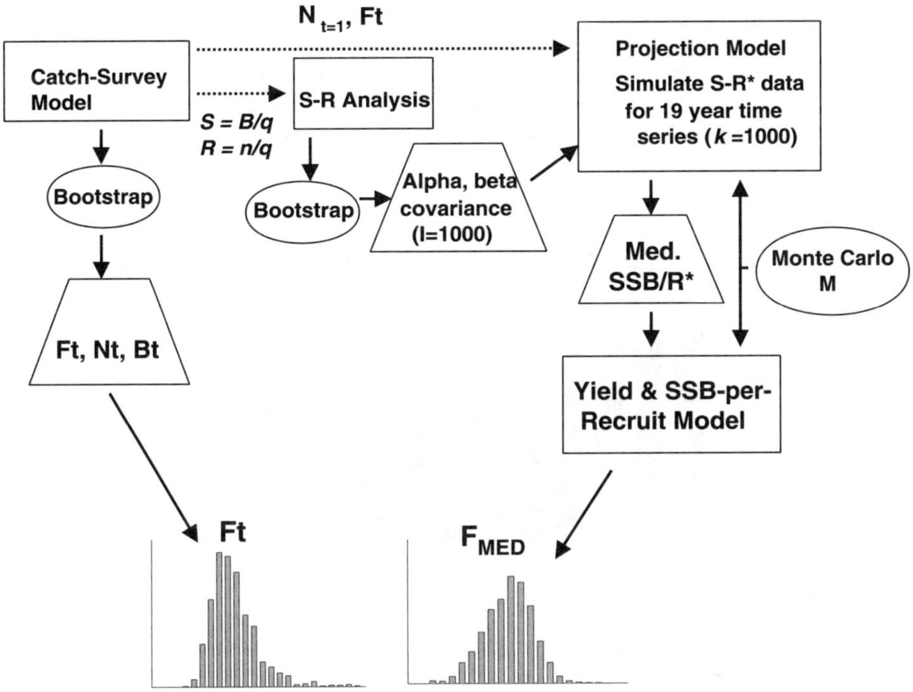

FIGURE 3. Conceptual structure of the uncertainty analysis applied to Delaware Bay blue crab. Endpoints in the analysis are empirical distributions of the terminal fishing mortality rate (F_t) and the replacement fishing mortality rate shown by frequency histograms and the analysis progresses through various assessment procedures. Rectangles represent assessment procedures, parallelograms are model outputs, and ovals indicate points in the analysis for which bootstrap and Monte Carlo techniques were applied.

Methods

The conceptual structure of the assessment methodology and decision framework for incorporating uncertainty into the blue crab assessment is shown in Figure 3. The rectangles represent standard assessment models, parallelograms are model outputs, and the ovals indicate points in the analysis for which bootstrap and Monte Carlo techniques were applied to characterize the uncertainty of model outputs. Since the endpoints of the analysis are empirical distributions of terminal fishing mortality rates and biological reference points, the analysis progresses through the various assessment procedures toward those final endpoints. The following description of methods consists of three major sections: Population Dynamics Model, Incorporation of Uncertainty into BRPs, and Development of F_{MED} Probability Distribution. In the analysis that follows, we first generate deterministic estimates of terminal F and various biological reference points using the standard assessment procedures. We then characterize the parent distributions of these estimators and incorporate these variances into the decision-based framework.

The stochastic decision-based framework attempts to capture the fact that there is uncertainty in both the estimates of terminal fishing mortality rate (F_t) and replacement fishing mortality rate (F_{MED}) by using the empirical distributions of these quantities generated from the uncertainty analysis. Specifically, we desired to provide a decision as to whether $F_t > F_{MED}$ in the form of a probability statement [i.e., $\Pr(F_t > F_{MED})$]. Probability statements of this type can be easily estimated using the empirical distribution of F_t and the deterministic (expected) value of F_{MED}, which is typically done in many assessments. However, uncertainty in F_{MED} remains uncharacterized. Therefore, we proposed a second tier probability statement that specifies the condition $\Pr(F_t > F_{MED})$ for a given decision confidence (%) calculated from the empirical distribution of F_{MED}. Figure 4 gives a conceptual illustration for calculating the $\Pr(F_t > F_{MED})$ for various decision confidence

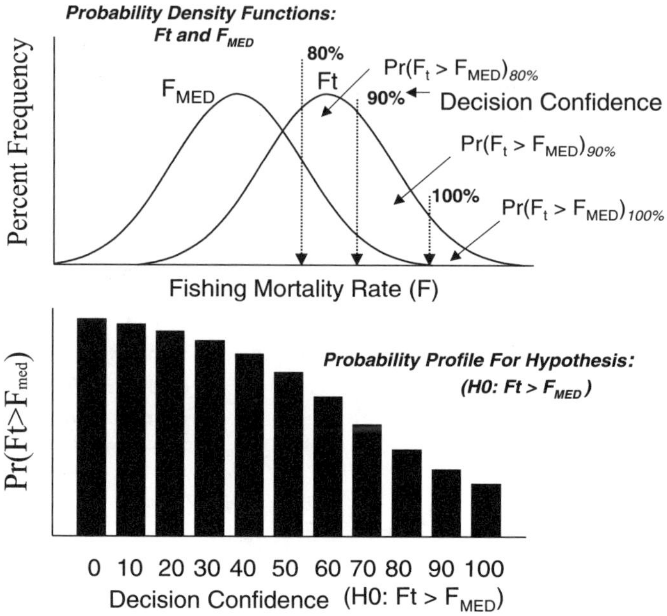

FIGURE 4. Conceptual illustration showing resulting probability density functions and probability profile from the uncertainty analysis for Delaware Bay blue crab. (A) Stylized probability density functions for terminal fishing mortality rate and the replacement fishing mortality rate (F_{MED}). Decision confidence levels show area under and to the right of the replacement fishing mortality rate probability density. (B) Stylized probability profile showing for each decision confidence level the probability that the terminal fishing mortality rate is greater than the replacement fishing mortality rate, $\Pr(F_t > F_{\text{MED}})$. The decision confidence level is the percentage of the area under the replacement fishing mortality rate probability density.

levels (%) from two hypothetical probability density functions of F_t and F_{MED} as well as the resulting probability profile. The decision confidence level can be thought of as a one-tailed α-probability from standard statistical hypothesis testing. Therefore, at the value of F along the domain of the x-axis (top graph) where the probability level $\alpha = 90\%$ (10% of the area under the F_{MED} PDF), the corresponding area under the F_t PDF gives the $\Pr(F_t > F_{\text{MED}})$. Thus, for each decision confidence level there is a corresponding $\Pr(F_t > F_{\text{MED}})$ that forms a probability profile (lower graph). For instance, a statement such as, there is an 80% confidence level that $\Pr(F_t > F_{\text{MED}}) = 0.80$ provides strong evidence of overfishing.

Population Dynamics Model

The catch-survey model (Collie and Sissenwine 1983; Conser and Idoine 1992) was applied to Delaware Bay blue crabs to estimate abundance and fishing mortality rates (Helser and Kahn 1999). Data employed included: 1) annual indices of population size (numbers) for each life history stage (i.e., recruit and fully-recruited size) estimated by research surveys, 2) annual fishery catch in numbers, 3) partial recruitment pattern of recruit size animals to the fishery (fully-recruited life stage), and 4) an estimate of instantaneous natural mortality. While stock assessment input data is not discussed here, the population dynamics model formulation and model results will be reiterated in as much as is necessary for subsequent development of the uncertainty analysis.

The catch-survey model is based on the first order difference equation

$$N(t+1) = [N(t) + R(t) - C(t)]e^{-M} \qquad (1)$$

which relates the fully-recruited stock size at the beginning of the year, $N(t+1)$, to the fully-recruited stock size at the beginning of the previous year, $N(t)$, plus recruitment in the previous year, $R(t)$, minus the catch, $C(t)$, all discounted for natural mortality, M. Equation (1) assumes that a recruit is any animal smaller than the minimum size vulnerable to the fishery at the beginning of the survey year whose size will be greater than or equal to the fully-recruited size by the end of the survey year. In other words, recruits must grow from a size not vulnerable to fully-vulnerable size within one year. In the formulation of the model, recruit and fully-recruited size groups were defined as blue crabs less than 120 mm and as blue crabs greater than or equal to 120 mm, respectively. This was somewhat of a natural construct for the catch-survey model for several reasons: 1) current conservation measures specify a 120 mm and 127 mm minimum carapace width for hard crabs in New Jersey and Delaware, respectively; 2) width frequencies indicate that the large majority of recruits, which appear in the research surveys during August-September, have grown to the fully-recruited size (120 mm) by the following survey (Helser and Kahn 1999); and 3) Kahn et al. (1998) demonstrated a significant positive correlation between indices of recruit crabs (<60 mm) in one year and indices of large crabs (defined as >120 mm) during the previous year. Kahn et al. (1998) also showed that the positive correlation between indices of recruits and fully-recruited crabs was not necessarily linear suggesting density-dependence between these life history stages.

Since approximately 70% of the blue crab catches are taken prior to the mid-point of the August-September research survey, we lagged indices of recruit and fully-recruited relative abundance to index population abundance at the beginning of the next calender year. In the fishery, median catches of blue crab are taken at a point 60% into the calender year. Therefore, the above difference equation can be formulated

$$N(t+1) = \left[(N(t) + R(t))\, e^{-0.60M} - C(t)\right] e^{-0.40M} \tag{2}$$

so that recruited blue crabs, $N(t)+R(t)$, experience 60% of the natural mortality ($e^{-0.60M}$), catch is removed, then the survivors, $[(N(t)+R(t))\ e^{-04M} - C(t)]$, experience the remaining 40% of natural mortality. For the analysis, natural mortality was estimated to be 0.75–0.8 based on an assumed longevity of approximately four years using the M = (3/maximum age) convention (Vetter 1985). While blue crab longevity has been reported as high as eight years (Rugolo et al. 1998), recent studies using lipofusin accumulation rates in blue crabs under natural conditions (Ju et al. 1999; In press) as an alternative to aging hard parts suggest that blue crabs grow faster than previously assumed; implying higher natural mortality and lower longevity.

Survey indices of abundance are related to absolute stock sizes estimated by the catch-survey model by a catchability coefficient which is estimated in the model,

$$n(t)' = q_n N(t) e^{\eta t} \tag{3}$$

and

$$r(t)' = q_r R(t) e^{\delta t} \tag{4}$$

where $r(t)'$ and $n(t)'$ are the observed research survey indices of recruit and fully-recruited blue crabs, and q_n and q_r are the catchability coefficients for recruit and fully-recruited blue crabs. The error terms in the above equations, $e^{\eta t}$ and $e^{\delta t}$, are log-normally distributed random variables which represent survey measurement errors for the recruit and fully-recruited indices, respectively. Substituting equations (3) and (4) into the difference equation, multiplying through by q_n and introducing a scaling parameter ($s_r = q_r/q_n$) that describes the relative selectivity of recruit indices to the fully-recruited blue crab indices we obtain:

$$n(t+1) = \left[\left(n(t) + \frac{r(t)}{s_r}\right) e^{-0.60M} - q_n C(t)\right] \times e^{-0.40M} e^{\varepsilon t} \tag{5}$$

where $e^{\varepsilon t}$ is the lognormally distributed process error. The process error is the difference between recursively calculated indices of fully-recruited animals (5) and the expected value of the fully-recruited indices predicted within the nonlinear least squares framework from the model. For the analysis of Delaware Bay blue crabs the relative selectivity was set equal to unity. The research survey uses a 1.5 inch mesh with a 0.5 inch codend liner (shrimp trawl) and it is unlikely that small blue crabs of recruit size observed in the survey would escape at a rate different from that of fully-recruited crabs. However, smaller crabs which inhabit the littoral areas of the estuary may be less available to the survey. The effect of unequal selectivity between recruit and fully-recruited crabs was evaluated in a sensitivity analysis and showed relatively minor effects on estimates of mortality (Helser and Kahn 1999).

Equation (5) is a statistically estimable function with $2Y$ parameters to be estimated; n for all T years, and r for all years except the last year, and q_n. Given estimates of n, r, and q_n from the nonlinear least squares minimization, and the value of relative selectivity, s_r, population sizes for the recruit and fully-recruited blue crabs are:

$$N(t) = \frac{\hat{n}(t)}{\hat{q}_n} \qquad R(t) = \frac{\hat{r}(t)}{s_r \hat{q}_n} \tag{6}$$

Instantaneous rates of total and fishing mortality for all crabs of recruit size and larger can be calculated as:

$$Z_{R+N}(t) = \log_e \left[\frac{N(t) + R(t)}{N(t+1)} \right] \tag{7}$$

$$F_{R+N}(t) = Z_{R+N}(t) - M \tag{8}$$

Fishing mortality can then be partitioned to recruit and fully-recruited sizes according to the average partial recruitment (p_R) of recruits over the survey year (typically a function of growth toward the size at which they become fully-recruited):

$$F_N(t) = \frac{F_{R+N}(t)[R(t) + N(t)]}{pR(t) + N(t)} \tag{9}$$

and

$$F_R(t) = p_R F_N(t) \tag{10}$$

Finally, to characterize the variance estimates of fishing mortality rates the conditioned nonparametric bootstrap, which resamples the residual pool of both recruit and fully-recruited measurement errors, was applied (Efron and Tibshirani 1993). The resampling procedure is conditioned because the pool of residuals are conditional upon the statistical model applied to the data; it is nonparametric because no distributional assumptions are made with regard to the residuals. Using the bootstrap, 500 realizations of terminal year fishing mortality rates were estimated and the variances were characterized using relative frequency histograms.

Incorporation of Uncertainty into BRPs

Management of fish stocks at sustained and productive levels requires the development of biological reference points that define limits beyond which stocks cannot be exploited or targets about which exploitation is expected to fluctuate. Biological reference points defined as limits and targets can be developed both in terms of fishing mortality rates or biomass levels. For the Delaware Bay blue crab stock, Helser and Kahn (1999) proposed a limiting fishing mortality rate threshold, known as the replacement fishing mortality rate (F_{REP}), to define the recruitment overfishing threshold. The fishing mortality rate F_{REP} defines the fishing pattern for which, on average, adequate recruitment was generated from the stock to replace the stock biomass lost due to mortality; therefore, stock biomass and recruitment theoretically stabilize. This concept was first developed by Sissenwine and Shepherd (1987) who showed that biologically meaningful reference points could be generated from combining spawning per recruit analysis and stock–recruitment data. Sissenwine and Shepherd (1987) proposed estimating F_{REP} from the median survival ratio from stock–recruitment (S-R) data and denoted it with the F_{MED}, which we refer to as the replacement fishing mortality rate.

We used this approach to generate the replacement fishing mortality rate for the threshold biological reference point for blue crab. Based on the original stock–recruitment data set, an estimate of F_{MED} can be derived by estimating median survival ratio R/S, taking its inverse (S/R) and finding a value of F corresponding to the SSB/R curve as a function of fishing mortality from an equilibrium spawning stock biomass per recruit analysis (SSB/R). This, however, only provides one realization of F_{MED} based on the original stock–recruitment data. Since we wanted to quantify the uncertainty in the estimate of F_{MED} corresponding to the median survival ratio through the simulation of multiple S-R data sets we employed a blue crab population projection model. The conceptual basis for this approach is that while uncertainty is incorporated into yield per recruit (Y/R) and SSB/R analyses through variable M, the projection model attempts to capture the uncertainty and inherently stochastic nature of the stock's dynamics through time. The SSB/R and S-R projection model were structurally linked through random values of natural mortality drawn from the triangular distribution and seeded into both analyses simultaneously. The output from the projection model contained 1000 data sets each with 19 pairs of simulated S-R observations, thus allowing estimation of 1000 values of the median survival ratio and corresponding values of S/R. Then, coupling these S-R realizations with standard SSB/R analysis, 1,000 realizations of F_{MED} could be generated.

Survival ratios from simulated S-R data

Survival ratios (R/S) are estimated from S-R data and when coupled with a SSB/R analysis, are used to derive a reference point known as the fishing replacement level (F_{MED}). This section deals specifically with the S-R data used in this analysis and how uncertainty was incorporated. Rather than simply bootstrapping the residuals from the original S-R values estimated from the catch-survey model to generate survival ratios, we chose to use a stochastic projection model for blue crab. This procedure attempted to capture the inherent stochastic nature of the process that originally generated the S-R data by incorporating uncertainty in the form of variable M (triangular distribution) and two sources of uncertainty associated with the stock–recruitment data. The projection model depended on input sets of Ricker model parameter estimates that were generated from a conditioned nonparametric bootstrap. First, the Ricker model (1954) was fit to the original blue crab S-R data estimated from the catch-survey model using nonlinear least squares in SAS (1985) with additive normal error structure. The bootstrap procedure was then applied by randomly resampling model residual errors with equal probability and adding these at random to the predicted values of recruitment from the Ricker S-R model to generate 1000 new psuedo data sets each consisting of 19 new stock and recruitment realizations. Finally, the Ricker model was then refit to the 1000 new psuedo-replicated data sets to generate 1000 realizations of the Ricker S-R parameters. The projection model used sets of 1000 random (a^* b^*) pairs of the Ricker model parameter estimates rather than single realizations from univariate distributions to maintain the covariance structure of the parameters.

For the stochastic projection model, the blue crab population abundance can be represented as the vector $\boldsymbol{N}(t)$,

$$N(t) = \begin{bmatrix} N_0(t) \\ N_1(t) \\ N_2(t) \\ N_3(t) \end{bmatrix} \tag{11}$$

where $N_a(t)$ was the number of age-a crabs alive at the beginning of the year. Here we assumed that crabs do not live past their third year and therefore a plus group was not used to accumulate survivors after that age. To describe annual survival through mortality, let M_a denote the constant instantaneous natural mortality rate and let $F_a(t)$ denote the instantaneous fishing mortality rate for age-a crabs in year t of the 19-year time horizon. For the projection analysis, a value of M was drawn randomly from a triangular distribution centered on $M = 0.80$ (range 0.60–1.0) and used for each of the 1000 19-year time trajectories. The population size at age in year t is given by

$$N_a(t) = N_{a-1}(t-1)\, e^{-M_{a-1}-F_{a-1}(t-1)} \quad \text{for } a = 0 \text{ to } 3 \tag{12}$$

Population sizes at age from initial year estimates (1979) of the catch-survey model were seeded into the projection model and served as the starting conditions. Since the catch-survey model provides only recruit (age-0) and fully recruited (age 1–3) stock size estimates, population sizes for age 1–3 were decomposed into individual ages using the above equation and the estimated F for that year and $M = 0.80$. The vector of year-specific Fs estimated from the catch-survey model was used for each subsequent year in the 19 year projection.

The numbers of new recruits that enter the fishery at age-0 (1 year after hatching) at the beginning of each year, are determined as a function of spawning biomass with some stochastic elements. The spawning stock biomass that produced recruits the following year is

$$SSB(t) = \sum_{a=0}^{7} W_{S,a} V_a N_a(t) e^{-p[M_a+F_a(t)]} \tag{13}$$

where: $SSB(t)$ is the sum over all age-classes of the weight of mature crabs at the midpoint of the spawning season; V_a is the fraction of age-a crabs that are mature; $W_{S,a}$ is the average weight of an age-a crab; and p is the average fraction of total annual mortality that occurs from January 1st to the mid-point of the spawning season.

The Ricker model was used for simulating recruitment, where survival to recruitment age is a function of SSB with a density-dependent or compensatory response, subject to stochastic variation. This stochastic variation incorporates two sources of uncertainty. The first arises from the fact that the functional form of the Ricker model is ill defined because there were only 19 original S-R data points for the blue crab stock with which to estimate the parameters. Therefore, random realizations of pairs of Ricker parameter estimates (a^* b^*) generated from the bootstrap were used to characterize the family of possible functional forms of the Ricker S-R model:

$$R_0(t) = a^* SSB(t-1)\, e^{-b^* SSB(t-1)} \tag{14}$$

We then simulated environmental disturbances, which added variability about the expected S-R relationship from each particular random set of parameters. This random noise component to the recruitment function was added using

$$N_0(t) = R_0(t) + N(\mu, \sigma^2) \tag{15}$$

where μ is a randomly drawn value from a normal distribution with mean equal to 0 and variance equal to the 58% of the mean square error (MSE) from the Ricker fit to the original S-R data. Since the original fit of the Ricker model explained 42% of the variability in R through SSB, we attributed the remaining 58% to environmental variability. We did not explicitly consider serial autocorrelation in fitting the stock–recruitment data or correct for it in projections, although weak evidence for it existed (i.e., Durbin-Watson test statistic = 1.796., $p < 0.10$). Positive autocorrelation in time series data produces nominal variances about parameter estimates that are smaller than they should be (Bence 1995). Its impact on the simulation results would lead to increased variability than expected. Finally, median survival ratios (R/SSB) were estimated for each of the simulated time series of spawning stock biomass and recruitment values and SSB/R estimates obtained from the reciprocal of the survival ratios.

Yield and SSB-per-recruit analyses

The equilibrium Y/R and SSB/R analyses were performed using the methods of Thompson and Bell (1934) and Gabriel et al. (1989), respectively. The structure of Y/R and SSB/R simulations included: 1) ages 0 through 3 with age-3 as the terminal year, 2) range of fishing mortality rates ranging from 0 to 2.0 by increments of 0.1, and 3) constant natural mortality and growth rates. Model input for these analyses consisted of a fishery partial recruitment pattern, mean weights at age in the fishery and stock, proportion mature at age, and a point estimate of natural mortality ($M = 0.8$) for the deterministic run. Details for the calculation and estimation of inputs for Y/R and SSB/R models for Delaware Bay blue crab can be found in Helser and Kahn (1999).

The estimate of natural mortality (M) is probably the most uncertain parameter used as input for Delaware Bay blue crab Y/R and SSB/R modeling. The effects of varying M on these analyses are somewhat predictable, at least in a directional sense; higher M results in a lower absolute value of Y/R, a Y/R curve which is flat-topped and a shift of reference points such as F_{MAX} to higher values of F. The effects of different values of M on the distributional properties of reference points are less predictable. Therefore, we chose to incorporate a Monte Carlo simulation into the structure of the model to evaluate the distributional properties of biological reference points generated from Y/R and SSB/R analyses. For this, we simply reran the analysis for each realization of a value of M that was generated randomly from a triangular distribution. From this underlying distribution, 1000 values of M were generated from a possible range of values approximately from 0.60 to 1.0 with a most likely value of 0.8. As a result of these simulations a probability distribution of either F_{max} or $F_{10\%}$ was derived instead of a single fixed value and upper and lower limits of SSB/R for different levels of F were estimated.

Development of F_{MED} probability distribution

Results of the SSB/R analysis were then coupled with S-R plots for Delaware Bay blue crab to generate biomass-based reference points needed for stock enhancement or replacement. Under a given fishing pattern (F and recruitment pattern to the fishery), the value of S/R from the SSB/R analysis can be superimposed on the standard S-R plot as a line that has a slope equal to the inverse of (S/R) and an intercept of zero (Shepherd 1982; Gabriel et al. 1989). A common reference point from this analysis is F_{MED} or F_{REP} (Sissenwine and Shepherd 1987), which, for a certain fishing mortality rate, generates a line that bisects the standard S-R plot (i.e., equal number of recruitment points on each side of the line). Other reference points of this type such as F_{LOW} and F_{HIGH} can also be calculated from this analysis; these define stock and recruitment plot lines that leave 90% and 10% of the points above the line, respectively. The output of the projection model contained 1000 data sets, each with 19 pairs of simulated S-R observations, thus allowing estimation of 1000 values of the median survival ratio and corresponding values of F_{MED}. The frequency distribution of F_{MED} was treated as a probability distribution of F_{MED} as a biological reference point, that could be compared with the probability distribution of the terminal fishing mortality estimate.

Results

Parameter estimates from the catch-survey model were reasonably precise and model residual errors showed no temporal patterns (Figure 5). The bootstrap estimates of coefficients of variation, which are the preferred variance measures of the model's parameter estimates, were generally smaller

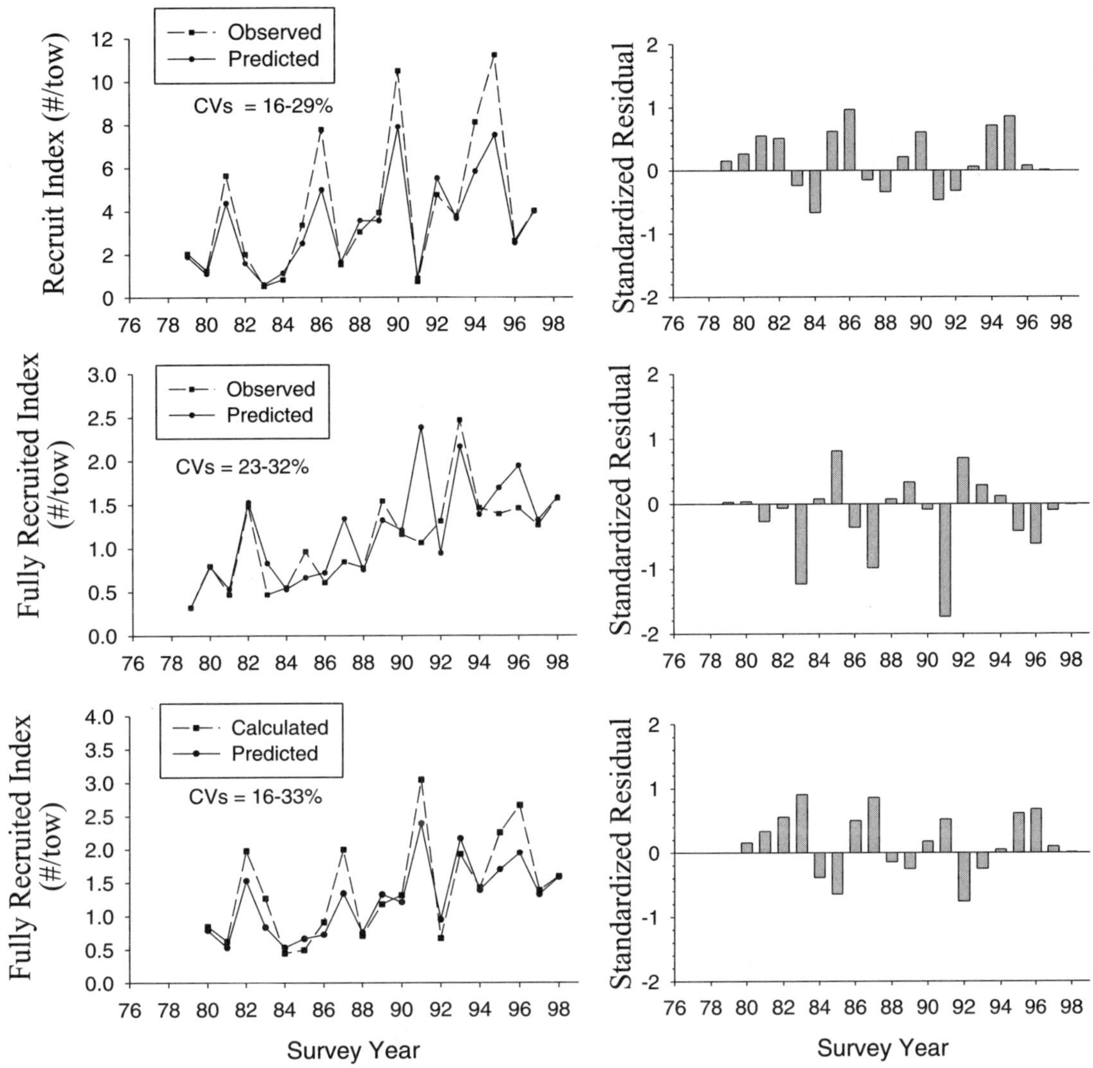

FIGURE 5. Summary of results from catch-survey model applied to Delaware Bay blue crab fishery. Shown are observed and predicted recruit and fully-recruited indices along with standardized residual plots.

than those from the nonlinear least squares (NLLS) routine. Bootstrap CVs for fully recruited stock sizes ranged from 23% to 32%, CVs for recruit stock sizes ranged from 16% to 29%, and CV for the catchability coefficient was 16%.

Fishing mortality rates on fully recruited blue crabs ranged from a low of 0.20 to a high of 1.45, and averaged 0.83 over the 1979–1997 period (Figure 6). Reflective of increases in fishing effort (thousands of crab pots) after 1985, fishing mortality rates increased compared with earlier years but have fluctuated without trend. Because of the instability of the terminal year in the catch-survey model and because Fs in the blue crab fishery have fluctuated greatly from year to year, we propose using the average of the three terminal years in the analysis to determine stock status. The average terminal fishing mortality rate (F_{95-97}) was estimated to be 0.93 for fully recruited blue crabs. Coefficients of variation associated with fishing mortality rates were estimated from five hundred bootstrap replicates, which were derived from randomly resampling measurement errors estimated from a base run of the catch-survey model. Coefficients of variation associated with annual fishing mortality estimates were highly variable in some years. In particular, CVs on Fs in 1983 and 1984 exceeded 75%, and therefore 80% confidence intervals encompass zero (Figure 6). Excluding these years

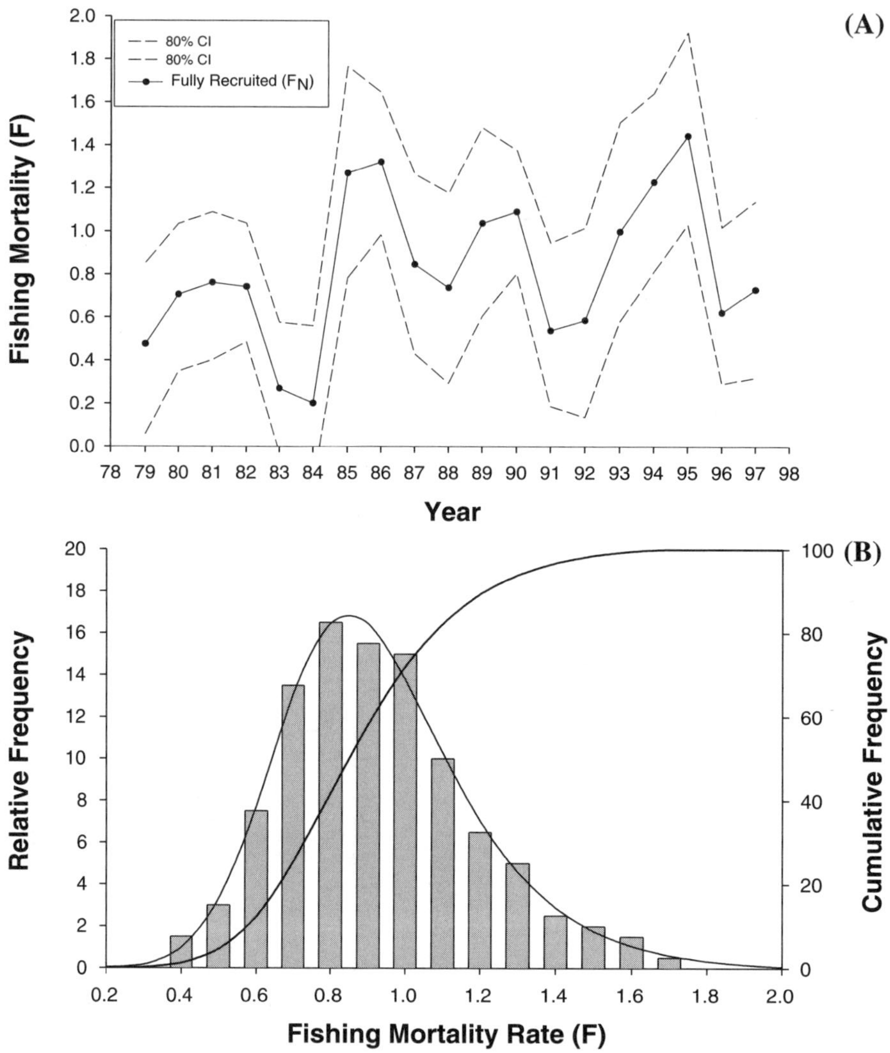

FIGURE 6. (A) Instantaneous fully recruited fishing mortality rates on Delaware Bay blue crab estimated from the catch-survey model. (B) Empirical distribution of average fishing mortality rates (95–97) estimated from the catch-survey model using the conditioned nonparametric bootstrap with 500 replications.

and the first year in the analysis, most CVs on estimated fishing mortality rates were less than 50%. Despite the variability associated with estimated fishing mortality rates for a number of years, CVs on the average terminal year Fs (F_{95-97}) were only 25%, indicating that these were relatively precisely estimated. Confidence intervals (80%) about the expected value of fishing mortality on fully recruited blue crabs ranged within 0.60–1.20.

Our approach to incorporating uncertainty into the calculation of the biological reference point, specifically F_{MED}, was to characterize the uncertainty associated with the stock and recruitment analysis and the value of natural mortality. It is apparent from Figure 7 that there is considerable variability in the blue crab stock and recruitment data; only 43% of the variability in recruitment was explained by variations in spawning stock biomass as suggested by the Ricker model. This empirical model suggests both a high degree of resilience based on the initial slope of the curve and significant compensation in recruitment with increasing levels of stock size. Because of the limited number of empirical data points the functional form of the S-R model may be

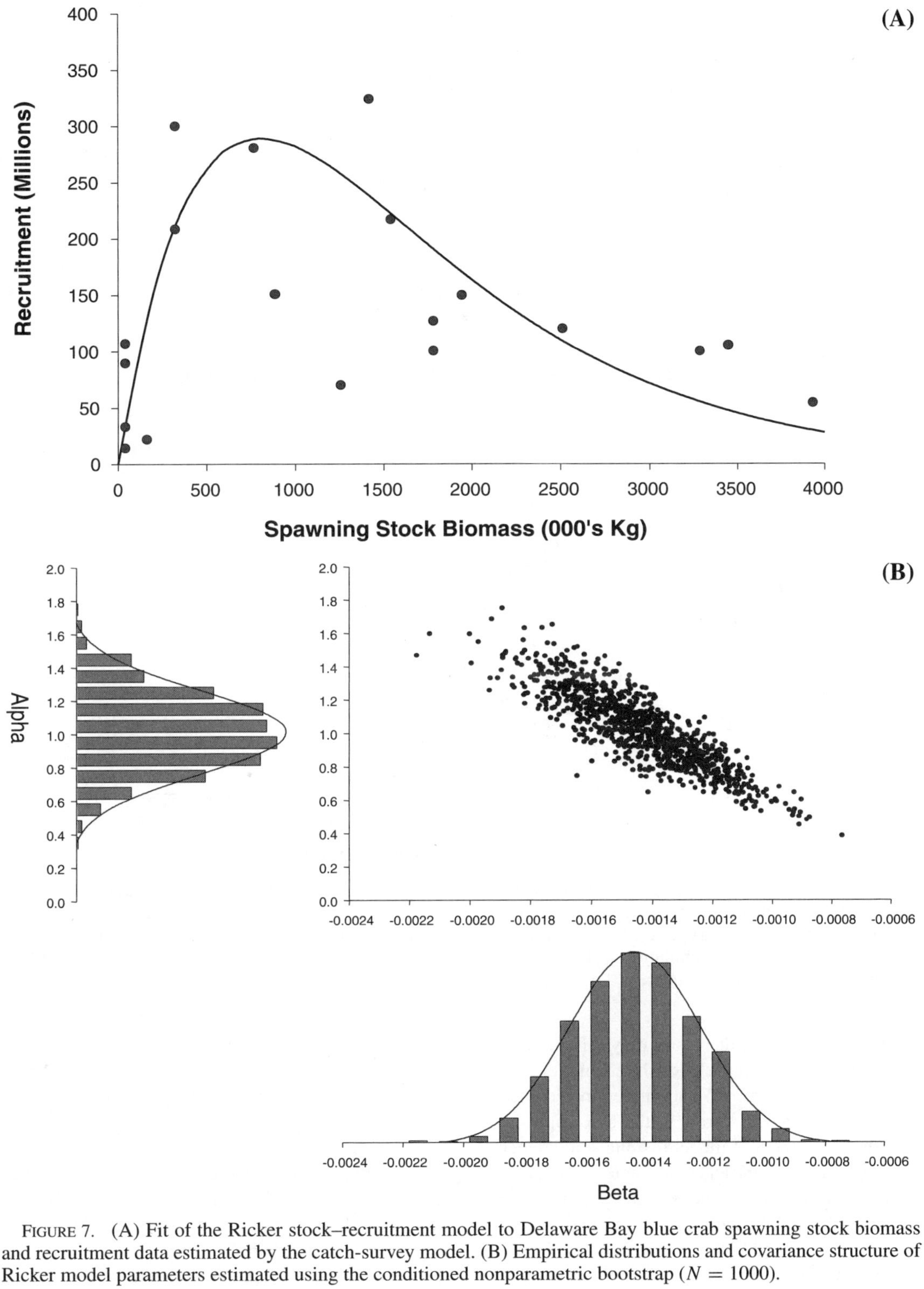

FIGURE 7. (A) Fit of the Ricker stock–recruitment model to Delaware Bay blue crab spawning stock biomass and recruitment data estimated by the catch-survey model. (B) Empirical distributions and covariance structure of Ricker model parameters estimated using the conditioned nonparametric bootstrap ($N = 1000$).

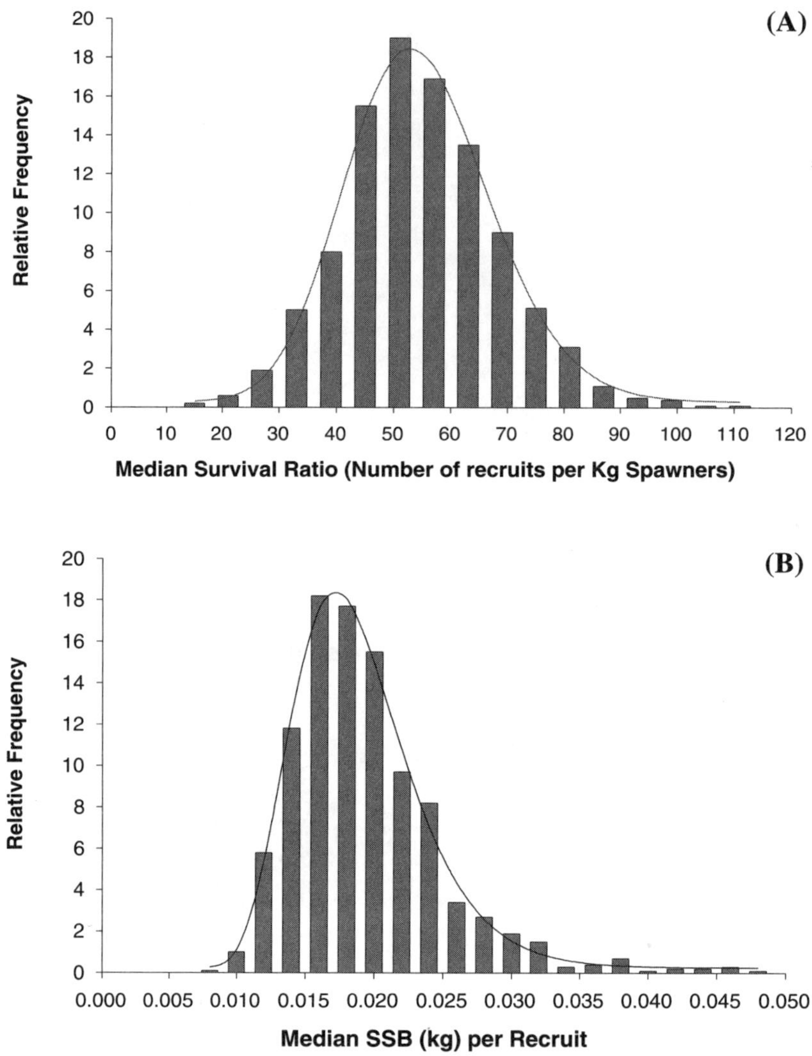

FIGURE 8. (A) Empirical distribution of the median survival ratio (number of recruits per kg spawners) generated from blue crab simulation of stock and recruitment. (B) Empirical distribution of the inverse of median survival ratio giving the median SSB (kg) per recruit generated from the blue crab simulation. Realizations of SSB-per-recruit generated from stochastic simulation will be superimpose on SSB-per-recruit analysis to derive realizations of the replacement fishing mortality rate.

poorly specified, and therefore, may differ from the true stock–recruitment relationship. This represents a type of process error. We attempted to account for this source of process error in the S-R analysis by bootstrapping the S-R model residuals and generating 1000 pairs of the Ricker parameters to characterize other possible realizations of the functional form of the true model. Based on the bivariate distribution of bootstrap derived Ricker parameters, the family of realizations of the Ricker models indicate a high degree of resilience, with alpha parameters ranging from 0.4 to 1.6 (×1000) recruits per kilogram of spawning stock (Figure 7). These realizations also reveal a range of compensatory effects, with beta ranging from −0.0022 to −0.0008.

The empirical probability distribution of median survival ratios (R/S) generated from 1000 simulated S-R scatter plots is shown in Figure 8. This distribution, as well as its inverse function (S/R), is clearly log-normally distributed (Figure 8). For the simulated S-R data, median survival ratios ranged from roughly 20–100 recruits/kg spawners, with a median

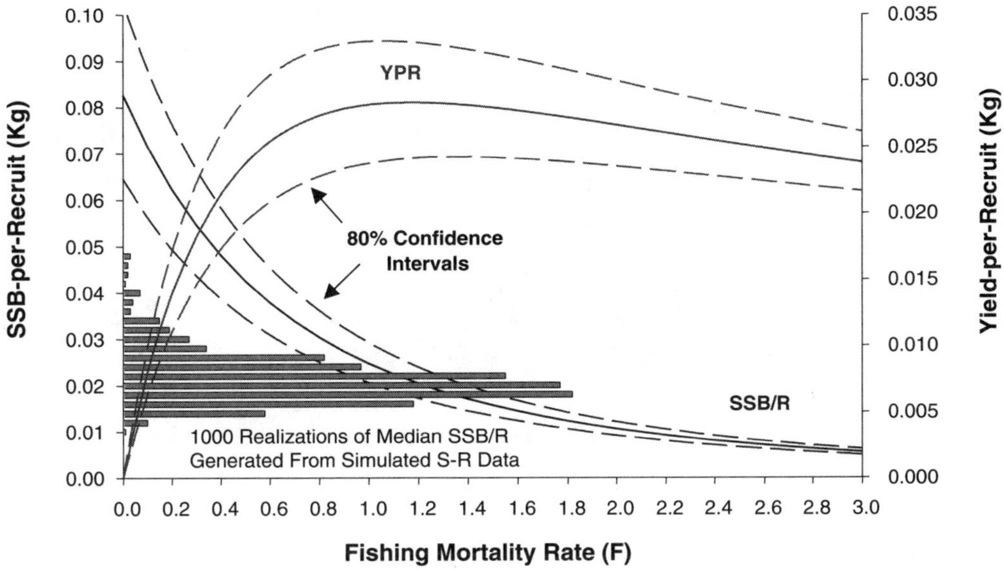

FIGURE 9. Results of uncertainty analysis showing variability about the expected yield and SSB-per-recruit with increasing levels of fishing mortality. Empirical distribution of SSB/R from simulated stock and recruitment realizations generated from the stochastic projection model is superimposed on the SSB-per-recruit scale.

value of 55 (Figure 8). When the inverse is taken, these survival ratios translate into median S/R values ranging from 0.010 to 0.045 kg S/R, with a median value of 0.020. Finally, coupling these realizations of S/R generated from the simulated S-R data with the SSB/R analysis, the median fishing mortality rate at replacement (F_{MED}) can be found for each realization. Figure 9 illustrates the full extent of uncertainty captured in this analysis showing the empirical distribution of S/R realizations from the simulated S-R data superimposed on the standard SSB/R graph. Variability in the estimate of natural mortality was incorporated into both the SSB/R analysis and in simulating the S-R data. Thus, confidence intervals around yield and SSB/R curves shown in Figure 9 illustrate the extent that uncertainty in M effect the model output. The point where the expected value of S/R from the S-R realizations intersects the expected value of S/R from the SSB/R analysis, is an estimate (not necessarily the expected value) of the replacement fishing mortality rate (F_{MED}).

The most cogent aspect of this analysis is not the point estimate but rather the empirical probability distribution of the replacement fishing mortality rate. The probability distribution of the replacement fishing mortality rate (F_{MED}) is provided in Figure 10 in concert with the terminal year fishing mortality rate (F_{95-97}). Considerable uncertainty exists in the estimation of F_{MED}, with the empirical probability distribution negatively skewed and values ranging from 0.5 to 2.1. In contrast, the probability distribution for the terminal year fishing mortality rate is positively skewed with values ranging from 0.4 to 1.7. If the deterministic approach were used to evaluate status of the stock by comparing median values, it would be concluded that F_t is substantially below F_{MED} and therefore the stock is not at risk. However, without characterizing the uncertainty in both estimates through development of their probability distributions (which overlap significantly) our level of confidence in that decision would be largely unquantifiable. Thus, the probability profile shown in Figure 10 calculates for each level of decision confidence (i.e., 90%, 80%, 70% based on the empirical F_{MED} PDF), the $\Pr(F_t > F_{\text{MED}})$. As an illustration, at the 80% decision confidence level, which is the value of F corresponding to the upper 20% of the F_{MED} PDF, the $\Pr(F_{95-97} > F_{\text{MED}}) = 0.03$. Thus, with high confidence (80%) it would be concluded that the Delaware Bay blue crab stock is not currently being overfished. As a point of comparison, we applied the same approach to evaluate the blue crab stock status using the same probability distribution of F_{MED} but to a period of years with relatively higher fishing mortality rates (1993–1995). Here, the probability distribution of the terminal years fishing mortality (F_{93-95}) shows a substantially greater overlap ($F = 0.8 - 2.4$) with the empirical probability distribution

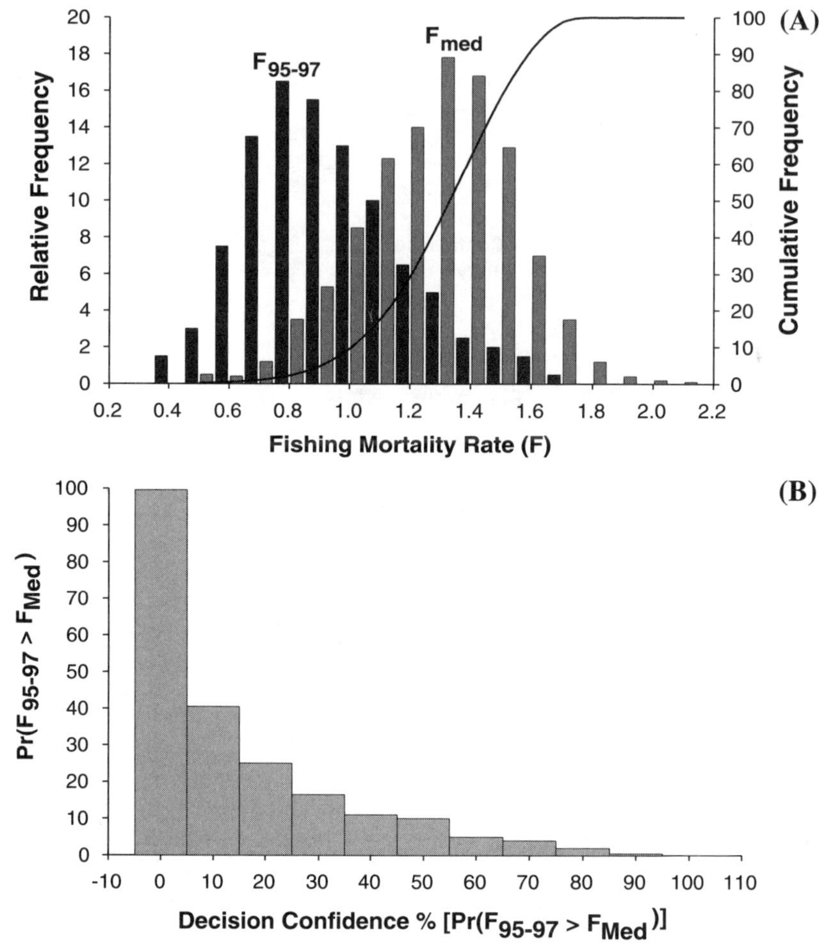

FIGURE 10. (A) Probability density functions for the average fully recruited fishing mortality rate for 1995–1997 (F_{95-97}) and the replacement fishing mortality rate (F_{MED}) for Delaware Bay blue crab. (B) Probability profile specifying the probability that recent fishing mortality rates are greater than the replacement fishing mortality rate, $\Pr(F_{95-97} > F_{\text{MED}})$ for various decision confidence levels along the x-axis.

of F_{MED}, and in some instances higher terminal fishing mortality rates (Figure 11). In contrast to the Fs for the later 1995–1997 period, $\Pr(F_t > F_{\text{MED}})$ during the 1993–1995 period were substantially greater for any decision confidence level. Using the 80% decision confidence level, the $\Pr(F_{93-95} > F_{\text{MED}}) = 33\%$.

Discussion

The aim of this paper was to characterize the uncertainty in both the blue crab population dynamics model that generated current fishing mortality rates and uncertainty in the biological reference point to which F is compared. We expressed uncertainty as empirical probability distributions of these quantities, and then developed a framework using a probability profile to evaluate the likelihood that the blue crab stock is currently in a state of overfishing. While numerous studies examined the uncertainty in stock assessment models (Restrepo et al. 1992; Conser 1994) and uncertainty in biological reference points (Smith et al. 1993; Cordue and Francis 1994) separately, few that we are aware of have attempted to integrate these uncertainty analyses into a single framework with which to evaluate the quality of decisions regarding the overfishing status of the resource of interest. The standard approach that has emerged in the northeast United States and Canada is to generate a probability profile that expresses uncertainty in the current estimate of F from a bootstrap analysis of the assessment model and presents

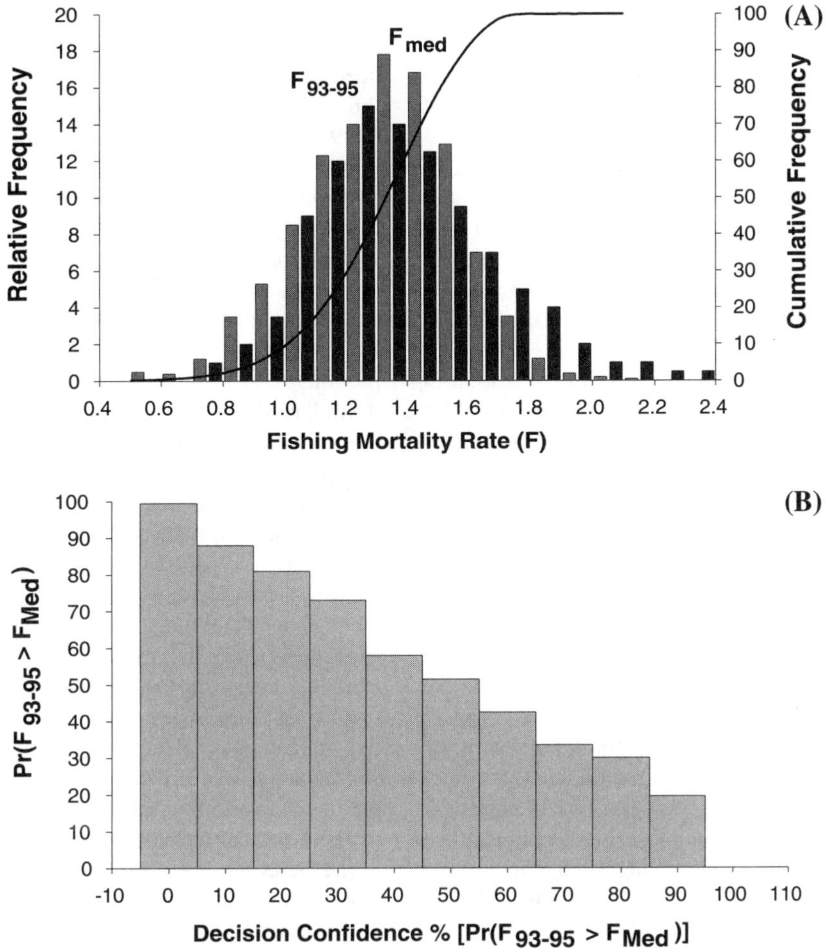

FIGURE 11. (A) Probability density functions for the average fully recruited fishing mortality rate for 1995–1997(F_{93-95}) and the replacement fishing mortality rate (F_{MED}) for Delaware Bay blue crab. (B) Probability profile specifying the probability that recent fishing mortality rates are greater than the replacement fishing mortality rate, $\Pr(F_{93-95} > F_{\mathrm{MED}})$ for various decision confidence levels along the *x*-axis.

advisory statements in the form "there is an X% probability that the current year fishing mortality rate is greater than F corresponding to the overfishing definition" (Rosenberg and Restrepo 1994). The intrinsic problem with this approach is that it fails to account for the uncertainty in estimation of the biological reference point. Fogarty et al. (1992, 1996) and Gabriel (1994) clearly recognized the inherent nature of uncertainty in both stock assessment models that generate terminal fishing mortality rates and the models that estimate biological reference points (F_{MED}), and characterized the empirical probability distributions of each for Georges Bank cod and haddock through the bootstrap. We applied this general philosophical approach of constructing a probability framework that accounts for uncertainty in both distributions for evaluating the status of Delaware Bay blue crab.

In our uncertainty analysis, we used a mixed Monte Carlo-bootstrap procedure and attempted to account for estimation error, process error, and to some degree model error. Since we used an observation-process error model (catch-survey model) (Collie and Sissenwine 1983) we applied the conditional nonparametric bootstrap, which resamples model residual errors to generate the empirical distribution of the current fishing mortality rate. Generation of the probability distribution for F_{MED} was a bit more complex and required accounting for uncertainty in both the SSB/R model and stock–recruitment (S-R) data. Instead of simply bootstrapping residual errors in the S-R data and assuming

SSB/R to be estimated without error as Fogarty et al. (1996) and Gabriel (1994) did, we incorporated uncertainty in the natural mortality rate through a Monte-Carlo simulation. While additional sources of uncertainty seem likely from other SSB/R input parameters, Jakobsen (1993) demonstrated that F_{MED} was relatively stable under variation in input parameters. However, based on our SSB/R recruit analysis for blue crab, variation in M did seem to impart significant variability in model outputs, probably due to the fact that blue crab have a relatively short life span (3–4 years).

For each trial, the same random realization of M that was seeded into the SSB/R analysis was also simultaneously seeded into the stochastic projection model that generated simulated stock and recruitment observations. Here we were specifically interested in accounting for process and model error with regard to the stock and recruitment data. Since only 19 stock and recruitment observations were estimated from the catch-survey model, we conceded that it might be easy to misspecify the functional form of the Ricker model. The observed empirical fit to the data clearly indicated significant compensation in the stock and recruitment dynamics of the blue crab population. Other investigators have suggested that compensatory processes are important in blue crab population dynamics, and have thus fit the Ricker-type model to data (Tang 1985; Lipcius and Van Engel 1990; Rugolo et al. 1998). In addition, information on blue crab life history suggests that crabs are cannibalistic (Hines and Ruiz 1995; Dittel et al. 1995) and therefore a compensatory mortality mechanism seems likely. But how strong is the compensatory response? Because of the lack of empirical observations, in general, and specifically for larger stock sizes (only 4 observations > 2 million kg), simulating stock and recruitment data from a possible family of functional forms of Ricker models defined by 1,000 realizations of alpha/beta pairs is a reasonable approach to account for model (misspecification) error. In addition, the projection model that incorporated variable M, uncertainty in the functional form of the S-R relationship and random environmental variation was developed to provide a realistic representation of the process error.

Under threshold management policies, control rules are developed to initiate management actions that immediately end overfishing when, for instance, a maximum overfishing threshold rate is surpassed. For instance, F_{MED} as an approximation of the replacement fishing mortality rate (F_{REP}) has been suggested as a recruitment overfishing threshold (Sissenwine and Shepherd 1987). This definition has been proposed as a maximum fishing mortality rate threshold for the blue crab fishery (Helser and Kahn 1999). We suggest that this recruitment overfishing threshold for Delaware Bay blue crab is inherently risk averse for two reasons. First, $F_{10\%}$ which is a commonly used recruitment overfishing threshold for other crustaceans with short life span and high fecundity, was estimated to be substantially higher than F_{MED} for Delaware Bay blue crab (Helser and Kahn 1999). The second reason is that F_{REP} (or approximated as F_{MED}) is a conservative estimate when the null model of density-independence is rejected in favor of compensation (Mace and Sissenwine 1993). It appeared to us that compensation was clearly important in the blue crab population dynamics. However, the need to balance a risk averse management strategy with regard to the biological health of the stock with the economic and social risks of restrictive management measures must be considered. If control rules specify that management actions be taken to immediately end overfishing on the stock when the fishing mortality rate exceeds F_{MED}, then restrictive regulations on the fishery could have potentially significant and severe economic and social impacts on the fishing community. This is clearly evidenced by the fact that blue crab support important fisheries in the Delaware Bay; commercial landings have averaged more than 10 million pounds annually with an annual estimated exvessel value of nearly $6 million in recent years (Cole 1998).

This uncertainty analysis for the blue crab fishery is only the first step in the construction of a formal risk assessment. Additional work to more fully characterize all important sources of uncertainty in the assessment process is needed, as well as the development of a utility or loss function that considers both biological and economic risk to the stock and fishing community, respectively. While certainly not the only approach possible, our formulation of a stochastic decision framework specifies the probability that overfishing is occurring $\Pr(F_t > F_{\text{MED}})$, but also provides, as a second tier in the assessment of the stock, the statistical level of confidence (i.e., $\alpha = 0.05$) associated with that decision. While the probability profile provides the overfishing status, $\Pr(F_t > F_{\text{MED}})$ for numerous decision confidence levels, we suggest that a simulation study is necessary to evaluate outcomes of using one decision level over another (e.g., 80% versus 90%) with associated probability states $\Pr(F_t > F_{\text{MED}})$ of the blue crab stock before an appropriate maximum threshold mortality rate decision rule can be proposed. For instance, realizations of fishing mortality rates could be selected with the probabilities as defined by the empirical distribution of F_t, conditional of the various confidence regions defined by the empirical distribution of F_{MED}, and seeded into the projection model already developed to evaluate population state variables through time. In concert, additional simulations would be conducted to determine the risks to the stock and fishery (through development of a formal utility function) that would result from various restrictive harvest scenarios to prevent overfishing. Finally, this approach and further simulation studies can be extended to decisions regarding control laws that specify both maximum fishing mortality rate and minimum biomass thresholds.

Acknowledgment

We thank Lisa Kline of the Atlantic States Marine Fisheries Commission and Jim Berkson and Don Orth from Virginia Tech who convened the Workshop Incorporating Uncertainty in Fishery Models where this work was first presented. We also thank David Eggleston and Martin Posey for the opportunity to present this work at the Blue Crab Symposium held in conjunction with the 2000 Annual Benthic Ecology meeting. Thanks are also extended to the two anonymous reviewers. This research was funded by the Atlantic Coastal Fisheries Cooperative Management Act, NOAA grant #NA96FG0319.

References

Bence, J. R. 1995. Analysis of short time series: correcting for autocorrelation. Ecology 76:628–639.

Cargo, D. G. 1954. Maryland commercial fishing gears. III. The crab gears. Chesapeake Biological laboratory, Education Series 36.

Churchill, E. P. 1919. Life history of the blue crab. Bull. U.S. Bureau of Fisheries for 1917–1918. 36:91–128.

Claytor, R. R., G. A. Nielsen, and P. A. Shelton. 1993. Using jackknife and Monte Carlo simulation experiments to evaluate forecast models for Atlantic salmon (*Salmo salar*). Pages 203–219 *in* S. J. Smith, J. J. Hunt, and D. Rivard, editors. Risk evaluation and biological reference points for fisheries management. Can. Spec. Publ. Fish. Aquat. Sci. 120.

Cole, R. W. 1998. Changes in harvest patterns and assessment of possible long term impacts on yield in the Delaware commercial blue crab fishery. Journal of Shellfish Research 17(2):469–474.

Cole, R. V., W. Whitmore, and D. M. Kahn. 1997. Recreational blue crab effort and harvest in the Delaware estuary. Delaware Division of Fish and Wildlife, Dover, Delaware.

Collie, J. S., and M. P. Sissenwine. 1983. Estimating population size from relative abundance data measured with error. Can. J. Fish. and Aquat. Sci. 40:1871–1879.

Conser, R. J. 1994. Stock assessment methods designed to support fishery management decisions in data-limited environments: development and application. Doctoral dissertation. University of Washington, Seattle.

Conser, R. J., and J. Idoine. 1992. A modified DeLury model for estimating mortality rates and stock sizes of American lobster populations. NEFSC Research Document, SAW 14/7.

Cordue, P. L., and R. I. C. C. Francis. 1994. Accuracy and choice in risk estimation for fisheries assessment. Canadian Journal of Fisheries and Aquatic Science 51:817–829.

Dittel, A. I., A. H. Hines, G. M. Ruiz, and K. K. Ruffin. 1995. Effects of shallow water refuge on behavior and density-dependent mortality of juvenile blue crabs in Chesapeake Bay. Bull. Mar. Sci. 57:902–916.

Efron, B., and R. J. Tibshirani. 1993. An Introduction to the Bootstrap. Chapman and Hall, New York.

Epifanio, C. B. 1995. Transport of blue crab (*Callinectes sapidus*) larvae in the waters off Mid-Atlantic states. Bull. Mar. Sci. 57:902–916.

Fischler, K. J. 1965. The use of catch-effort, catch sampling and tagging data to estimate populations of blue crabs. Transactions of the American Fisheries Society 94: 287–310.

Fischler, K. J., and C. H. Walburg. 1962. Blue crab movements in coastal South Carolina. Transactions of the American Fisheries Society 91:275–278.

Fogarty, M. J., R. K. Mayo, L. O'Brien, F. M. Serchuk, and A. A. Rosenberg. 1996. Assessing uncertainty and risk in exploited marine populations. Reliability Engineering and System Safety 54:183–195.

Fogarty, M. J., A. A. Rosenberg, and M. P. Sissenwine. 1992. Fisheries risk assessments: sources of uncertainty. Environ. Sci. Tech. 26:440–447.

Francis, R. I. C. C. 1993. Monte Carlo evaluation of risks for biological reference points used in New Zealand fishery assessments. Pages 221–230 *in* S. J. Smith, J. J. Hunt, and D. Rivard, editors. Risk evaluation and biological reference points for fisheries management. Can. Spec. Publ. Fish. Aquat. Sci. 120.

Francis, R. I. C. C. 1992. Use of risk analysis to assess fishery management strategies: a case study using orange roughy on the Chatham Rise, New Zealand. Canadian Journal of Fisheries and Aquatic Science 49:922–930.

Gabriel, W. L. 1994. A simple method for estimating uncertainty associated with F_{MED}. ICES C. M. 1994/D: 5., 1994.

Gabriel, W. L, M. P. Sissenwine, and W. J. Overholtz. 1989. Analysis of spawning stock biomass per recruit: An example for Georges Bank haddock. North American Journal of Fisheries Management 9:383–391.

Helser, T. E., and D. M. Kahn. 1999. Stock assessment of Delaware Bay blue crab (*Callinectes sapidus*) for 1999. Department of Natural Resources and Environmental Control, Delaware Division of Fish and Wildlife, Document 40-05/99/02/01, Dover, Delaware.

Hilborn, R., E. K. Pikitch, and R. C. Francis. 1993. Current trends in including risk and uncertainty in stock assessments and harvest decisions. Canadian Journal of Fisheries and Aquatic Science 50:874–880.

Hines, A. H., and G. M. Ruiz. 1995. Temporal variation in juvenile blue crab mortality: nearshore shallows and cannibalism in Chesapeake Bay. Bull. Mar. Sci. 57: 884–901.

Jakobsen, T. 1993. The behavior of F_{LOW}, F_{MED}, and F_{HIGH} in response to variation in parameters used for their estimation. Pages 119–125 *in* S. J. Smith, J. J. Hunt, and D. Rivard, editors. Risk evaluation and biological reference points for fisheries management. Can. Spec. Publ. Fish. Aquat. Sci. 120.

Ju, S. J., D. H. Secor, and H. R. Harvey. 1999. Use of extractable lipofuscin for age determination of blue crabs (*Callinectes sapidus*). Mar. Ecol. Prog. Ser. 185:171–179.

Ju, S. J., D. H. Secor, and H. R. Harvey. In press. Growth rate variability, and lipofuscin accumulation rates in the blue crab. Mar. Ecol. Prog. Ser.

Kahn, D. M., R. W. Cole, S. F. Michels, and W. H. Whitmore. 1998. Development of life-stage-specific indices of relative abundance and stock-recruitment relationships for the Delaware Bay blue crab stock. J. Shellfish Research 17(2):529–541.

Kennish, M. J., J. J. Vouglitois, D. J. Danila, and R. A. Lutz. 1984. Shellfish. Pages 170–200 *in* M. J. Kennish and R. A. Lutz, editors. Ecology of Barnegat Bay, New Jersey. Springer-Verlag, New York.

Lipcius, R. N, and W. A. Van Engel. 1990. Blue crab population dynamics in Chesapeake Bay: variation in abundance (York River, 1972–1988) and stock-recruit functions. Bull. Mar. Sci. 46:180–194.

Mace, P. M., and M. P. Sissenwine. 1993. How much spawning per recruit is enough. Pages 101–118 *in* S. J. Smith, J. J. Hunt, and D. Rivard, editors. Risk evaluation and biological reference point for fisheries management. Can. Spec. Publ. Fish. Aquat. Sci. 120.

Mohn, R. K. 1993. Bootstrap estimates of ADAPT parameters, their projection in risk analysis and their retrospective patterns. Pages 173–184 *in* S. J. Smith, J. J. Hunt, and D. Rivard, editors. Risk evaluation and biological reference point for fisheries management. Can. Spec. Publ. Fish. Aquat. Sci. 120.

Perry, H. M., J. Warren, C. Trigg, and T. Van Devender. 1998. The blue crab fishery of Mississippi. J. Shellfish Research 17:425–433.

Perry, H. M. 1975. The blue crab fishery in Mississippi. Gulf Research Repts. 5:39–57.

Restrepo, V. R., J. M. Hoenig, J. E. Powers, J. W., and S. C. Turner. 1992. A simple simulation approach to risk and cost analysis, with application to swordfish and cod fisheries. Fish. Bull. U.S. 90:736–748.

Ricker, W. E. 1954. Stock and recruitment. J. Fish. Research Board. Can 11:559–623.

Rosenberg, A. A., and V. R. Restrepo. 1994. Uncertainty and risk evaluation in stock assessment advice for U.S. marine fisheries. Canadian Journal of Fisheries and Aquatic Science 51:2715–2720.

Rothschild, B. J., and A. F. Sharov. 1996. Abundance estimation and population dynamics of the blue crab in the Chesapeake Bay. Chesapeake Biological Laboratory. Reference No. UMCEES[CBL]95–207.

Rugolo, L. J., K. S. Knotts, A. M. Lange, and V. A. Crecco. 1998. Stock assessment of Chesapeake Bay blue crab (*Callinectes sapidus*). Journal of Shellfish Research 17(2):493–517.

Shepherd, J. G. 1982. A versatile new stock-recruitment relationship of fisheries and construction of sustainable yield curves. J. Cons. Int. Explor. Mer. 40:67–75.

Sissenwine, M. P., and J. G. Shepherd. 1987. An alternative perspective on recruitment overfishing and biological reference points. Canadian Journal of Fisheries and Aquatic Science 44:913–918.

Smith, S. J., J. J. Hunt, and D. Rivard, editors. 1993. Risk evaluation and biological reference points for fisheries management. Can. Spec. Publ. Fish. Aquat. Sci. 120.

Tagatz, M. E. 1968a. Growth of juvenile blue crabs, *Callinectes sapidus* Rathbun, in the St. John's River, Florida. U.S. Fish. Wild. Bull. 67:281–288.

Tagatz. M. E. 1968b. Biology of the blue crab, *Callinectes sapidus* Rathbun, in the St. John's River, Florida. Fish. Bull. 67:17–33.

Tang, Q. 1985. Modification of the Ricker stock recruitment model to account for environmentally induced variation in recruitment with particular reference to the blue crab fishery in Chesapeake Bay. Fish. Res. 3:13–21.

Thompson, W. F., and F. H. Bell. 1934. Biological statistics of the Pacific halibut fishery. 2. Effect of changes in intensity upon total yield and yield per unit of gear. Rep. Int. Fish. (Pacific Halibut). Comm. 6:108 p.

Uphoff, J. H. 1998. Stability of the blue crab stock in Maryland's portion of Chesapeake Bay. J. Shellfish Research 17(2):519–528.

Van Engle, W. A. 1958. The blue crab and its fishery in Chesapeake Bay. Part 1: Reproduction, early development, growth and migration. Comm. Fish. Rev. 20:6–17.

Van Engle, W. A. 1982. Factors affecting the distribution and abundance of the blue crab in Chesapeake Bay. Pages 177–209 *in* S. K. Majumdar, L. W. Hall, Jr., and H. M. Austin, editors. Contaminant problems and management of living Chesapeake Bay resources. Pennsylvania Academy of Sciences, Philadelphia, Pennsylvania.

Vetter, E. F. 1985. Estimation of natural mortality in fish stocks: a review. Fish. Bull. U.S. 86(1): 25–43.

Walters, C., and Ludwig, D. 1994. Calculation of Bayes posterior probability distributions of key population parameters. Canadian Journal of Fisheries and Aquatic Science 51:713–722.

Williams, A. B. 1974. The swimming crabs of the genus *Callinectes* (Decapoda: Portunidae). Fish. Bull. 77:685–798.

American Fisheries Society Symposium 27:83–112, 2002

Consideration of Uncertainty in Stock Assessment of Atlantic Menhaden

D. S. VAUGHAN, M. H. PRAGER, AND J. W. SMITH
National Marine Fisheries Service, NOAA, Beaufort Laboratory, 101 Pivers Island Road Beaufort, North Carolina 28516, USA

Abstract.—Stock assessments of Atlantic menhaden are conducted annually for the Atlantic States Marine Fisheries Commission, as required by the recently updated Fishery Management Plan, adopted in 1992. Uncertainties in stock assessments have been explored over the years from many perspectives. Two general areas of analysis are considered here. The first area is largely deterministic and concerns the virtual population analysis (VPA), including development and coherence of the catch-at-age matrix; historical retrospective problems; implications of assuming constant M at all ages analyzed; and reliability of recruitment estimates relative to fishery-independent juvenile abundance indices when used for calibrating the VPA. The second area of consideration comprises stochastic analyses, including stochastic projections based on biological benchmarks determined from yield-per-recruit and spawning-stock-biomass-per-recruit models; bootstrapped application of a surplus-production model; and projections from that production model. Nonetheless, the largest uncertainty in assessment of the stock stems not from modeling considerations, but is a biological question: Can the high stock levels observed in the 1950s be regained by reducing fishing mortality? Projections based on production modeling assume that they can, but if exogenous forces (for example, habitat loss or pollution) have affected the stock, it may be that they cannot. If the recent pattern of lower fishing mortality rates in response to social and economic factors continues, the fishery will in essence conduct an experiment that may answer the question.

Introduction

The purpose of this study is to illustrate areas of uncertainty that have been or are being addressed within the context of Atlantic menhaden stock assessments and to suggest additional areas for future consideration. By "uncertainty," we mean imprecision or bias in data, model choice, and estimation methods, as well as in the resulting assessment results, specifically in benchmarks used for management. Much of the information contained in this paper is provided annually to the Atlantic Menhaden Advisory Committee (AMAC) of the Atlantic States Marine Fisheries Commission.

Natural history and exploitation

Atlantic menhaden *Brevoortia tyrannus* is a euryhaline clupeid found in coastal and intertidal waters from West Palm Beach, Florida, to Nova Scotia, Canada (Reintjes 1969). They form surface schools off Florida, Georgia, and the Carolinas in April and May as the waters warm, and move slowly northward, stratifying by age and size during summer; older and larger fish are generally found farther north. Southward migration begins in early fall, with surface schools last seen in late December or early January off the Carolinas. Spawning occurs principally at sea, from May to October off southern New England and the mid-Atlantic states and from October to April off the southeastern United States. Eggs hatch at sea and larvae are moved into estuaries by ocean currents (Nelson et al. 1977), where they metamorphose into juveniles. In late fall and early winter, juveniles leave the estuaries and move into large bays or open waters. Adult menhaden are filter feeders feeding primarily on phytoplankton and, in turn, support predatory fishes and sea birds.

Sampling of juvenile Atlantic menhaden by the National Marine Fisheries Service began in 1955. In the 1970s, sampling activities culminated in extensive coastwide trawl surveys, conducted through 1978 (Ahrenholz et al. 1989). A stream survey (two streams each in North Carolina and Virginia) was continued through 1986. Ahrenholz et al. (1989) found no significant correlations between relative juvenile abundance estimates and later fishery-dependent estimates of year-class strength. However, recent investigation with state data sets (described later) have shown more promise (Vaughan and Smith 1999), but more work is needed to better separate a small number of age-1+ menhaden from juveniles (age-0).

As noted in Ahrenholz et al. (1987), fishing on Atlantic menhaden has occurred since colonial times, but the purse seine fishery for reduction began in New England about 1850. Landings and nominal effort (measured as number of weeks a vessel unloaded during the fishing year, vessel-weeks) are available since 1940 (Figure 1). Landings rose during the 1940s (from 167,000 to 376,000 t), peaking during the 1950s (high of 712,000 t in 1956), and then declined during the 1960s (from 576,000 t in 1961 to 162,000 t in 1969). During the 1970s the stock rebuilt (landings rose from 250,000 t in 1971 to 376,000 t in 1979), and then maintained intermediate levels during the 1980s (varying between 238,000 t in 1986 when fish meal prices were extremely low to 418,600 t in 1983). Landings during the 1990s have varied between 245,900 t in 1998 and 401,200 t in 1990. Through analysis of extensive catch-effort data bases, Clark and Mangel (1979) demonstrated that catchability is frequently density dependent and thus catch per unit effort (CPUE) and nominal fishing effort are poor measures of population abundance and fishing mortality rate in purse seine fisheries. Schaaf (1979) demonstrated density-dependent catchability in the Atlantic menhaden stock, with a statistical relationship that fits remarkably well ($R^2 = 0.77$).

The number of fish reduction plants, the main users of the resource, has declined from more than twenty during the late 1950s to two in 2000; one in Reedville, Virginia, and one in Beaufort, North Carolina. Similarly, the number of purse-seine vessels in the reduction fishery has declined from more than one hundred thirty vessels during the late 1950s to twelve during 2000.

Assessment and management

Annual estimation of six "trigger" variables is required by the Atlantic menhaden fishery management plan [Atlantic Menhaden Advisory Committee (AMAC) 1992] (Table 1). These variables were developed by the AMAC for evaluation of the health of the Atlantic menhaden stock (AMAC 1992). The first three variables are estimated directly from fishery statistics: (1) reported landings in weight from the reduction fishery, (2) proportion of age-0 Atlantic menhaden in the landings by number, and (3) proportion of adult (age-3+) Atlantic menhaden in the landings by number. Three remaining trigger variables are obtained from virtual population analysis (VPA), and have less precision associated with current year estimates. These variables include (4) recruitment to age 1 in numbers, (5) spawning stock biomass (weight of mature females in metric tons estimated

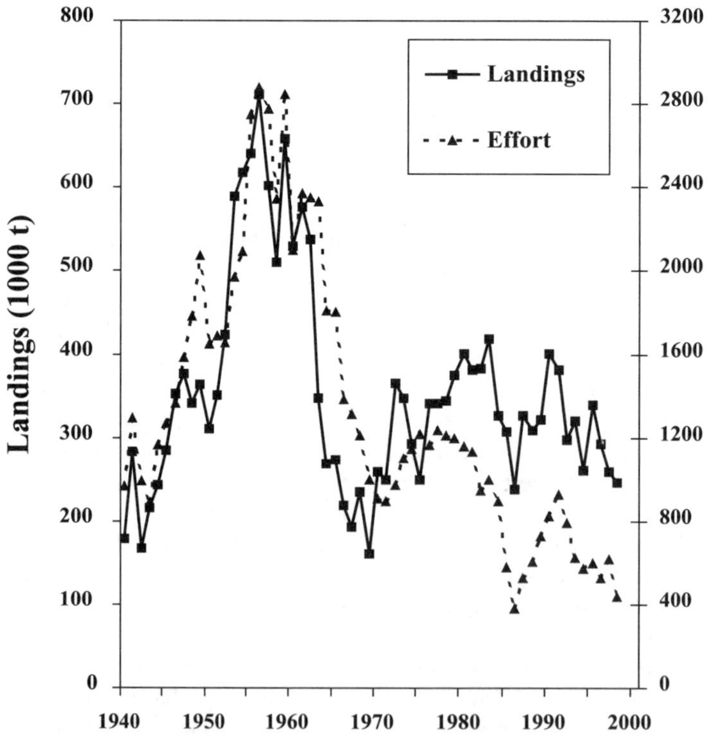

FIGURE 1. Atlantic menhaden landings and nominal fishing effort (vessel-weeks).

TABLE 1. Annual estimated values of six Atlantic menhaden triggers. (Boldface indicates years in which respective trigger would have been exceeded.)

Year	Landings[a]	P0[b]	P3+[c]	Recruitment[d]	SSB[e]	SPR[f]
1940	179.0	–	–	–	–	–
1941	283.1	–	–	–	–	–
1942	167.4	–	–	–	–	–
1943	215.0	–	–	–	–	–
1944	243.5	–	–	–	–	–
1945	285.6	–	–	–	–	–
1946	351.8	–	–	–	–	–
1947	376.4	–	–	–	–	–
1948	341.3	–	–	–	–	–
1949	363.4	–	–	–	–	–
1950	311.2	–	–	–	–	–
1951	351.3	–	–	–	–	–
1952	423.6	–	–	–	–	–
1953	589.2	–	–	–	–	–
1954	617.9	–	–	–	–	–
1955	641.4	24.4	20.1	3.1	327.0	13.8
1956	712.1	1.0	15.5	5.7	258.7	6.6
1957	602.8	8.5	7.1	7.3	133.2	6.7
1958	510.0	3.9	4.4	3.3	88.7	16.1
1959	659.1	0.2	8.4	15.1	173.7	8.6
1960	529.8	2.6	7.7	2.2	123.3	24.1
1961	575.9	0.0	**48.6**	3.0	360.3	13.3
1962	537.7	2.5	**33.3**	2.2	200.0	4.9
1963	346.9	5.5	13.3	2.2	65.3	3.1
1964	269.2	17.5	6.8	**1.7**	30.8	**2.4**
1965	273.4	17.1	6.2	**1.9**	20.8	**1.7**
1966	**219.6**	**26.1**	2.7	**1.4**	**9.1**	3.3
1967	**193.5**	0.7	8.0	**1.9**	20.9	5.5
1968	**234.8**	13.4	6.7	**1.2**	**16.8**	**2.1**
1969	**161.6**	18.2	6.2	**1.7**	**14.1**	5.4
1970	259.4	1.5	2.6	2.6	**16.2**	6.6
1971	250.3	7.5	11.2	**1.3**	28.1	6.6
1972	365.9	2.9	11.3	3.4	48.0	**2.0**
1973	346.9	3.0	2.5	2.7	**12.5**	**1.3**
1974	292.2	15.9	2.6	3.0	**12.1**	**1.5**
1975	250.2	13.8	2.6	3.7	**13.6**	**1.9**
1976	340.5	8.4	1.7	6.8	**15.6**	2.8
1977	341.1	13.2	2.8	5.1	25.6	4.3
1978	344.1	14.8	9.5	4.7	44.5	3.7
1979	375.7	**38.6**	3.9	4.2	40.4	6.4
1980	401.5	2.6	9.2	6.7	58.0	4.6
1981	381.3	**29.8**	7.2	4.7	42.4	5.0
1982	382.4	3.6	12.7	6.4	48.8	3.1
1983	418.6	24.5	4.2	2.5	35.8	3.8
1984	326.3	**36.5**	9.5	3.8	55.3	**1.7**
1985	306.7	21.1	2.9	5.0	18.8	**2.5**
1986	**238.0**	5.1	3.5	4.5	**15.7**	7.5
1987	327.0	1.9	7.8	3.4	37.5	7.8
1988	309.3	15.7	17.6	3.0	58.9	5.1
1989	322.0	5.7	6.4	5.5	38.3	5.2
1990	401.2	14.3	7.6	2.2	35.3	7.7
1991	381.4	**27.8**	9.6	3.5	59.6	**2.8**
1992	297.6	19.5	6.3	3.5	29.8	8.6
1993	320.6	4.3	10.3	3.2	39.3	10.1
1994	260.0	5.9	16.1	2.4	62.9	13.0
1995	339.9	3.5	23.2	2.9	74.7	9.5
1996	292.9	3.1	15.6	2.1	54.9	13.9
1997	259.1	2.5	**30.0**	2.0	102.1	14.0
1998	**245.9**	7.6	20.5	**1.3**	73.2	11.5

TABLE 1. (continued)

Year	Landings[a]	P0[b]	P3+[c]	Recruitment[d]	SSB[e]	SPR[f]
Median[g]	324.1	13.6	7.1	3.4	26.9	4.1
25%	250.3	3.6	3.7	2.2	15.7	2.1
75%	365.9	18.2	10.3	4.7	42.4	5.5
Trigger	<250.0	>25.0	>25.0	<2.0	<17.0	<3.0

[a] Landings in thousands of metric tons.
[b] Percent by numbers of age 0's in landings.
[c] Percent by numbers of adults (ages 3+) in landings.
[d] Estimated numbers of recruitment to age 1 in billions.
[e] Estimated mature female biomass (spawning stock biomass or SSB) in thousands of metric tons.
[f] Estimated equilibrium maximum spawning potential based on egg production (for estimated F vs $F = 0$) in percent (includes F at age 0).
[g] Median, 25th, and 75th percentiles based on fishing years from 1965 through 1990, except for P3+ which is based on fishing years 1955 through 1990.

on March 1), and (6) percent maximum spawning potential (or static spawning potential ratio, static SPR), estimated as the ratio between egg production with and without fishing, under equilibrium conditions. The triggers are considered annually along with other information on stock status to determine the need for management.

Detection of a series of poor year classes was a major concern when the triggers were developed. By the time the first trigger variable (landings) is reduced below the warning level, possibly because of poor recruitment, the recent recruiting year classes to the stock that support the fishery are one or two years in the past. The second trigger variable (proportion of age-0 menhaden in the landings) may be indicative of a good recruit year-class, but in many cases it seems to indicate good weather late in the year instead. The third trigger variable (proportion of adults, age-3+, in the landings) was designed to indicate whether one or more moderate to strong year classes are followed by several weak year classes (poor recruitment). The fourth trigger variable (recruitment to age-1) is a direct estimate of recruitment, and is particularly sensitive to retrospective error. The fifth trigger variable (spawning stock biomass) is indicative of whether there are sufficient spawners available to improve chances for adequate subsequent recruitment, but the weak spawner-recruit relationship for Atlantic menhaden gives no guarantee that moderate to high recruitment will come from high spawning stock (and may suggest the reverse). The sixth trigger variable (% maximum spawning potential) compares an index of egg production by mature females (summed over all mature ages) under equilibrium assumptions and the estimated F to the same index with $F = 0$. Thus, the sixth trigger variable is a measure of fishing mortality, or the rate at which stock abundance is being reduced annually because of fishing activity, and is not a measure of spawning stock biomass.

Virtual population analysis is used to reconstruct the fish population and fishing mortality rates by age and year, typically assuming a constant rate of natural mortality (M). In most stock assessments, constant values for M are obtained from life history analogies (e.g., maximum age, growth rates), with the methods of Pauly (1979) and Hoenig (1983) among the most widely used. Estimates of M in the early literature on Atlantic menhaden vary, although only moderately (Ahrenholz et al. 1991). Schaaf and Huntsman (1972) estimated $M = 0.37$/year based on an ad-hoc approach regressing total mortality rate (Z) on fishing effort. Nearly 438,000 juvenile Atlantic menhaden were tagged coastwide, 1970–1986; estimates were $M = 0.52$/year from a preliminary tag-recovery analysis (Dryfoos et al. 1973), and $M = 0.50$/year from a more extensive tag-recovery analysis (Reish et al. 1985). The mean of the range ($M = 0.45$/year) has been used routinely in Atlantic menhaden assessments beginning with Ahrenholz et al. (1987).

Errors in VPA-based quantities are usually greatest in the most recent year analyzed. Estimates for previous years quickly converge to true values, provided that catch at age and natural mortality are well estimated and that fishing mortality is at least moderate (Jones 1961; Tomlinson 1970; Pope 1972, Ulltang 1977; Megrey 1989). Thus, VPAs in successive years often give revised estimates of stock status in what was the final year of a previous VPA. For some stocks, such revised estimates display patterns suggestive of bias (i.e., current-year estimates consistently appear to have been too high or too low). The susceptibility of an assessment procedure to retrospective errors can be evaluated using *retrospective analysis,* which recreates an historical series of VPAs using a specified assessment procedure (Sinclair et al. 1990). The remaining three trigger variables are subject to this source of potential error.

Analyses that support advice to management contain additional uncertainty from various sources. Such uncertainty in analyses of Atlantic menhaden has been investigated in several contexts, including effects of sampling error (Dorval 1998) and the retrospective problem (Cadrin and Vaughan 1997). Simulation of the population through time has been used to examine alternative management scenarios (Vaughan 1977a, 1993).

Analyses presented in this study build on the above areas of uncertainty. Two general areas of analysis are described. The first area is largely deterministic and encompasses virtual population analysis and related work. Uncertainty associated with development of the catch-at-age matrix (or catch matrix) is briefly discussed. Some discussion and updating of retrospective and historical analyses are presented. Typically natural mortality (M) is assumed to be constant over all ages and years; here, we explore the sensitivity of our analyses to age-varying M. Fishery-independent indices of recruitment (juvenile abundance surveys) that have become available recently are compared with lagged estimates of recruitment to age-1 from the VPA. Although they have not yet proved useful for calibrating VPA, these indices have been helpful in confirming recruitment patterns obtained from VPA.

The second area of analysis includes stochastic projections and an alternative modeling approach, which itself includes a form of stochasticity. Population models such as yield per recruit, spawning stock biomass per recruit, and various stock–recruitment relationships were developed from annual, age-specific estimates of population size and fishing mortality rates. Various biological reference points can be obtained from these modeling approaches (e.g., $F_{0.1}$, F_{max}, and $F_{x\%}$, where x% represents a specified level of SPR). Population projections under alternate levels of fishing mortality can be made, after assuming a specific relationship between spawning stock and subsequent recruitment. The stock–recruitment relationship is probably the greatest source of uncertainty in any attempted population projection (Vaughan 1977b). The event-tree approach used here is based on conditional probabilities developed from the historical pattern of recruitment and spawning stock biomass, rather than from an assumed mathematical relationship (Vaughan 1993). Finally, application of different modeling approaches may provide additional insight. Hence, results from an alternative approach based on surplus production models are compared with those from VPA.

Methods

The Beaufort Laboratory (NMFS) has conducted coastwide biological sampling of Atlantic menhaden for length and weight at age since 1955, and the result is one of the longest time series of biological data on an exploited fish stock in the nation. The reduction fishery is sampled for fish size and age according to a two-stage cluster design (first stage consisting of purse-seine sets and second stage consisting of individual fish from each sampled set), conducted over the range of the fishery, both temporally and geographically (Chester 1984). Number of fish sampled in the second cluster was reduced during the early 1970s from 20 fish to 10 fish to increase sampling intensity of the first cluster (number of sets). This sampling program provided the basic data for VPA and subsequent analyses.

Virtual population analysis (VPA)

This section includes analyses conducted on the catch-at-age matrix used by the VPA, output from the VPA, including consideration of retrospective error and comparison of VPA with age-fixed M to those with age-variable M. In addition, VPA estimates of recruitment to age-1 are compared with a coastwide juvenile abundance index developed from state indices covering Florida to southern New England.

Catch-at-age matrix and VPA results.—The catch-at-age matrix was developed on a plant-week basis from detailed landings and biological sampling data. An analysis of internal consistency of this matrix is based on pairwise correlations between lagged catches from a cohort. This approach permits one to assess the ability of the catch matrix to follow a cohort through its lifetime. One should be able to follow a strong cohort (or a weak cohort) from full recruitment (at least age-2) through its eventual disappearance from the stock. However, there is an implicit assumption that F is relatively constant over time, as changes in F could reduce the correlations.

Our VPA methods generally followed those described in Ahrenholz et al. (1987), Vaughan and Smith (1988), Vaughan (1990, 1993), Vaughan and Merriner (1991), and Cadrin and Vaughan (1997). Using the Murphy (1965) VPA approach applied separately to each cohort, annual age-specific estimates of fishing mortality rate (F) and population numbers are obtained. A constant value of M (0.45/year) was used. Starting values for final F for complete cohorts were estimated from the catch curve for the cohort. A separable VPA approach was applied to recent years (1995–1998) in developing starting fishing mortality rates on oldest ages for incomplete cohorts (Clay 1990). Median values with their interquartile range (25th and 75th percentiles) were computed to provide historical perspective on estimates of the last three trigger variables (1955–1998).

A detailed retrospective analysis for Atlantic menhaden is presented in Cadrin and Vaughan (1997) and is not repeated here. The retrospective analysis was accomplished using separable virtual population analysis with catch-at-age data for 1970 through a variable final year (1984–1995). Cadrin and Vaughan (1997) also presented an historical analysis based on the series of trigger reports to AMAC, which is updated in this paper. The estimates in Cadrin and Vaughan (1997) updated here are from the revised Atlantic menhaden fishery management plan (FMP) (AMAC 1992) and annual trigger reports to AMAC (Vaughan 1994–1999). Comparisons are based on proportional error between estimates from the terminal year (1998) with estimates from prior years (1990 and 1992–1997). Proportional error is calculated for prior year estimates by subtracting and then dividing by terminal year estimates.

Age-specific estimates of M.—With very few exceptions, stock assessments routinely assume a constant M across years and ages. Assessments for Atlantic menhaden have followed this approach. However, the usual assumption of constant M over all ages is probably not realistic, especially when young ages (i.e., age 0 and age 1) are included in the analysis. An empirical relationship devised by Boudreau and Dickie (1989) was to be used to obtain annual, age-specific estimates of M as a function of mean weight at age:

$$\hat{M} = 2.88(592W)^{-0.33}, \qquad (1)$$

where W is weight (in pounds) at age. Thus, as the fish grows, M declines with age. Assuming decreasing M with increased size (or age) seems biologically reasonable (e.g., Gulland 1987), but this assumption has not been widely adopted in marine stock assessments. For example, the Boudreau and Dickie (1989) method was used in assessments of Atlantic weakfish *Cynoscion regalis* (Seagraves 1992), but was later rejected by a review committee (SARC 1998). The basis for the SARC (1998) rejection was in specifying the functional relationship of M with age, rather than the concept that M varies (declines) with age. Proportional error is used to present the general magnitude of difference between estimates of recruitment, SSB, and static SPR for the age-variable M and age-fixed M. Proportional error is calculated from the age-variable M estimates by subtracting and then dividing by age-fixed M estimates. The analysis presented here is for the purpose of exploring the sensitivity of our VPA to age-varying estimates of M and the effect on management benchmarks.

Juvenile abundance indices.—Recently, extant data sets of predominantly juvenile menhaden have been made available by states along the U.S. Atlantic coast; including South Carolina (SEAMAP), North Carolina, Virginia, Maryland, Connecticut, and Rhode Island. When averaging indices within or between states, all indices are first standardized by subtracting the series mean and dividing by the series standard deviation.

Menhaden data collected by the SEAMAP program were provided by South Carolina Department of Natural Resources. Trawl sampling was conducted from coastal North Carolina (primarily south of Cape Fear River) to northern Florida in 1989–1998 during spring (April–May), summer (June–August), and fall (September–October). Standardized indices are calculated for each season and averaged across seasons for each year to form a SEAMAP juvenile abundance index (1989–1998).

Four juvenile indices were provided by the North Carolina Division of Marine Fisheries: (1) Alosid seine index (1972–1998), (2) Albemarle striped bass nursery trawl index (1982–1998), (3) Pamlico Sound survey index (1987–1998), and (4) NC juvenile trawl survey index (1987–1998). These indices were standardized and averaged across the four indices by year to form a North Carolina juvenile abundance index (1972–1998).

Juvenile Atlantic menhaden CPUE for 1959–1998 was obtained from annual striped bass seine surveys conducted by the Maryland Department of Natural Resources. Indices of Atlantic menhaden abundance were computed, based on arithmetic means for each of the four main regions (head of the Chesapeake Bay and the Choptank, Nanticoke, and Potomac rivers) from the Maryland survey. Standardized indices for each region were calculated, and then averaged across the four regions. A similar index of juvenile Atlantic menhaden for tributaries to the Virginia portion of Chesapeake Bay (1980–1998) was provided by the Virginia Institute of Marine Sciences (VIMS) based on samples from the lower portion of rivers where Atlantic menhaden are found. Standardized Maryland and Virginia indices were averaged to form a Chesapeake Bay juvenile abundance index (1959–1998).

Juvenile Atlantic menhaden indices were also obtained from the Connecticut Department of Environmental Protection (indices from fall 1984–1998 trawl survey and from 1987 to 1998 Connecticut River alosid survey) and from the Rhode Island Department of Fish and Wildlife (indices from 1990 to 1998 juvenile finfish survey of Narragansett Bay and 1979–1998 trawl survey of Rhode Island coastal waters). Each index was first standardized and then averaged to form statewide Connecticut and Rhode Island indices. These, in turn, were again standardized and averaged to form a southern New England juvenile abundance index (1979–1998).

A coastwide juvenile abundance index (1972–1998) was developed as a weighted average of standardized indices for SEAMAP (weight = 0.15), North Carolina (weight = 0.35), Chesapeake Bay (weight = 0.45), and southern New England (weight = 0.05). Statistical weights by regions are based on estimates of relative juvenile menhaden density × area[1]. Only North Carolina and Chesapeake Bay indices were combined for 1972–1978, southern New England was included for 1979–1988, and all four regions were included for 1989–1998.

Stochastic projections and alternate approach

In this section, we develop biological reference points that serve as the basis for subsequent event tree simulated projections. An alternative approach based on surplus production models is also considered.

Biological reference points for projections.— Yield per recruit and spawning stock biomass per recruit (static SPR) were estimated based on partial recruitment for 1995–1998, using the numerical yield-per-recruit (YPR) approach of Ricker (1975). Unlike the analytical approach of Beverton and Holt (1957), which assumes knife-edged recruitment to the fishery, Ricker's approach allows for partial recruitment at ages not fully recruited (e.g., ages 0 and 1 compared with age 2). Static SPR (the sixth trigger variable used for Atlantic menhaden) compares spawning stock biomass per recruit calculated for different levels of fishing mortality to spawning stock biomass per recruit calculated with $F = 0$ under an equilibrium (static) assumption (Gabriel et al. 1989; Mace et al. 1996). Thus, the comparison is not to "virgin" spawning stock biomass, but to a theoretical spawning stock biomass under the assumption that life history parameters are not affected by fishing. Benchmark values of instantaneous fishing mortality rates (F) were developed from yield per recruit ($F_{0.1}$ and F_{max}) and from static spawning potential ratio (F_{10} and F_{20}) approaches for Atlantic menhaden.

Event tree projections of recruitment and SSB.— Event-tree techniques were used to replace the usual functional stock–recruit model with an empirically derived probabilistic model of recruitment conditioned on spawning stock biomass or other measure of spawning capacity. Simulations were made consisting of 25-year population projections (replicated 100 times) using the event tree approach described in Vaughan (1993) as the only source of uncertainty. To define the event tree, conditional probabilities of recruitment to age 0 given spawning stock biomass were estimated from historical estimates of spawning stock biomass and subsequent recruitments to age 0, as estimated by VPA for 1955–1998. A set of 100 replicates of 25-year projections was accomplished for each level of fishing mortality (described below). In these projections natural mortality ($M = 0.45$/year) was assumed constant. The estimated population age structure for 1997 (from VPA) was used as the initial age structure for projections. Parameters from annual von Bertalanffy growth equations and fitted weight-length equations, averaged for 1995–1998 (the most recent four-year period), were used to represent growth. Output of particular interest from these simulations are recruitment to age 1 and spawning stock biomass (trigger variables 4 and 5 represented in Figure 2A and B).

Current fishing mortality was obtained from the mean estimates of age-specific F for the period 1995–1998 (when only Virginia and North Carolina plants have been active in the reduction fishery) from the most recent Murphy VPA. Mean age-specific F forms the basis of partial recruitment (relative mortality at age) used for varying estimates

[1] by Dr. D. Ahrenholz, NOAA Center for Coastal Fisheries and Habitat Research, Beaufort, North Carolina (personal communication)

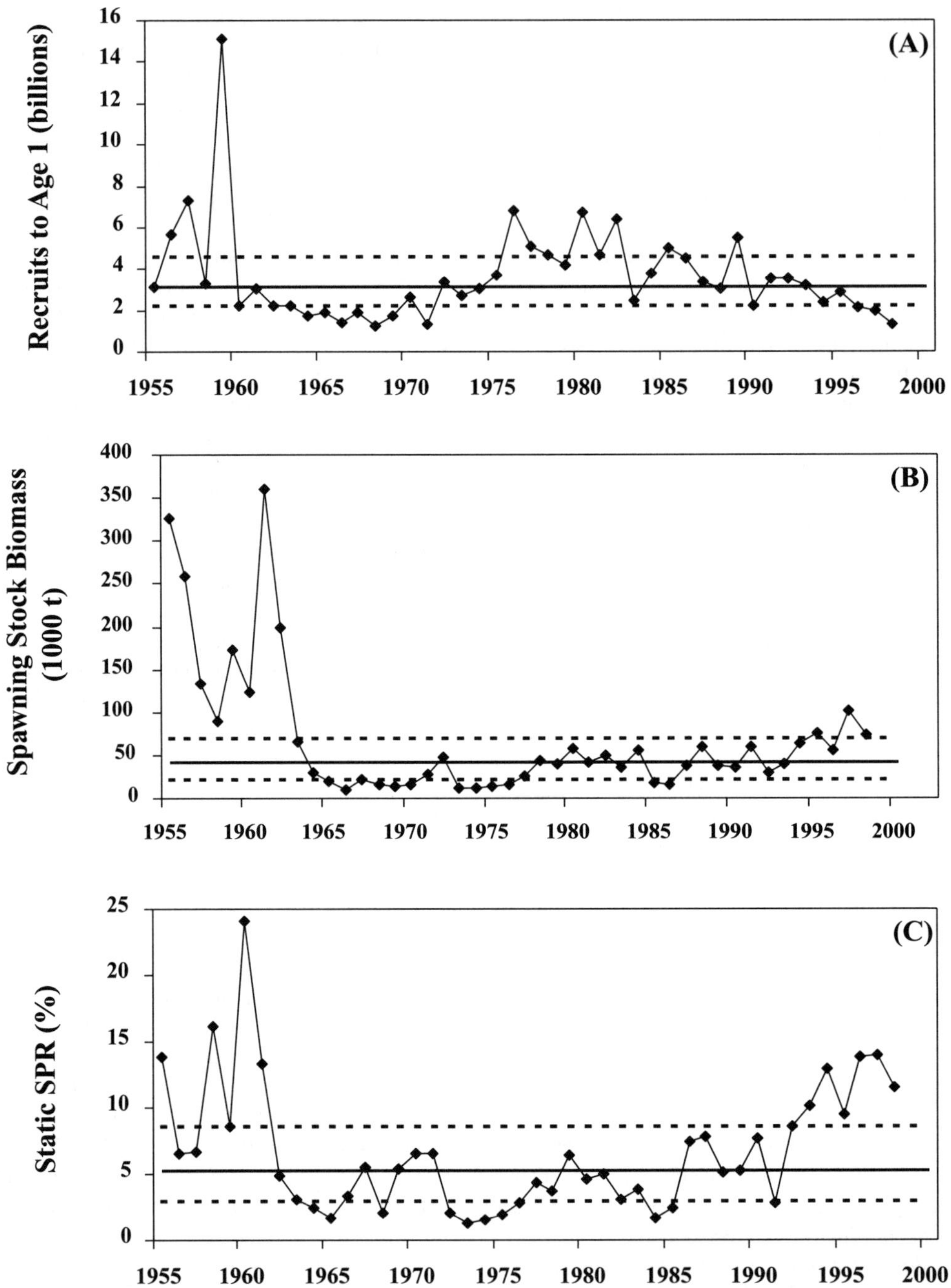

FIGURE 2. Historical plots for (A) recruitment to age-1 (R_1), (B) spawning stock biomass (SSB), and (C) static spawning potential ratio (SPR) for Atlantic menhaden with their temporal median and interquartile range.

of age-specific F according to the biological reference points developed from yield per recruit and spawning stock biomass per recruit analyses. Projections are then based on the fishing mortality associated with two biological reference points obtained from yield-per-recruit analysis ($F_{0.1}$ and $F_{\max}$), from static spawning potential ratio analysis (F_{10} and F_{20}), and from F_{rep}. The reference point F_{rep} is based on spawning stock biomass per recruit, uses the median of plotted R/SSB, and compares the inverse (median SSB/R) to the theoretical curve of SSB/R (values in Figure 2C are proportional to SSB/R) (Sissenwine and Shepherd 1987). Because estimates of F form the basis of the projections and the final trigger variable (static SPR) is calculated directly from F, there would be no uncertainty associated with this variable. Hence, output from these projections include only the first five trigger variables.

Surplus production models.—Surplus production modeling was used in addition to VPA to obtain an additional perspective on population dynamics. In this, we used data on total annual landings in weight. Annual data on either relative abundance or relative fishing mortality rate were also required; we used estimates of stock biomass (ages 0+) derived from the VPA. Biomass was used rather than recorded data on fishing effort rate because our production model assumed constant catchability q, and density-dependent catchability has been demonstrated in this species (Schaaf 1979). Using biomass estimates also allowed us to set $q = 1$, thus reducing the number of estimated parameters by one.

Production models assume that the stock's production of new biomass B per unit time t is a function solely of current stock biomass; that is,

$$\frac{dB}{dt} = f(B) \tag{2}$$

Under this fundamental assumption, biomass can be removed from the stock at this same rate dB/dt while leaving the stock size unchanged; thus the time rate of production as given above is often denoted surplus production, equilibrium yield, or sustainable yield.

The simplest, logistic form of production model was described by Lotka (1924) as a model of population growth; Schaefer (1954, 1957) applied it to fisheries problems, while adding a term for fishing mortality; Pella (1967) showed how to fit the model without assuming that observed yields are equilibrium yields; and Prager (1994) introduced several refinements, including the use of bootstrapping to estimate nonparametric confidence intervals. Under the logistic model, the production function is a parabola, which implies that the curve of production as a function of biomass is symmetrical. Using the parameterization of Fletcher (1978),

$$\frac{dB_t}{dt} = 4m\frac{B_t}{K} - 4m\left(\frac{B_t}{K}\right)^2, \tag{3}$$

where B_t is stock biomass at time t, m is maximum sustainable yield, and K is the maximum population size, or carrying capacity, of the stock. In the Pella-Tomlinson generalized production model (Pella and Tomlinson 1969), the production curve could be skewed in either direction. In the improved parameterization of Fletcher (1978),

$$\frac{dB_t}{dt} = 4g\frac{B_t}{K} - 4g\left(\frac{B_t}{K}\right)^n, \tag{4}$$

where n is a parameter determining the shape of the production function, and g (used for notational convenience) is the following function of n:

$$g = \frac{n^{n/n-1}}{n-1}. \tag{5}$$

It appears that few data sets on real populations provide enough information to obtain useful estimates of n in equation (4) (Hilborn and Walters 1992).

In fitting, we used an observation-error estimator, conditioned on the observed yield and estimated in logarithmic transformation, as implemented by an extended version of the ASPIC computer program (Prager 1995). We fit both logistic and generalized production models and made projections based on model results and several assumed levels of future landings.

Results

Virtual population analysis

Catch-at-age matrix and VPA results.—Coherence of the catch matrix is presented in Table 2. Significant correlations ($P < 0.05$) were obtained for all lagged catches through age 6, with the exception of nonsignificant correlation between catches at age 2 with age 6 four years later ($P = 0.165$). Although this does not imply that the catch matrix is without error, it does suggest an exceptionally high degree of precision to the catch matrix.

Estimates of the last three triggers for 1955–1998, derived from the most recent assessment (Vaughan and Smith 1999), are summarized in Table 3. Estimates of historical recruitment to age-1 range between 1.2 billion in 1968 and 15.1 billion

TABLE 2. Coherence of Atlantic menhaden catch-at-age matrix based on pair-wise correlations of subsequent catches from each cohort, 1955–1998. For each age, the Pearson correlation coefficient is shown in the first row, and the probability of exceeding this coefficient under the null hypothesis that the coefficient equals zero is in the second row.

Age	1	2	3	4	5	6
1	1.0	0.59	0.77	0.80	0.76	0.37
	0.0	0.0001	0.0001	0.0001	0.0001	0.0222
2	–	1.0	0.55	0.47	0.53	0.22
		0.0	0.0001	0.0016	0.0003	0.1652
3	–	–	1.0	0.92	0.71	0.37
			0.0	0.0001	0.0001	0.0166
4	–	–	–	1.0	0.75	0.58
				0.0	0.0001	0.0001
5	–	–	–	–	1.0	0.83
					0.0	0.0001

in 1959 (1958 cohort). The estimate for the most recent year (1998) is 1.3 billion. The historical median is 3.1 billion, and the interquartile range is between 2.2 and 4.6 billion. Historical estimates of spawning stock biomass have ranged between 9,100 t in 1966 and 360,300 t in 1961 (when the 1958 cohort was 3 years of age). The estimate for the most recent year is 73,200 t in 1998. The historical median is 41,400 t and the interquartile range is between 20,900 t and 69,300. Similarly, historical estimates of static SPR have ranged between 1.3 in 1973 and 26.0 in 1960 (when the 1958 cohort became fully recruited to the fishery). The estimate for the most recent year is 11.5% in 1998. The historical median is 5.3% and the interquartile range is between 2.9% and 8.6%. The historical period covered by this catch-at-age matrix includes a period when the stock was extremely large (during the late 1950s), a period when it was seriously depressed (1960s and early 1970s), and one when it apparently had recovered (late 1970s and 1980s). In recent years, there has been declining recruitment, increasing spawning stock biomass, and historically high values for static spawning potential ratio (reduced fishing mortality).

TABLE 3. Current estimates of VPA-generated indices ("triggers") and juvenile abundance indices, compared to their long-term median and quartiles. Estimates for 1998 that fall within the interquartile range are not considered markedly different from the long-term median.

Variable	n	Current estimate (1998)	Historical 25th pctl	Historical median estimate	Historical 75th pctl
Population-Based (VPA) Triggers:					
Fixed M:					
R_1 (billions)	44	1.3	2.2	3.1	4.6
SSB (1000 t)	44	73.2	20.9	41.4	69.3
SPR (%)	44	11.5	2.9	5.3	8.6
Variable M:					
R_1 (billions)	44	4.6	2.5	3.4	5.0
SSB (1000 t)	44	65.5	19.9	37.9	62.1
SPR (%)	44	10.1	2.2	4.0	6.3
Standardized Juvenile Abundance Indices for:					
SEAMAP	10	1.36	−0.98	−0.08	1.23
NC Combined	27	0.33	−0.80	−0.10	0.46
VA Seine	19	−0.77	−0.74	−0.35	0.44
MD Seine	40	−0.87	−0.94	−0.32	0.83
CT Combined	15	0.73	−0.59	−0.31	0.23
RI Combined	20	1.68	−0.39	−0.39	−0.37
Coastwide	27	0.02	−0.43	0.13	0.64

Historical and retrospective analysis of the Atlantic menhaden VPA revealed substantial inconsistency in estimates of management variables in the last year of stock assessments (Cadrin and Vaughan 1997). Retrospective error of management variables in terminal years was positively skewed but not biased. Cadrin and Vaughan (1997) found that historical estimates of management variables converged to similar values within three years. An updated historical analysis using proportional error based on Murphy VPA runs is summarized for the VPA-based triggers (Figure 3), to show recent performance of current estimates of these trigger values. As a cohort appears in the landings for more years, its year-class strength is better estimated. Initial estimate of recruitment to age 1 is based directly on fishing mortality estimated by separable VPA on some recent short time period (this process is used for incomplete cohorts). Once the cohort is represented by age 4 or age 5 in the catch matrix, little if any change in recruitment to age 1 occurs, because uncertainty in starting (final) F converges quickly for Atlantic menhaden. Recently, initial estimates of recruitment to age 1 for a given cohort have been low (often below the trigger warning level of 2 billion). As the cohort is fished for additional years, progressively increasing estimates of recruitment have been more commonly obtained.

Age-specific estimates of M.—Sensitivity of VPA-based output was explored using the Boudreau and Dickie (1989) approach of equation (1), yielding estimates of M averaged across 1955–1998 as follows: age 0, $M = 0.82$/year; age 1, $M = 0.58$/year; age 2, $M = 0.46$/year; and age 3 and older, $M = 0.34$/year. The estimate of M averaged over all ages (0–8) and all years was 0.46, similar to the constant M of 0.45/year that has been used historically in Atlantic menhaden VPAs. The VPA-based triggers for 1955–1998 were re-estimated with age-specific M. Estimates of recruitment to age 1 ranged from 1.3 billion in 1968 to 16.3 billion in 1959 (1958 cohort). Proportional error suggests slightly higher estimates of recruitment to age-1 were obtained with age-variable M compared with age-fixed M (Table 3; Figure 4). However, estimated recruitment in 1998 is much higher, relative to the estimate based on age-fixed M.

Historical estimates of spawning stock biomass range between 8,400 t in 1966 and 325,000 t in 1961 (when the 1958 cohort was 3 years of age). The estimate for the most recent year is 65,500 t in 1998. The historical median is 37,900 t and the interquartile range is between 19,900 t and 62,100 t. Generally lower estimates of spawning stock biomass were obtained with age-variable M compared with age-fixed M. Similarly, historical estimates of static SPR range from 1.0 in 1973 to 19.2 in 1960 (when the 1958 cohort became fully recruited to the fishery). Based on variable M, static SPR for Atlantic menhaden has never been above 20% including a period of very robust stock size. The estimate for the most recent year is 10.1% in 1998. The historical median is 4.0% and the interquartile range is between 2.2% and 6.3%. Generally lower estimates of spawning stock biomass were obtained with age-variable M than with age-fixed M. Again, this information suggests that in recent years there has been declining recruitment (though with apparent increase in 1998), increasing spawning stock biomass, and historically high values of static spawning potential ratio (i.e., reduced fishing mortality).

Juvenile abundance indices.—There was limited consistency among the juvenile abundance indices by region that were combined to create the coastwide index (Table 4). The SEAMAP index was marginally correlated ($P < 0.1$) with the Chesapeake Bay combined juvenile abundance index (VA and MD). The North Carolina and New England indices were uncorrelated with each other as well as with the other regional indices. As might be expected, the SEAMAP, North Carolina and Chesapeake Bay juvenile abundance indices were all highly correlated with the Coastwide index ($P < 0.01$). The southern New England indices, which had minimal (5%) contribution to the coastwide index, was uncorrelated with the coastwide index. There were highly significant pairwise correlations found for MD–VA and CT–RI. Finally, the coastwide and Chesapeake Bay juvenile abundance indices were highly correlated ($P < 0.01$) with estimates of lagged recruitment (Figure 5).

The most recent values of these juvenile abundance indices are compared with their median and interquartile range (Table 3). Only the Virginia seine index is below its 25th percentile. The Maryland seine and coastwide indices fall between the 25th percentile and median. The North Carolina combined index falls between the median and 75th percentile. The SEAMAP, Connecticut combined, and Rhode Island combined indices all fall above the 75th percentile of their historical time series. We suggest that the interquartile range is an appropriate measure of historical performance against which to compare recent performance. Hence, values outside the

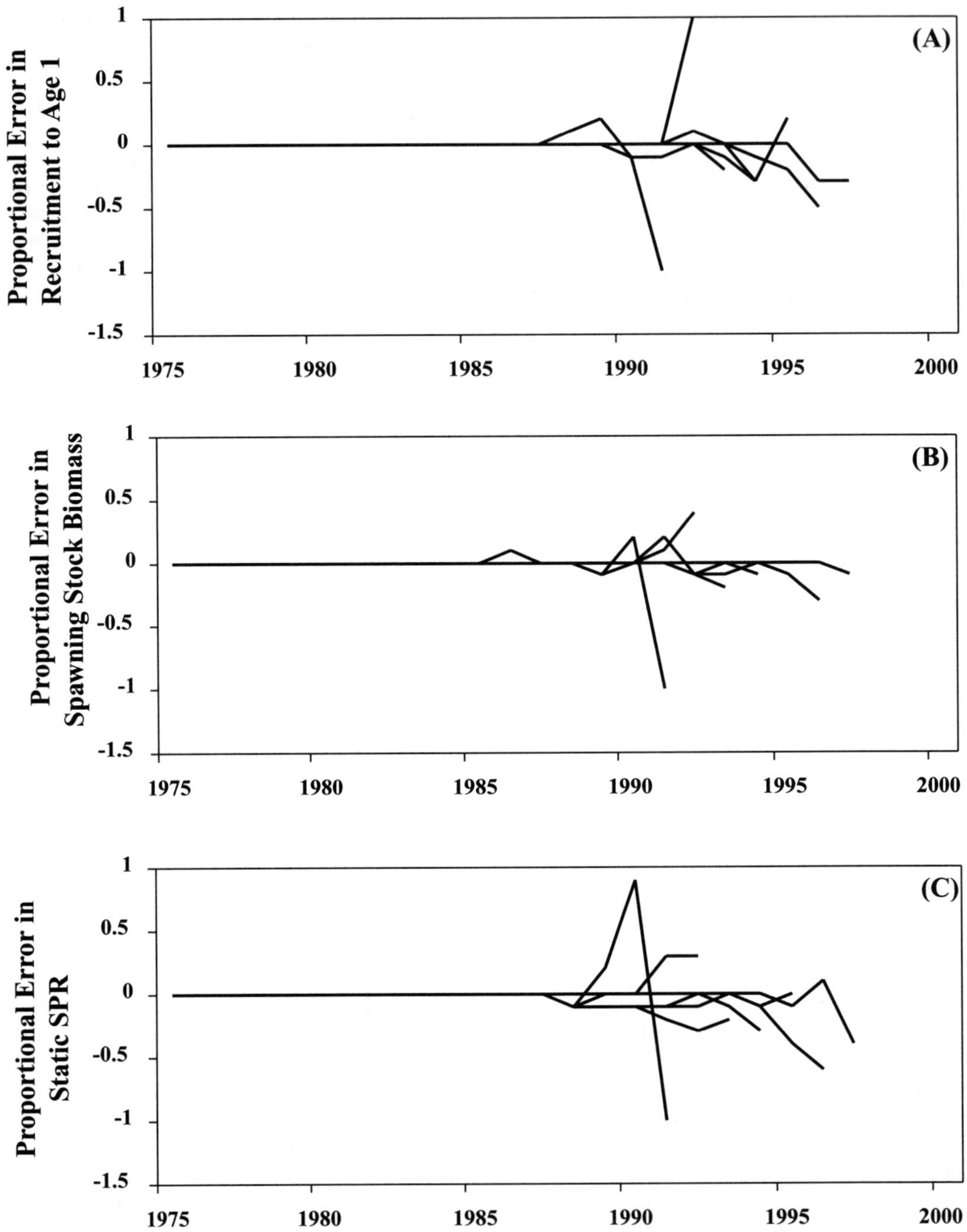

FIGURE 3. Retrospective error expressed as proportional error relative to the terminal year for (A) recruitment to age 1 (R_1), (B) spawning stock biomass (SSB), and (C) static spawning potential ratio (SPR) for Atlantic menhaden.

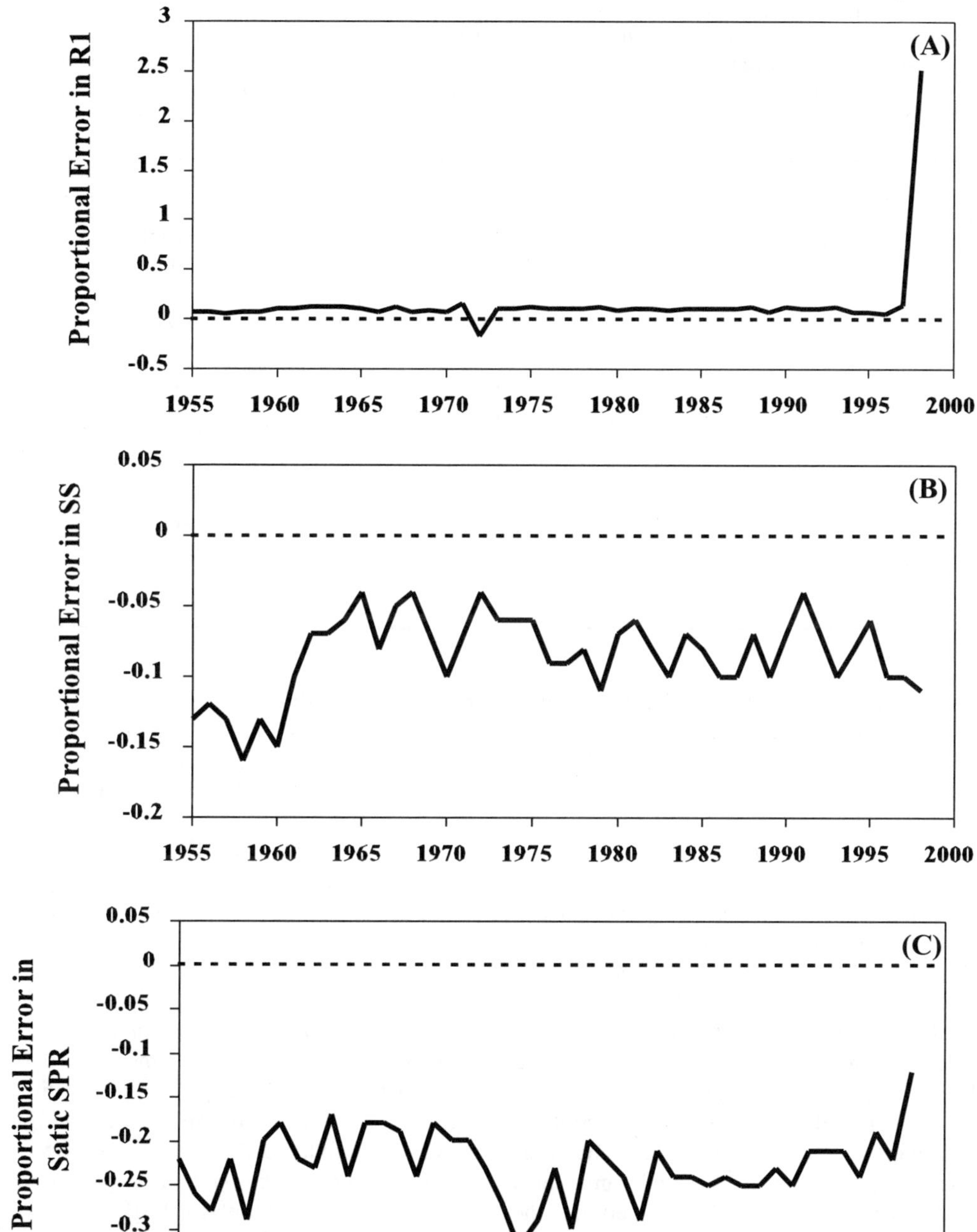

FIGURE 4. Sensitivity of virtual population analysis (VPA) based on age-variable M expressed as proportional error relative to VPA analysis based on age-fixed M for (A) recruitment to age 1 (R_1), (B) spawning stock biomass (SSB), and (C) static spawning potential ratio (SPR) for Atlantic menhaden.

TABLE 4. Coherence among regional juvenile abundance indices and with lagged recruits to age 1. For each index, the Pearson correlation coefficient is shown in the first row, the probability of exceeding this coefficient under the null hypothesis that the coefficient equals zero is in the second row, and sample size in the third row. Regions represented are U.S. south Atlantic bight (SEAMAP), North Carolina (NC), Chesapeake Bay (CB), southern New England (NE), and coastwide (Coast). Significant values ($P < 0.1$) indicated by asterisk, highly significant values ($P < 0.01$) indicated by double asterisk.

Index	SEAMAP	NC	CB	NE	Coast	Lagged R_1
SEAMAP	1.0	0.51	0.59*	0.04	0.80**	0.38
	0.0	0.1334	0.0719	0.9089	0.0051	0.3111
	10.0	10.0	10.0	10.0	10.0	9.0
NC	–	1.0	0.24	0.31	0.75**	0.23
		0.0	0.2205	0.1841	0.0001	0.2529
		27.0	27.0	20.0	27.0	26.0
CB	–	–	1.0	−0.33	0.82**	0.68**
			0.0	0.1490	0.0001	0.0001
			27.0	20.0	27.0	26.0
NE	–	–	–	1.0	−0.11	−0.10
				0.0	0.6573	0.6902
				20.0	20.0	19.0
Coast	–	–	–	–	1.0	0.58**
					0.0	0.0019
					27.0	26.0

interquartile range can be interpreted as low or high relative to their historical performance, while values within the interquartile range cannot be interpreted as different from the median value.

Stochastic projections, production models

Biological reference points for projections.— Current F is given by the vector ($F_0 = 0.02$, $F_1 = 0.21$, $F_{2+} = 0.95$) based on selectivity for the reference period of 1995–1998. The biological reference points described below were estimated based on the same selectivity of F on ages 0 and 1 relative to age 2 and older (ages at full F). Biological reference points traditionally obtained from the yield-per-recruit approach include F_{max} and $F_{0.1}$ (Figure 6). The reference point F_{max} represents the level of fishing mortality that in theory maximizes biomass return to the fishery, and $F_{0.1}$ is a modification of F_{max} based on an economic-return argument. The reference point $F_{0.1}$ is that level of F that occurs where the slope of the curve of yield-per-recruit versus F is 10% of its slope at the origin. This reference point was proposed as providing nearly the same yield in biomass as F_{max}, but with less expenditure of fishing effort and less chance of overfishing (Gulland and Boerema 1973). For Atlantic menhaden, F_{max} occurs at a full F (mean F over ages 2–8) of about 1.0, while $F_{0.1}$ occurs at a full F of about 0.5. Biological reference points based on static SPR are given as F subscripted with the level desired for static SPR (Figure 6). Hence, F_{20} refers to the level of F that would produce a theoretical ratio of 20% SPR. The reference point F_{20} occurs at full F of about 0.7, and F_{10} at about 1.1. In calculating F_{rep}, we obtained the median SSB/R for the historical Atlantic menhaden data (12.4 expressed as 1,000 t per million recruitment). The corresponding F from the SSB/R curve generated for the partial recruitment during 1995–1998 was interpolated at 0.9 (F_{rep}).

Event tree projections of recruitment and SSB.— Conditional probability diagrams were developed that relate the historical pattern in recruitment to age 0 or age 1 given the spawning stock biomass (SSB) that produced them (Table 5). Low, moderate and high are based on the interquartile range of the historical time series for SSB and recruitment (as in Vaughan 1993). The historical pattern of recruitment to age 1 (1956–1998) conditioned on SSB (1955–1997) suggests that low SSB (below the 25th percentile) has about a 36% chance of producing low recruitment, a 45% chance of producing a moderate recruitment, and only 18% chance of producing high recruitment. Thus, there is a tendency towards low recruitment when SSB is low (i.e., SSB < 20,900 t). Probabilities of recruitment to age-1 also were obtained for moderate and high SSB. Moderate SSB is somewhat more likely to produce high

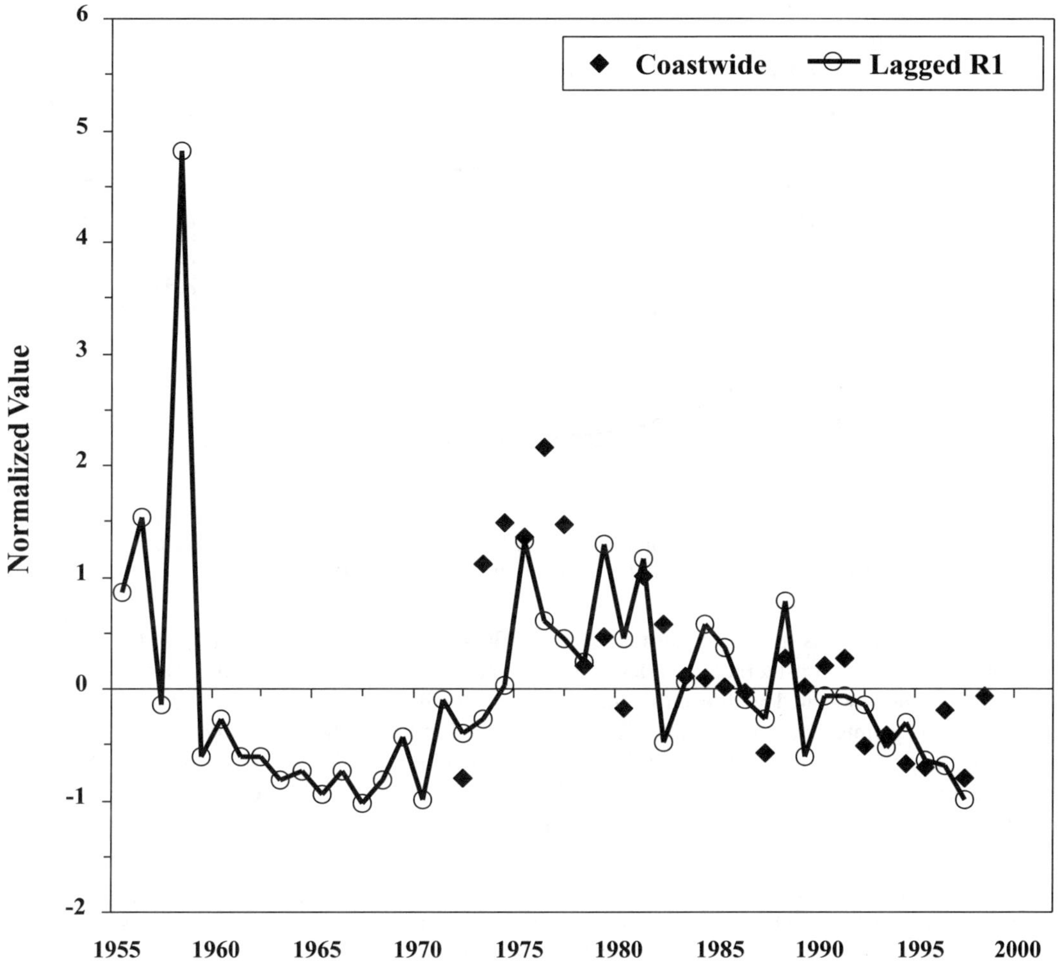

FIGURE 5. Comparison of coastal juvenile abundance index with lagged recruitment to age-1 (R_1) for Atlantic menhaden.

recruitment (23%) than low recruitment (27%); while high SSB is as likely to produce low (30%) as high recruitment (30%). This analysis suggests that maintaining SSB above approximately the 25th percentile for the long-term data series (20,900 t) will give a higher probability of good recruitment.

The conditional probabilities of recruitment to age-0 given SSB were used in the population projections to relate current spawning stock to subsequent recruitment (Vaughan 1993) (Table 5). Annual survival at age in these projections were based on constant M and age-specific F for the biological reference points developed from yield per recruit and spawning stock biomass per recruit analyses. Because static SPR is directly related to F, static SPR was not simulated in these projections. The 25-year projections are replicated 100 times with underlying error (estimated from historical data). The annual median and interquartile range of the fourth and fifth trigger variables are plotted for the event tree simulation approach (Figures 7–11). The results of these projections are for comparison only and are not intended as predictions of future values. The projections suggest no significant difference in recruitment to age-1 whether fishing at $F_{0.1}$ ($F = 0.5$) or F_{10} ($F = 1.1$). On the other hand, spawning stock biomass is expected to be higher when fishing at $F_{0.1}$ ($F = 0.5$) compared with F_{10} ($F = 1.1$).

The cumulative probability (or risk) of decline in recruitment to age 1 is developed from the 100 replicates of the projections in their 10th year (2007) (Figure 12). Risk is defined as the cumulative

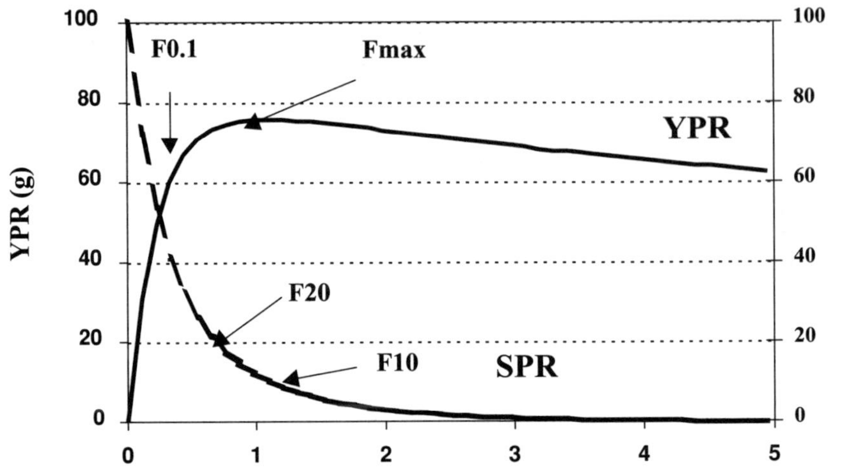

FIGURE 6. Critical values of instantaneous fishing mortality rates (F) from yield per recruit ($F_{0.1}$ and F_{max}) and from static spawning potential ratio (F_{10} and F_{20}) for Atlantic menhaden. The estimate of present F is for an age-at-entry of 0.5 years and mean and partial recruitment based on 1995–1998.

probability of a value less than or equal to a given value of recruitment. Considerable overlap exists among the curves associated with each of the biological reference points. Furthermore, all risk curves shown suggest moderate probability of falling below the current warning level of this trigger variable (2 billion age-1 recruits). These values range from 30% from F_{max} to 39% for F_{rep} and F_{20}, with intermediate values of 36% and 37% for $F_{0.1}$ and F_{10}, respectively.

Similarly, the cumulative probability (or risk) of decline in spawning stock biomass (SSB) is developed from the 100 replicate projections in their 10th year (2007) (Figure 13). Risk is defined as

TABLE 5. Conditional probabilities of recruitment to age 0 and to age 1 from spawning stock biomass stratified by interquartile range for 1955–1998. Conditional probabilities for recruitment to age 0 from spawning stock biomass used in population projections. Recruitment to age 0 and age 1 are in billions and spawning stock biomass is in thousands of metric tons (t).

	Spawning stock biomass		
Recruitment to:	Low (<20.9)	Moderate (20.9–69.2)	High (>69.2)
		Age 0	
Low (<2.79)	0.273	0.182	0.300
Moderate (2.79–6.12)	0.454	0.591	0.400
High (>6.12)	0.273	0.227	0.300
		Age 1	
Low (<2.22)	0.364	0.227	0.300
Moderate (2.22–4.57)	0.454	0.500	0.400
High (>4.57)	0.182	0.273	0.300

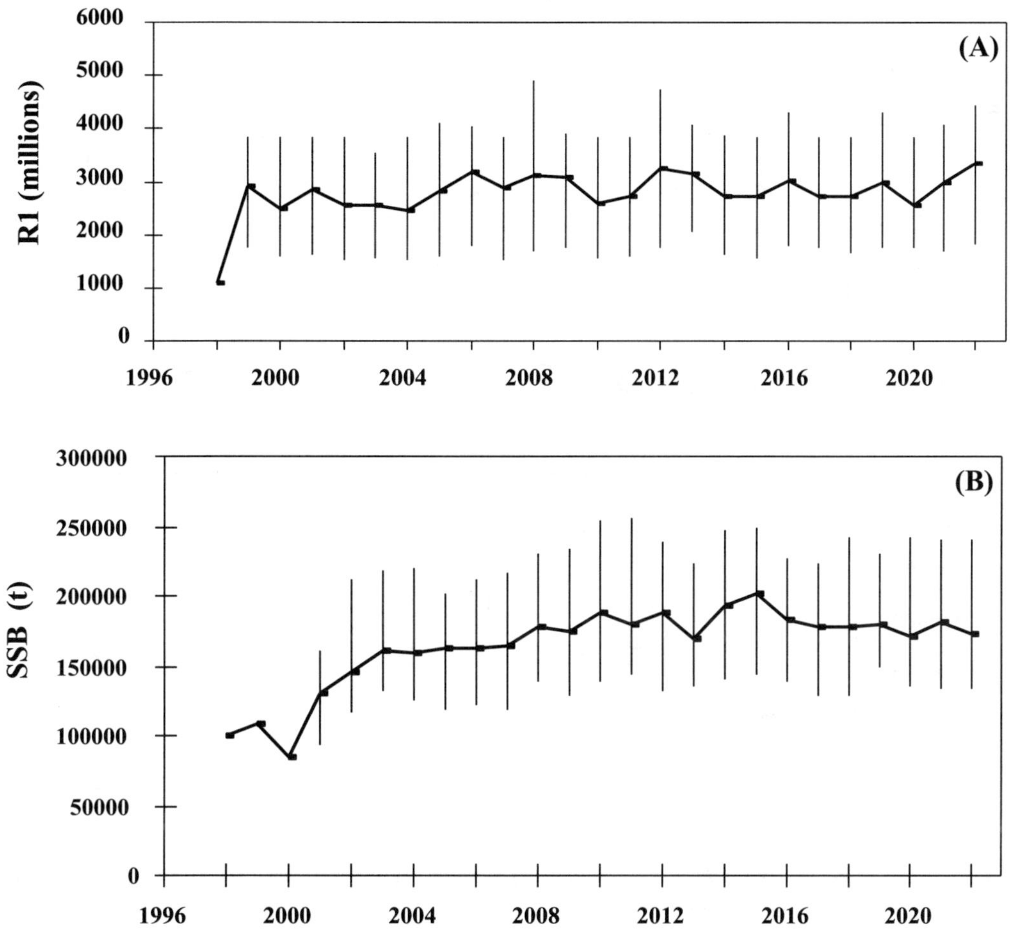

FIGURE 7. Projected values for $F = 0.5$ ($F_{0.1}$) of (A) recruitment to age 1, and (B) spawning stock biomass of Atlantic menhaden (interquartile range displayed).

the cumulative probability of a value less than or equal to a given value of SSB. SSB associated with lower values of F give lower risk of being below a specified value of SSB. That is, the probability of falling below 40,000 t, varies between 0% from for $F_{0.1}$, F_{20}, and F_{rep}, and 11% for F_{10}, with an intermediate value for F_{max} (6%). However, all risk curves shown exhibit only a minuscule probability of falling below the current warning level of this trigger variable (17,000 t).

We also examined the statistical properties of recruitment of Atlantic menhaden (to age 1) for different ranges of spawning stock biomass (Table 6). In addition to mean values, nonparametric statistics are calculated that are less sensitive to nonnormal (highly skewed) distribution of R_1 attributable to the remarkable strength of the 1958 year-class. These statistics include the median (or 50th percentile) and the 25th and 75th percentiles. The interquartile range, defined by the 25th and 75th percentiles, is a nonparametric analog to the standard deviation. Although the mean value of R_1 trends upward with increasing minimum spawning stock biomass, this is a result of the reduced sample size and the increasing effect of the recruitment to age 1 in 1959 (1958 year-class). It does not demonstrate increased likelihood of improved recruitment. No trend is evident in the median value (the lowest median value shown is for SSB values in excess of the 75th percentile of the historical time series for SSB). There is some indication of improvement in the upper limit of the interquartile range (75th percentile) with increasing spawning stock biomass at or above its historic median level (41,400 t).

Surplus-production models.—Results of fitting logistic and generalized models were quite similar;

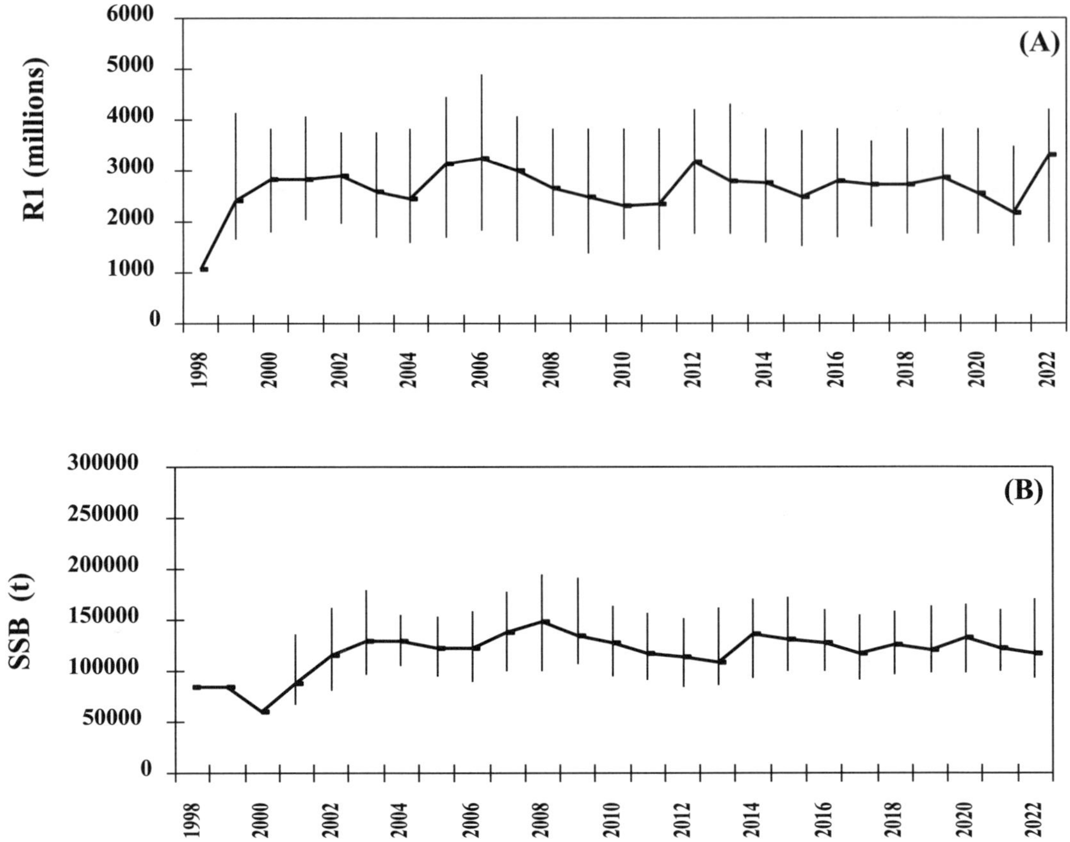

FIGURE 8. Projected values for $F = 0.7$ (F_{20}) of (A) recruitment to age-1, and (B) spawning stock biomass of Atlantic menhaden (interquartile range displayed).

here, we report primarily on the logistic model. This model fit the data reasonably well (Figure 14A); fit of the generalized model was quite similar. The largest discrepancies between observed and estimated CPUEs are in the years of peak abundance centered around 1960, where the production model, perhaps because it lacks age structure, is unable to replicate rapid changes in stock size. Estimates of benchmarks (Table 7) seem reasonable given the history of the stock. Estimated time trajectories of biomass and fishing mortality rate relative to their respective benchmarks B_{MSY} and F_{MSY} depict a stock that has been below the benchmark B_{MSY} since about 1960, and fishing mortality rate that has been above F_{MSY} for most of the time period (Figure 14B).

The generalized production model estimated that MSY was slightly larger than the logistic model and about the same stock status (Table 7). The shape exponent of the generalized model was estimated as $n = 2.76$, which skews the curve to the left so that $B_{\text{MSY}} = 0.81\,K$ (whereas in the logistic model, $B_{\text{MSY}} = 0.5\,K$). As the generalized model with its free exponent is more complex than the logistic, an F−ratio test was performed testing H_0: $n = 2$ against the alternative H_a: $n \neq 2$. The test failed to reject H_o ($P = 0.44$), suggesting the logistic model as a better choice for this stock. We note that the usefulness of the F-ratio test in this context has not been studied.

Projections under constant-catch scenarios estimate that, if the level of yield recorded in 1999 (Figure 15B) or estimated for 2000 (Figure 15C) is taken in 1999–2003, stock biomass will increase to greater than B_{MSY}. If the yield recorded in 1998 is taken, stock biomass will recover, but not to B_{MSY} by 2003. Estimates of fishing mortality rate from the same projections indicate that F will drop below F_{MSY} under all three scenarios. Nonparametric 80% confidence intervals on B and F are extremely narrow, and we emphasize that they include only one source of uncertainty, namely lack of fit of the logistic model to the observed data, which include VPA biomass estimates. To assess the reasonableness of

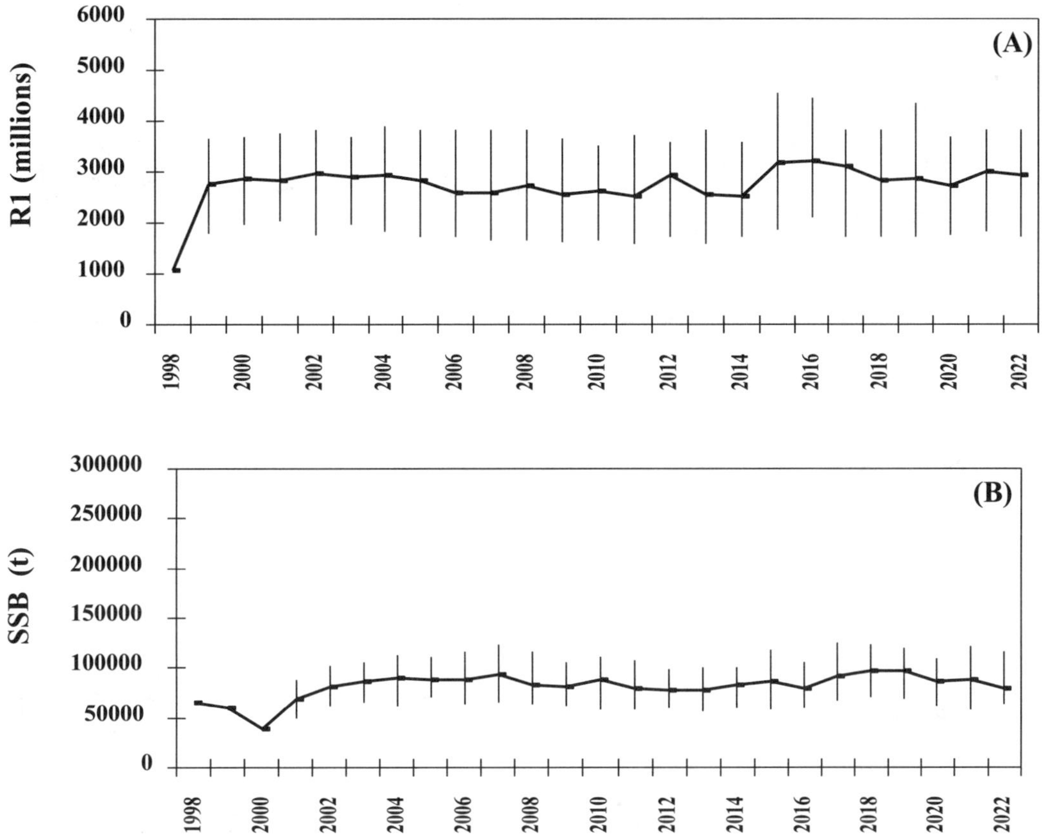

FIGURE 9. Projected values for $F = 0.9$ (F_{current}, F_{rep}) of (A) recruitment to age 1, and (B) spawning stock biomass of Atlantic menhaden (interquartile range displayed).

the assumption $q = 1.0$, we refit the logistic model, estimating q; the resulting estimate was $q = 1.02$, which suggests that the assumption conflicts with neither the data nor the model.

Discussion

Age-structured models

The quality of the catch matrix for the reduction fishery on the Atlantic menhaden is one of the best in the United States. This is partly because of the relatively few reduction plants (especially in recent years), and also because a statistically sound sampling program has been in place for over 40 years. The catch matrix is remarkably consistent and follows cohorts well through time. However, there are several areas where improvement will be needed. The growing bait fishery for menhaden along the Atlantic coast is a source of additional fishing mortality that should be better characterized. Preliminary assessments including bait landings have been made (Vaughan and Smith 1998). Another source of mortality is bycatch in the shrimp trawl fishery. Because of the poor quality of data available, little has been done to quantify this source of mortality, but the evidence so far suggests that it is minimal.

Occasional criticism has been leveled at the application of untuned virtual population analysis. When applied to poor quality catch matrices, such criticisms are justified. Because of the historic lack of appropriate abundance indices, tuned VPAs have not been applied to menhaden. Attempts to tune VPAs to recently obtained juvenile abundance indices have not been successful to date (Vaughan and Smith 1998, 1999), as model convergence was not achieved. However, this avenue continues to be explored. Regardless, we believe the quality of the Atlantic menhaden catch matrix permits reasonable confidence in the output variables. Because the Atlantic menhaden database is one of the longest in the United States (and the longest in the southeast), these variables can be viewed with reasonable

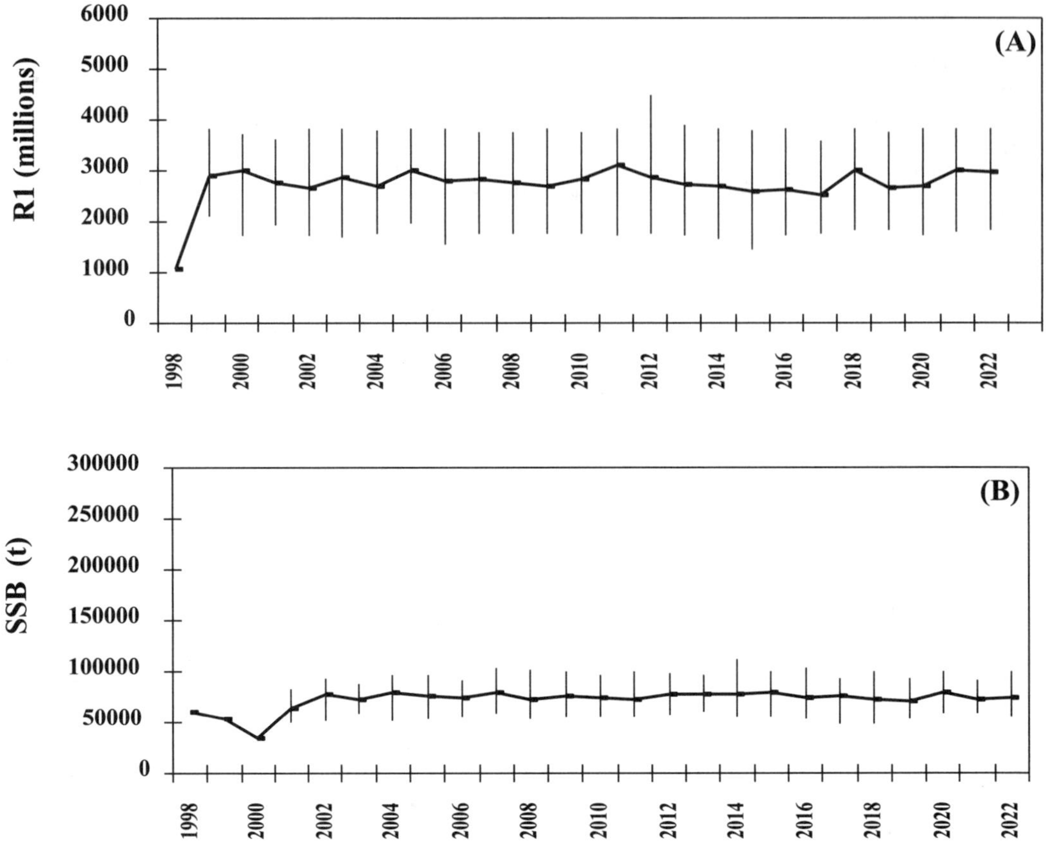

FIGURE 10. Projected values for $F = 1.0$ ($F_{\max}$) of (A) recruitment to age 1, and (B) spawning stock biomass of Atlantic menhaden (interquartile range displayed).

historical perspective. Hence, comparison of recent estimates to the median and interquartile range allows reasonable judgments as to current status of these variables. Thus, recent recruitment is probably low, while recent SSB and static SPR are probably high relative to these historical criteria. The only difference in these historic perspectives when a variable-aged M is used concerns recruitment in the most recent year. This latter approach suggests a recent higher value for recruitment in 1998. Confidence in estimates, especially recruitment, is lacking in the most recent few years, as suggested in Cadrin and Vaughan (1997), and updated historical analyses. Recent estimates of recruitment have been low initially, but have increased for the same year-class in subsequent assessments.

The recently developed coastwide juvenile abundance index compares favorably with our estimates of lagged recruitment to age 1. The level of agreement between the coastwide index and lagged recruits to age 1 serves as support for both the potential usefulness as a juvenile abundance index, but also supports the validity of the VPA-estimates of recruitment of Atlantic menhaden to age 1. The oldest values of this index are from the Maryland seine index. In more recent years, data from SEAMAP (U.S. South Atlantic Bight), North Carolina, Virginia, Connecticut, and Rhode Island have contributed to this coastwide index. Recent index values, when compared with their respective interquartile range, suggest low values of juvenile abundance from Chesapeake Bay, with moderate to high values outside of this region (Table 3).

A range of biological reference points is presented, based on YPR and static SPR analyses. These reference points form the basis of projections to investigate their utility as specified fishing mortality for management. Very little gain in recruitment to age 1 is suggested by these projections (Figure 12), while significant gains in SSB are evident with decreasing fishing mortality (Figure 13). However, the range of projected spawning stock biomass is consistently

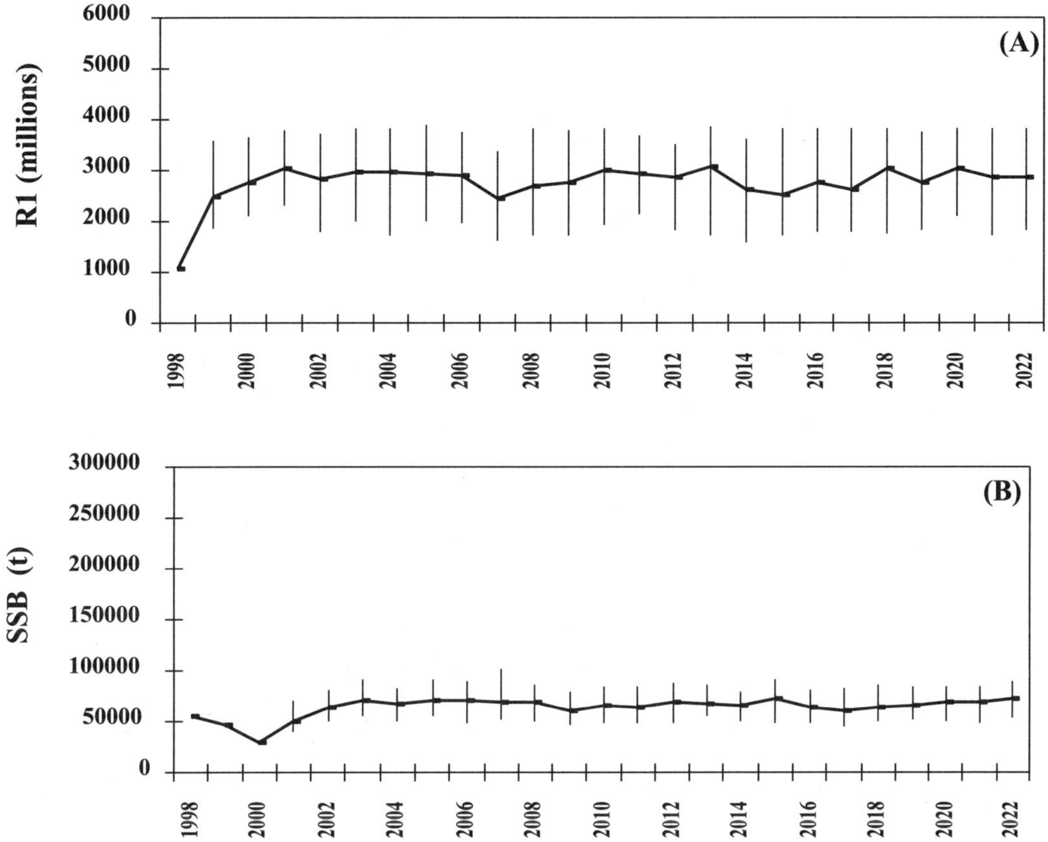

FIGURE 11. Projected values for $F = 1.1$ (F_{10}) of (A) recruitment to age 1, and (B) spawning stock biomass of Atlantic menhaden (interquartile range displayed).

above the current warning level for spawning stock biomass (17,000 t) or above the 25th percentile for the historic period (20,900 t) for all of the biological reference points for F. Furthermore, Myers et al. (1994) have suggested using the expected biomass that would produce half the maximum recruitment from a spawner-recruit relationship as a biological reference point for spawning stock biomass. The Ricker curve for Atlantic menhaden suggests that a spawning stock biomass of 25,300 t produces half the theoretical maximum recruitment to age 1 (2.6 billion). Although the Ricker spawner-recruit curve explains little of the underlying variability observed between recruitment to age 1 and spawning stock biomass, we do not intend to imply that there is no spawner-recruit relationships. If we apply the approach of Myers and Barrowman (1996), we would pose the following three questions: 1) did the greatest recruitment occur when SSB was high (yes, the 1958 cohort was produced when SSB was 89,000 t, the 82nd percentile); 2) did the lowest recruitment occur when SSB was low (yes, the 1967 cohort was produced when SSB was 21,000 t, the 27th percentile); and 3) was recruitment greater when SSB was above the median than when SSB was below the median (yes, 4.1 billion when above, 3.1 billion when below). Hence, conserving SSB should be a management concern.

Given the high variability in recruitment to age 1 at all levels of SSB, levels of recruitment are similar over the range of fishing mortalities that were simulated. This conclusion is further supported by the analysis on recruitment conditioned on different ranges of SSB (Table 6). These analyses suggest that only "improving" environmental conditions can **assure** improved recruitment. However, as noted earlier, there appears to be some real benefit to recruitment to age-1 from maintaining SSB above about the 25th percentile of its historic range (20,900 t). One caution should be noted. The risk curves themselves have associated uncertainty which has not been formally investigated in this study, although previous

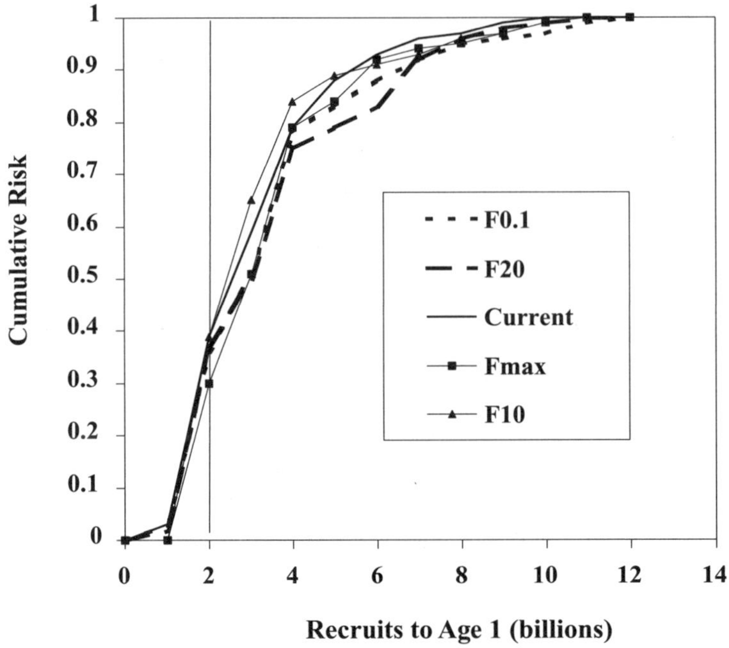

FIGURE 12. Cumulative probability (or risk) of decline in Atlantic menhaden recruitment to age 1 in 2007 for the event tree relationships with projection instantaneous fishing mortality rates based on 5 biological reference points ($F_{0.1}$, $F_{\max}$, F_{10}, F_{20}, F_{rep}). The vertical line represents a "trigger value" (2.0 billion fish) used in management.

experience (Vaughan 1993) would suggest this is relatively small.

Surplus-production models

Production-model estimates depict a stock that has been heavily exploited, perhaps excessively so, since at least the 1960s. As those estimates are made under the rather sweeping assumptions of production modeling, they should be considered together with other information on the stock. Nonetheless, they are consistent with the observations of an earlier study (Schaaf 1979), which held that as early as the mid-1950s, "the harvesting rate exceeded the growth rate of the population." If in future years fishing mortality is diminished, as in the projections (Figure 15, Figure 16), the population might again attain the levels of the 1950s. However, the projections implicitly assume that fishing mortality and density dependence are the only significant determinants of population size. Whether or not that is the case will in large part determine whether a reduction in fishing mortality will indeed result in those large population sizes.

Uncertainty is understated in most fisheries modeling contexts for several reasons: the variance of real-world processes appears to be increasing (Steele and Henderson 1984); our data sets are usually quite short; most fishery models are more or less misspecified. The last point is particularly true of the simplest models, such as production models. Uncertainty in production modeling was addressed here through bootstrapping, fitting two model formulations, and verifying the reasonableness of the assumption $q = 1.0$. Bootstrapping provided estimates of confidence intervals on benchmarks and other quantities; examining the generalized model provided insight into uncertainty related to model specification. Nonetheless, estimated confidence intervals in Figure 15 and Figure 16 are probably optimistic, in part because the input "data" include biomass estimates derived from VPA. In such quantities, some of the inherent variability in raw data are lost.

Comparison of approaches

Comparison of VPA and production modeling is comparison of dissimilar things. Virtual population analysis is a more detailed and precise approach, while production models are usually simpler and require less from the user. A more fundamental

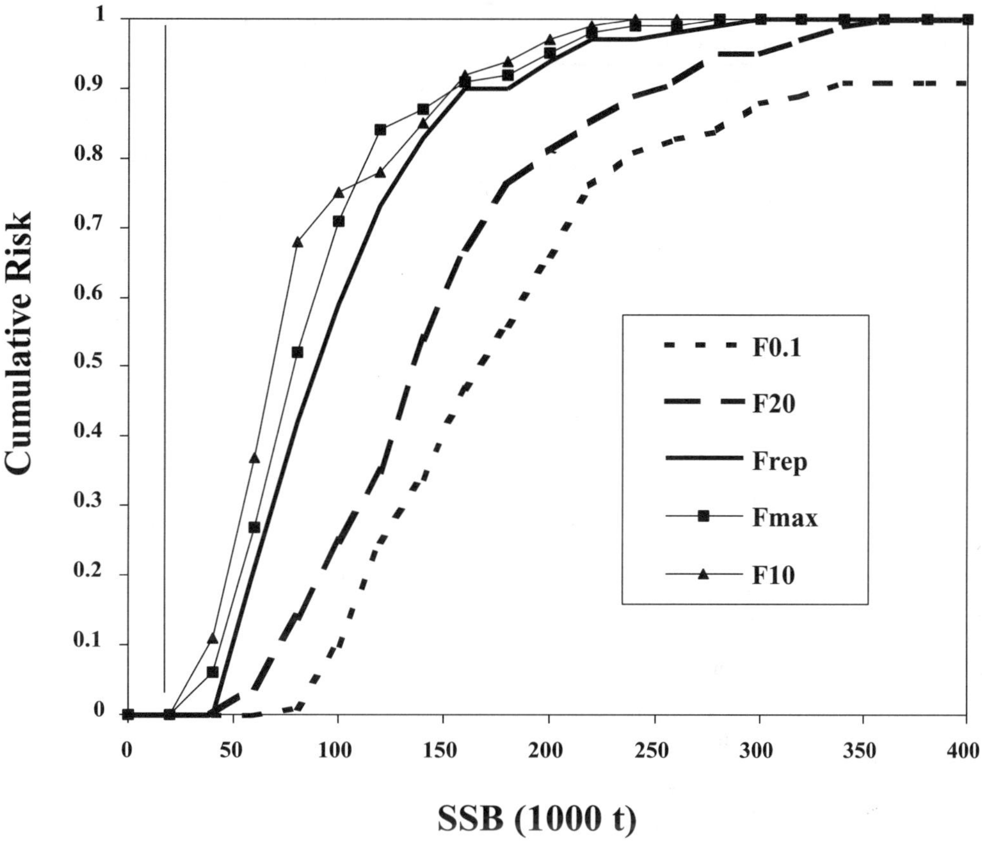

FIGURE 13. Cumulative probability (or risk) of decline in Atlantic menhaden stock biomass in 2007 for the event tree relationships with projection instantaneous fishing mortality rates based on 5 biological reference points ($F_{0.1}$, $F_{\max}$, F_{10}, F_{20}, F_{rep}). The vertical line represents a "trigger value" (17,000 t) used in management.

difference between the two models is that a production model includes an implicit recruitment function, while a VPA does not. Because of this distinction, benchmarks involving sustainability and optimality (e.g., MSY) can be estimated directly from a production model, while VPA estimates must be combined with auxiliary information or models to estimate such benchmarks. Each assessment tool is valuable and provides its own view of the data; neither can be judged "better" than the other.

In the context of the Atlantic menhaden assessment, the two modeling approaches are linked and complementary. They are linked because the input data for production modeling were in fact estimates from VPA. They are complementary because, as mentioned above, the two types of model are useful in estimating different quantities. Here, VPA was used to estimate population quantities; the production model to estimate maximum sustainable yield and stock status relative to optimal values.

A limited comparison of models can be made by considering what each says about the current state of the stock. The production model estimates that stock biomass is substantially below B_{MSY}, the level that can produce maximum sustainable yield (Figure 14B, Table 7). This reflects a close model fit to biomass estimates (ages 0+) from the VPA (Figure 14A), which indicate that the stock has long been below the remarkably high levels of the 1950s. The production model, however, says nothing specifically about the spawning component of the stock, which the VPA estimates as at high levels in the last few years analyzed (Table 1).

Comparison of estimated fishing mortality rates is more difficult. The production model estimates that the current fishing mortality rate is about 20% higher than the level that could produce MSY (Figure 14B, Table 7). Direct VPA estimates are

TABLE 6. Statistical properties of recruitment to age-1 Atlantic menhaden (R_1, in billions) dependent on the spawning stock biomass (SSB, t) that produced them.

Range of SSB (t)	Recruitment to age-1 (billions)				
				Percentiles	
	n	Mean	Median	25th	75th
All	44	3.6	3.1	2.2	4.6
Greater than:					
5,300[a]	44	3.6	3.1	2.2	4.6
17,000[b]	34	3.7	3.1	2.2	4.7
20,850[c]	32	3.8	3.1	2.2	4.7
25,300[d]	31	3.9	3.2	2.2	4.7
41,400[e]	21	4.1	3.0	2.2	5.0
69,250[f]	10	4.4	2.6	2.2	5.7
88,000[g]	9	4.7	3.0	2.2	5.7
Less than:					
5,300	–	–	–	–	–
17,000	10	3.3	3.1	1.9	3.7
20,850	12	3.2	3.1	1.8	4.1
25,300	13	3.1	3.0	1.7	3.7
41,400	23	3.2	3.1	1.9	3.8
69,250	34	3.4	3.1	2.2	4.5
88,000	35	3.4	3.1	1.9	4.5

[a] SSB that produces one-half maximum R_1 (Beverton-Holt)
[b] Current warning level for spawning stock biomass (SSB)
[c] Twenty-fifth percentile for historical SSB (1955–98)
[d] SSB that produces one-half maximum R_1 (Ricker).
[e] Median SSB for period 1955–98.
[f] Seventy-fifth percentile for historical SSB (1955–98)
[g] Minimum SSB during 1955–62 (period of historically high levels of SSB).

not compared with any measure of optimality, but it seems that static SPR is currently at relatively high levels (Table 3). The question of stock status, again, depends on whether the high levels of the 1950s can be taken as indicators of what might be possible now. The production model assumes that they can, while an approach based solely on observed percentiles, as in Table 3, tends to discount them as unusual events.

Summary on uncertainty

In our analyses of Atlantic menhaden, we considered uncertainty in a variety of areas, ranging from purely biological to purely statistical, with most uncertainties somewhere in between. Analyses were conducted under alternative assumptions on the natural mortality rate M, one of the most difficult parameters to estimate precisely in any stock assessment. The standard assumption, M invariant with age, is used in most fisheries work; the alternative assumption considered here, M decreasing with age but having the same average value, is probably more realistic, but has not been widely accepted. We found sensitivity of the assessment to this factor to be minimal.

The retrospective analysis of Cadrin and Vaughan (1997) provides insight into retrospective patterns expected in assessment of this stock by current methods. Unfortunately, understanding of the retrospective problem in general is incomplete; however, the conclusions of Pope (1972) and other investigators are still true: estimates from catch-at-age (VPA) methods are least reliable in the final years. Use of tuning indices has potential to ameliorate this situation somewhat; indices of the Atlantic menhaden stock are in relatively early stages of development.

Event-tree analysis offers a more empirical method of conducting stock simulations than more common methods that use parametric stock–recruitment models. Event-tree projections were used here to examine probable outcomes, in terms of recruitment and spawning-stock size, of various management measures. Simulations using a parametric recruitment model were not reviewed

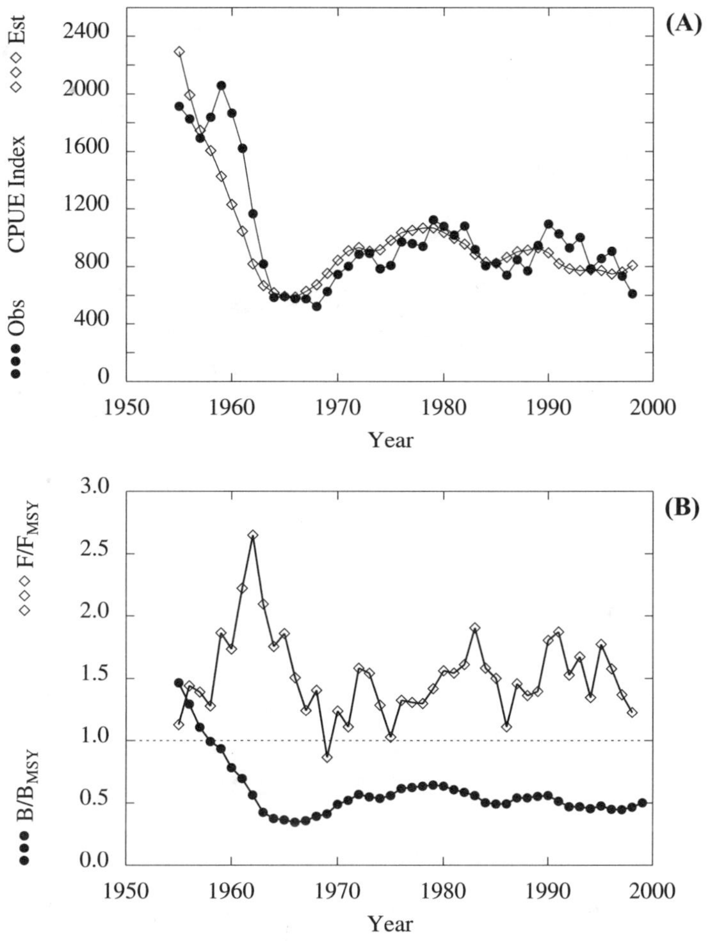

FIGURE 14. Logistic stock-production model of Atlantic menhaden. (A) Observed and estimated CPUE (ages 0+). (B) Estimated time trajectories of age 0+ stock biomass B and annual fishing mortality rate F, each relative to its management benchmark (B_{MSY} and F_{MSY}).

here, but were compared with event-tree analysis by Vaughan (1993).

Use of a production model in addition to the age-structured model (VPA) addresses another area of uncertainty: appropriate model structure. By using both types of model, this assessment examined a wide range of model structure and complexity. Results of the two types of model were by no means incompatible, but of course the production model depended heavily on VPA results. It may be possible in future assessments to make the methods more independent, perhaps by estimating a density-dependent catchability coefficient within the production model.

TABLE 7. Estimates of three management benchmarks for Atlantic menhaden, estimated by logistic and generalized production modeling. Bias-corrected 80% confidence intervals, derived from bootstrapping, are given for logistic model. For technical reasons, bootstrap was not applied to generalized fit.

Model	MSY (kt/yr) Point estimate	MSY (kt/yr) BC 80% interval	B_{99}/B_{MSY} Point estimate	B_{99}/B_{MSY} BC 80% interval	F_{98}/F_{MSY} Point estimate	F_{98}/F_{MSY} BC 80% interval
Logistic	414	391–443	0.50	0.41–0.63	1.23	1.03–1.45
Generalized	433		0.48		1.20	

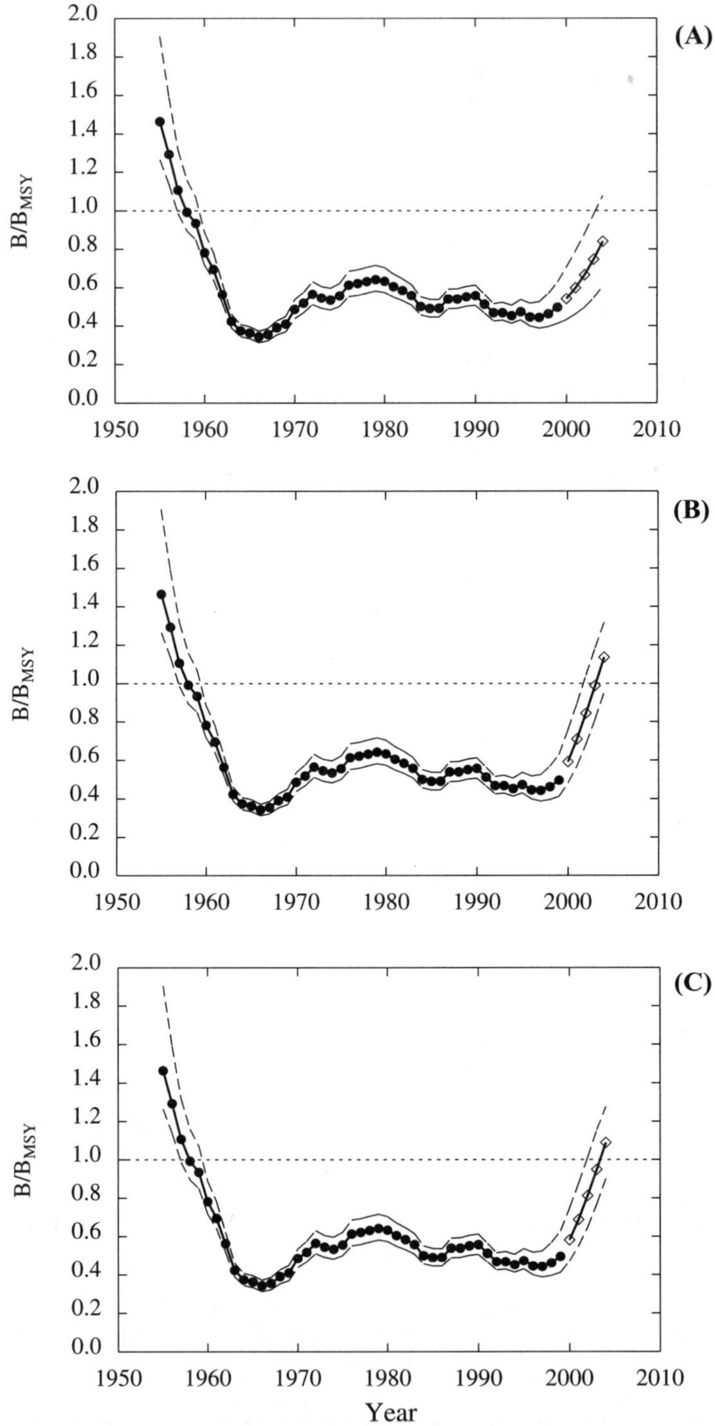

FIGURE 15. Projections of Atlantic menhaden stock based on logistic production-model results. Trajectory of biomass relative to B_{MSY}, assuming yields in 1999–2003 of (A) 245.9 kt/year, the recorded yield in 1998; (B) 171.2 kt/year, the recorded yield in 1999; or (C) 185.0 kt/year, the estimated yield in 2000. Fitted years shown as solid circles; projected years as hollow diamonds; 80% confidence interval as dashed lines.

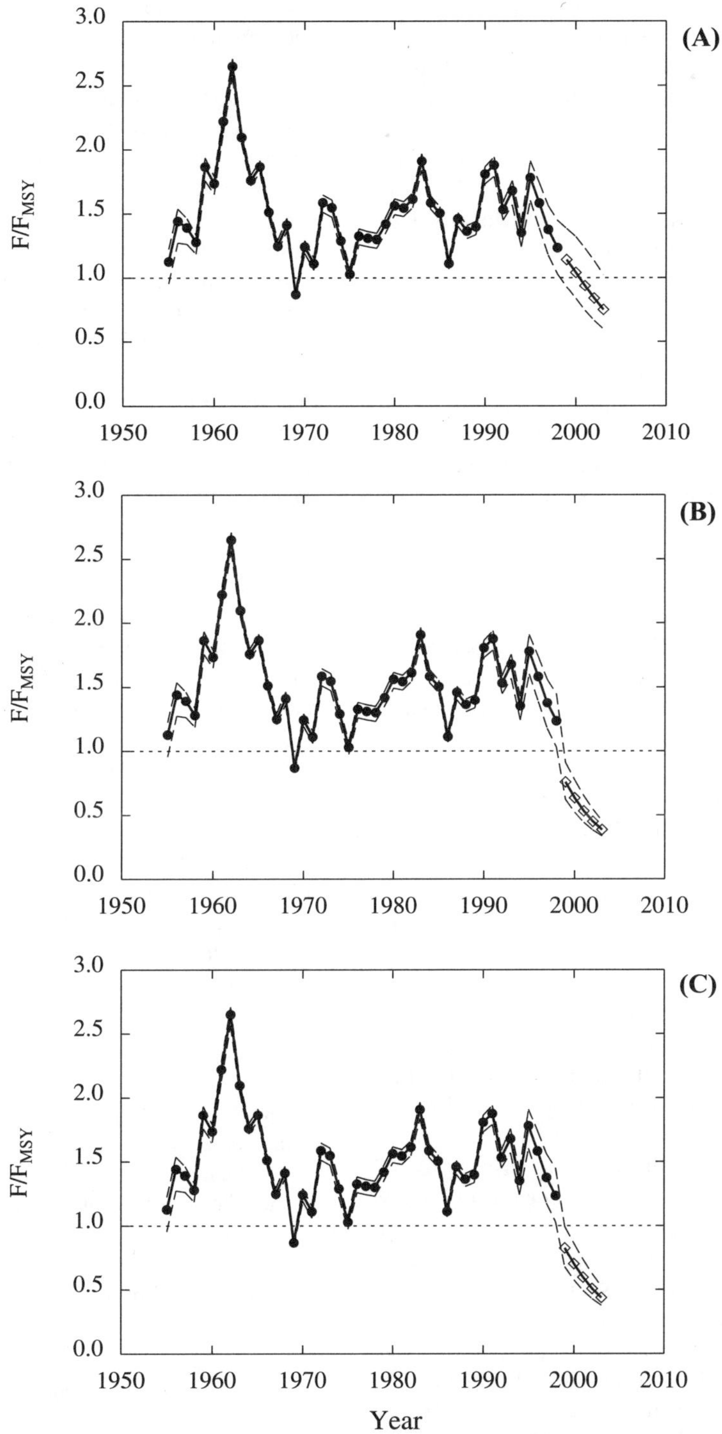

FIGURE 16. Projections of Atlantic menhaden stock based on logistic production-model results. Trajectory of fishing mortality rate relative to F_{MSY}, assuming yields in 1999–2003 of (A) 245.9 kt/year, the recorded yield in 1998; (B) 171.2 kt/year, the recorded yield in 1999; or (C) 185.0 kt/year, the estimated yield in 2000. Fitted years shown as solid circles; projected years as hollow diamonds; 80% confidence interval as dashed lines.

Uncertainties in the production model analysis itself were addressed through both bootstrapping and alternative runs. The bootstrap was used for non-parametric estimates of variability for parameters and related quantities; sensitivity runs were used to examine effects of varying model form, and the validity of the assumption $q = 1.0$. We did not attempt to translate these uncertainties directly into management terms, except to observe that all sensitivity runs estimate essentially the same present stock status.

Projections based on the production model emphasize perhaps the greatest uncertainty about this stock: whether or not the high stock levels of the 1950s can be regained by reducing fishing mortality. If fishing has been the main force regulating the stock size, it should be possible to regain those larger stock sizes with less fishing. However, if external factors—for example, habitat loss or changes in water quality—have contributed markedly to depressing stock sizes, reduced fishing mortality will not result in a remarkably large stock. None of the work here has attempted to answer that question, but if recent reductions in fishing mortality rate and yield continue, an unplanned experiment will take place that may reveal the answer.

Despite the variety of methods used, we believe that uncertainty is routinely underestimated in this and other stock assessments. A variety of assumptions used in modeling serve to reduce apparent uncertainty, yet those assumptions are not entirely valid.

- Autocorrelation is present in most fishery time series; this serves to reduce the information content of the data, and the best method of handling this situation is unclear. In modeling, we tend to assume that this is inconsequential.
- Specification error is the technical term for incorrect formulation of a model. Because all fishery models are highly simplified, specification error always exists. We do not suggest an increase in model complexity, as it is usually accompanied by an increase in variance and can have other undesirable properties, e.g., need for excessive biological data. However, it is important for all concerned to understand that bias from specification error is a likely component of stock assessments.
- Many methods of modeling assume that errors are normally distributed, either in the original data or through transformations. When maximum-likelihood methods are used for fitting (and many least-squares methods are also maximum-likelihood), model properties are assured only when distributional assumptions are met. Furthermore, some methods of constructing confidence intervals rely on distributional properties. Some results are robust to violations of distributional assumptions, while in other cases, bias or increased variance can result. Only one independent set of estimates is made in each year's assessment, and the effect of high variance can be large errors in those estimates, not just large confidence intervals.
- Finally, fishery scientists universally acknowledge that recruitment depends on unknown environmental factors, yet our lack of knowledge requires us in most cases to model recruitment with little or no environmental influence. The variance of the environment seems to be increasing with time (Steele and Henderson 1984), which can serve to invalidate even previously well-known environmental–biological relationships. This variance in the environment is probably the largest and most poorly understood component of uncertainty in stock assessments, and may elude our grasp indefinitely.

Given all the issues raised here, is stock assessment worthwhile? We believe that it is, provided that it is used together with precautionary principles and with a clear understanding that uncertainty is not just a buzzword, but a consequence of nature. Models will inevitably fail at times, and the measure of any management policy is whether it can succeed in the long term given the profound uncertainties of nature.

Acknowledgments

This work was supported by the U.S. National Oceanic and Atmospheric Administration through the Southeast Fisheries Science Center, NMFS; and the Center for Coastal Fisheries and Habitat Research in Beaufort, North Carolina. We thank D. Ahrenholz, S. Cadrin, J. Waters, and an anonymous reviewer for reviewing the manuscript and for their valuable suggestions.

References

Ahrenholz, D. W., D. L. Dudley, and E. J. Levi. 1991. Overview of mark-recovery studies on adult and juvenile Atlantic menhaden, *Brevoortia tyrannus*, and gulf menhaden, *B. patronus*. Mar. Fish. Rev. 53(4):20–27.

Ahrenholz, D. W., J. F. Guthrie, and C. W. Krouse. 1989. Results of abundance surveys of juvenile Atlantic and gulf menhaden, *Brevoortia tyrannus* and *B. patronus*. NOAA Technical Report, NMFS 84. Seattle, Washington.

Ahrenholz, D. W., W. R. Nelson, and S. P. Epperly. 1987. Population and fishery characteristics of Atlantic menhaden, *Brevoortia tyrannus*. Fish. Bull, U.S. 85:569–600.

Atlantic Menhaden Advisory Committee (AMAC). 1992. Fishery Management Plan for Atlantic Menhaden, 1992 Revision. Atlantic States Marine Fisheries Commission, Fisheries Management Report 22. Washington, D.C.

Beverton, R. J. H., and Holt, S. J. 1957. On the dynamics of exploited fish populations. Reprinted 1993 by Chapman and Hall, London.

Boudreau, P. R., and L. M. Dickie. 1989. Biological model of production based on physiological and ecological scaling of body size. Canadian Journal of Fisheries and Aquatic Science 46:614–623.

Cadrin, S. X., and D. S. Vaughan. 1997. Retrospective analysis of virtual population estimates for Atlantic menhaden stock assessment. Fish. Bull. U.S. 95:445–455.

Chester, A. J. 1984. Sampling statistics in the Atlantic menhaden fishery. NOAA Technical Report, NMFS 9. Seattle, Washington.

Clark, C. W., and M. Mangel. 1979. Aggregation and fishery dynamics: a theoretical study of schooling and the purse seine tuna fisheries. Fish. Bull. U.S. 77:317–337.

Clay, D. 1990. TUNE: A series of fish stock assessment computer programs written in FORTRAN for microcomputers (MS DOS). Int. Comm. Conserv. Atl. Tunas, Coll. volume Sci. Papers 32:443–460.

Dorval, E. 1998. Effect of sampling errors on estimates of recruitment and fishing mortality from separable virtual population analysis. Thesis, Old Dominion University, Norfolk, Virginia.

Dryfoos, R. L., R. P. Cheek, and R. L. Kroger. 1973. Preliminary analyses of Atlantic menhaden, *Brevoortia tyrannus*, migration, population structure, survival and exploitation rates, and availability as indicated from tag returns. Fish. Bull. U.S. 71:719–734.

Fletcher, R. I. 1978. On the restructuring of the Pella–Tomlinson system. Fish. Bull. U.S. 76:515–521.

Gabriel, W. L., M. P. Sissenwine, and W. J. Overholtz. 1989. Analysis of spawning stock biomass per recruit: An example for Georges Bank haddock. North American Journal of Fisheries Management 9:383–391.

Gulland, J. A. 1987. Natural mortality and size. Marine Ecology—Progress Series 39:197–199.

Gulland, J. A., and L. K. Boerema. 1973. Scientific advise on catch levels. Fish. Bull. U.S. 71: 325–335.

Hilborn, R., and C. J. Walters. 1992. Quantitative fisheries stock assessment: choice, dynamics, and uncertainty. Chapman and Hall, New York.

Hoenig, J. M. 1983. Empirical use of longevity data to estimate mortality rates. Fish. Bull. U.S. 82:898–903.

Jones, R. 1961. The long term effects of changes in gear selectivity and fishing effort. Mar. Research Dep. Agricul. Fish. Scotland, Ser. 2.

Lotka, A. J. 1924. Elements of physical biology. Reprinted 1956 as "Elements of mathematical biology." Dover Press, New York.

Mace, P. M., D. Gregory, N. Ehrhart, M. Fisher, P. Goodyear, R. Muller, J. Powers, A. Rosenberg, J. Shepherd, and D. Vaughan. 1996. An evaluation of the use of SPR levels as the basis for overfishing definitions in Gulf of Mexico finfish fishery management plans. Report to Gulf of Mexico Fishery Management Council, Tampa, Florida.

Megrey, B. A. 1989. Review and comparison of age-structured stock assessment models from theoretical and applied points of view. Pages 8–48 *in* E. F. Edwards and B. A. Megrey, editors. Mathematical analysis of fish stock dynamics. American Fisheries Society, Symposium 6, Bethesda, Maryland.

Murphy, G. I. 1965. A solution of the catch equation. J. Fish. Research Board Can. 22:191–201.

Myers, R. A., and N. J. Barrowman. 1996. Is fish recruitment related to spawner abundance? Fish. Bull. U.S. 94: 707–724.

Myers, R. A., A. A. Rosenberg, P. M. Mace, N. Barrowman, and V. R. Restrepo. 1994. In search of thresholds for recruitment overfishing. ICES J. March. Sci. 51:191–205.

Nelson, W. R., M. C. Ingham, and W. E. Schaaf. 1977. Larval transport and year-class strength of Atlantic menhaden, *Brevoortia tyrannus*. Fish. Bull., U.S. 75:23–41.

Pauly, D. 1979. On the interrelationships between natural mortality, growth parameters, and mean environmental temperature in 175 fish stocks. J. Cons. Int. Explor. Mer 39:175–192.

Pella, J. J. 1967. A study of methods to estimate the Schaefer model parameters with special reference to the yellowfin tuna fishery in the eastern tropical Pacific Ocean. dissertation, University of Washington, Seattle.

Pella, J. J., and P. K. Tomlinson. 1969. A generalized stock production model. Bull. Inter-American Trop. Tuna Comm. 13:419–496.

Pope, J. G. 1972. An investigation of the accuracy of virtual population analysis using cohort analysis. Int. Comm. N. Atl. Fish. Research Bull. 9:65–74.

Prager, M. H. 1994. A suite of extensions to a nonequilibrium surplus–production model. Fishery Bulletin, U.S. 92:374–389.

Prager, M. H. 1995. User's manual for ASPIC: A stock-production model incorporating covariates, program version 3.6x. NMFS Southeast Fisheries Science Center, Miami Laboratory Document MIA–92/93–55, 4th edition Available from M. H. Prager.

Reintjes, J. W. 1969. Synopsis of biological data on Atlantic menhaden, *Brevoortia tyrannus*. U.S. Fish. Wildl. Serv. Circ. 320.

Reish, R. L., R. B. Deriso, D. Ruppert, and R. J. Carroll. 1985. An investigation of the population dynamics of Atlantic menhaden (*Brevoortia tyrannus*). Canadian Journal of Fisheries and Aquatic Science 42:1371–1379.

Ricker, W. E. 1975. Computation and interpretation of biological statistics of fish populations. Fisheries Research Board of Canada Bulletin 191, 1–382.

Schaaf, W. E. 1979. An analysis of the dynamic population response of Atlantic menhaden, *Brevoortia tyrannus*, to an intensive fishery. Rapp. P.-v. Réun. Cons. int. Explor. Mer 177:243–251.

Schaaf, W. E., and G. R. Huntsman. 1972. Effects of fishing on the Atlantic menhaden stock, 1955–1969. Transactions of the American Fisheries Society 101:290–297.

Schaefer, M. B. 1954. Some aspects of the dynamics of populations important to the management of the commercial marine fisheries. Bull. Inter-American Trop. Tuna Comm. 1(2):27–56.

Schaefer, M. B. 1957. A study of the dynamics of the fishery for yellowfin tuna in the eastern tropical Pacific Ocean. Bull. Inter-American Trop. Tuna Comm. 2:247–268.

Seagraves, R. J. 1992. Weakfish Fishery Management Plan, Amendment #1. Atlantic States Marine Fisheries Commission, Fish. Manage. Rep. No. 20.

Sinclair, A., D. Gascon, R. O'Boyle, D. Rivard, and S. Gavaris. 1990. Consistency of some northwest Atlantic groundfish stock assessments. North Atl. Fish. Org., SCR Doc. 90/96. Dartmouth, Nova Scotia.

Sissenwine, M. P., and J. G. Shepherd. 1987. An alternative perspective on recruitment overfishing and biological reference points. Canadian Journal of Fisheries and Aquatic Science 44:913–918.

Smith, J. W., and Population Dynamics Team. 1999. Forecast for the 1999 gulf and Atlantic menhaden purse-seine fisheries and review of the 1998 fishing season. NOAA Beaufort Laboratory, Beaufort, North Carolina.

Steele, J. H., and E. W. Henderson. 1984. Modeling long-term fluctuations in fish stocks. Science 224:985–987.

(SARC) Stock Assessment Review Committee. 1998. Consensus Summary of Assessments. 26th Northeast Regional Stock Assessment Workshop (26th SAW), Northeast Fisheries Science Center Ref. Doc. 98-03, Woods Hole, Massachusetts.

Tomlinson, P. K. 1970. A generalization of the Murphy catch equation. J. Fish. Research Board Can. 27:821–825.

Ulltang, O. 1977. Sources of errors in and limitations of virtual population analysis (cohort analysis). J. Cons. int. Explor. Mer 37:249–260.

Vaughan, D. S. 1977a. A stochastic analysis of the stability of the Atlantic menhaden fishery. dissertation, Graduate School of Oceanography, University of Rhode Island, Narragansett.

Vaughan, D. S. 1977b. Confidence intervals on mortality rates based on the Leslie matrix. Pages 128–150 *in* W. Van Winkle (editor.), Assessing the Effects of Power-Plant-Induced Mortality on Fish Populations, Pergamon, New York.

Vaughan, D. S. 1990. Assessment of the status of the Atlantic menhaden stock with reference to internal waters processing. NOAA Tech. Memo. NMFS SEFSC 262. Beaufort, North Carolina.

Vaughan, D. S. 1993. A comparison of event tree risk analysis to Ricker spawner-recruit simulation: An example with Atlantic menhaden. Can. Spec. Publ. Fish. Aquat. Sci. 120:231–241.

Vaughan, D. S. 1994–1999. Trigger variables for Atlantic menhaden. Annual reports to the Atlantic Menhaden Advisory Committee, Atlantic States Marine Fisheries Commission, NOAA Beaufort Laboratory, Beaufort, North Carolina.

Vaughan, D. S., and J. V. Merriner. 1991. Assessment and management of Atlantic menhaden, *Brevoortia tyrannus*, and gulf menhaden, *B. patronus*, stocks. Mar. Fish. Rev. 53(4):49–57.

Vaughan, D. S., and J. W. Smith. 1988. Stock assessment of the Atlantic menhaden, *Brevoortia tyrannus*, fishery. NOAA Tech. Rep. NMFS 63. Seattle, Washington.

Vaughan, D. S., and J. W. Smith. 1998. Supplemental analysis of the status of the Atlantic menhaden stock. Report to the Atlantic Menhaden Advisory Committee, Atlantic States Marine Fisheries Commission, NOAA Beaufort Laboratory, Beaufort, North Carolina.

Vaughan, D. S., and J. W. Smith. 1999. Supplemental analysis of the status of the Atlantic menhaden stock. Report to the Atlantic Menhaden Advisory Committee, Atlantic States Marine Fisheries Commission, NOAA Beaufort Laboratory, Beaufort, North Carolina.

American Fisheries Society Symposium 27:113–136, 2002

Bayesian Approaches to the Analysis of Uncertainty in the Stock Assessment of Pacific Halibut

ANA M. PARMA[1]
International Pacific Halibut Commission, P.O. Box 95009
Seattle, Washington 98145-2009, USA

Abstract.—The uncertainty associated with estimates of stock size is increasingly acknowledged in the provision of management advice. Estimated variances, however, are usually small compared with the variability of abundance estimates for any given year produced by successive assessments, especially when changes in the assessment methods are introduced. Of all the different kinds of uncertainty, uncertainty in the specification of the model structure is often the most significant source of assessment errors in some closely monitored fisheries. Recent changes in Pacific halibut assessments illustrate this problem. A separable catch-at-age model used since the mid-1980s performed very poorly in retrospective analyses, initially overestimating biomass and then underestimating it in the 1990s. The latter has been attributed to trends in catchability at age associated with a remarkable decrease in halibut growth rate over the last 15 years. A new model was developed which replaced the assumptions of constant catchability and selectivity made in the old model by a more flexible and realistic treatment of observation and process variability. The change in model structure resulted in estimates of present biomass more than double the previous estimates. While retrospective performance of the new model is much improved, major uncertainties still remain. In particular, the relative importance of size and age effects in determining catchability of the setline surveys is difficult to discern from the data. Two extreme models, one based on the assumption that survey selectivity is a function of size and the other based on the assumption that survey selectivity *at age* is constant, have been formulated to incorporate this uncertainty. In this paper, Bayesian methods are used to evaluate the uncertainty around abundance estimates and short-term forward projections, including the uncertainty due to alternative possible model structures. The posterior distributions of parameters of interest under each of the two models are approximated by Markov Chain Monte Carlo methods, and the support given to the models by the data is evaluated by computing their integrated likelihoods. While far from a complete representation of all sources of model uncertainty, the analysis illustrates how uncertainty in model choice, in addition to the standard parameter uncertainty, can be incorporated in risk computations.

Introduction

Fish stock assessments have come under close scrutiny in recent years, after a number of well-publicized collapses and fisheries crises have pointed to the limitations of the available data and management approaches. A positive outcome of this increased awareness is a more open acknowledgement of the assessment uncertainty in the provision of management advice. The precautionary approach (FAO 1995) has gained acceptance as a basis for fisheries management (Restrepo et al. 1998; Richards and Maguire 1998), which implies moving away from providing just point recommendations to fully evaluating consequences of alternative management actions in terms of the probabilities of achieving management goals. While increasingly sophisticated techniques for evaluating uncertainty and risks have found use in fish stock assessments (e.g., Kinas 1996; Punt and Hilborn 1997; Cordue 1998), these techniques commonly rely on the structure of the chosen statistical model being correct (but see Patterson 1999 for an exception). Yet, it is the problem of model misspecification that appears to be the most significant source of assessment errors, and the one that is most difficult to address and quantify, beyond standard sensitivity analyses.

Uncertainty in the estimates of stock size (or fishing mortality) imply that fishing rates are only imperfectly controlled, a fact that needs to be taken into account when evaluating alternative harvesting strategies or control rules. Whether management is based on catch quotas or effort controls, assessment errors translate into *implementation uncertainty*, compounding other management problems such as discards or poor enforcement, and substantially degrading policy performance (Walters and Parma 1996; NRC 1998). In this context, the uncertainty about model structure, often referred to as *structural uncertainty*, is particularly harmful; not only does it dramatically increase overall uncertainty, but model misspecification may also lead to persistent bias in the implementation of harvest

[1]Current address: Centro Nacional Patagónico, 9120 Puerto Madryn, Chubut, Argentina.

rules or in forecasts used for planning. The case of Northern cod is perhaps the best-known example of a persistent overestimation of stock size (Walters and Maguire 1996), which had devastating management consequences. Many other, less dramatic examples also show that structural uncertainty is usually substantial, as data used in stock assessments are frequently insufficient, and assumptions made to fill data gaps are never uniquely determined.

The recent history of assessments of Pacific halibut *Hippoglossus stenolepis* illustrates these problems. The Pacific halibut stocks have supported a longline commercial fishery in Canadian and U.S. waters for more than a 100 years, with catches fluctuating between 20 and 70 million pounds. Since 1923, stocks have been studied and managed by the International Pacific halibut Commission (IPHC; Sullivan and McCaughran 1995.) In spite of a wealth of fishery data available for Pacific halibut, stock assessments produced by IPHC have suffered from serious problems of model misspecification. A separable catch-at-age model used since the mid-1980s (Deriso et al. 1985) performed very poorly in retrospective analyses, initially overestimating biomass (Parma 1993) and then underestimating it in the 1990s (Clark et al. 1999). The latter has been attributed to a decrease in the catchability at age of the commercial fishery, associated with a remarkable reduction in halibut growth rate over the last 15 years. The decrease in growth rate was most dramatic in the Gulf of Alaska, at the center of the distribution of Pacific halibut, following a strong increase in recruitment in the 1980s. While this supports a hypothesis of density-dependence, direct environmental forcing cannot be discarded. Identifying the causes of the growth decline is an area of current research at IPHC. Whichever the explanation for the change, the decrease in growth had serious implications in the assessment performance.

Until 1994, annual assessments relied exclusively on fishery-dependent data, and no regular surveys were conducted during 1987–1992. So, while the halibut database is indeed very good in many respects, no consistent series of abundance indices was available when problems with the standard assessment model became evident. Model performance in retrospective analyses was used to guide the development of new assessment methods. Surveys were resumed in 1993 and survey-based indices are now used regularly in the assessments. The implementation of these methodological changes resulted in major changes in biomass estimates and catch quotas, as well as in our perceptions about stock-productivity relationships (Clark et al. 1999). While performance of the new models is much improved, some substantial uncertainties remain. As in other data-rich situations, confidence intervals around estimates of stock size derived ignoring structural uncertainty are deceivingly narrow, and most of the uncertainty revolves around the choice of the assessment model.

In this paper, I provide some background on the evolution of assessment methods used for Pacific halibut, mainly to stress the importance of structural uncertainty. I discuss the uncertainties present in the recent assessments and use Bayesian Markov Chain Monte Carlo (MCMC) methods to quantify them, including uncertainty in the structure of the model. I illustrate how the approach can be applied in risk analyses by conducting short-term projections under a constant harvest rate, the policy currently followed by IPHC. While a full policy evaluation is beyond the scope of this paper, the extension of the approach to contrast the performance of different candidate policies is straightforward.

Results below correspond to the assessment of Pacific halibut in IPHC Regulatory Area 3A, in the central Gulf of Alaska, which best illustrates the effects of the trends in size-at-age on model performance and uncertainty.

Model Misspecification and Retrospective Errors in Past Halibut Assessments

Standard catch-at-age models fall into two main classes: VPA-type and statistical catch-at-age models (Hilborn and Walters 1992). The latter have been the basis for the Pacific halibut assessments over the last 15 years. In particular, a separable model known as CAGEAN (Deriso et al. 1985) was the primary assessment tool from 1985 until 1995. In contrast to VPA models, statistical catch-at-age models have tended to prioritize the importance of errors in the catch-at-age observations, while making rather rigid assumptions about the population dynamics, and generally ignoring process stochasticity altogether, beyond of course variability in recruitment and fishing mortality. Separable models are extreme in this sense, as the schedule of relative age-specific fishing mortality (here referred to as *selectivity*) is assumed to be constant over time so that fishing mortality at age and time $F_{a,t}$ is separable into an age-dependent factor and a year-dependent factor as

$$F_{a,t} = s_a F_t \tag{1}$$

The year-dependent factor F_t, or full-recruitment fishing mortality, is related to the amount of effective fishing effort exerted in a given year. In the halibut assessments, F_t was assumed to be proportional (in expectation) to standardized fishing effort (E_t):

$$F_t = q E_t \varepsilon_t, \tag{2}$$

where q is the catchability coefficient assumed constant, and ε_t is random noise.

Both assumptions—constant selectivity and constant catchability—are questionable in the face of improvements in technology and changing fishing practices. Therefore, likely violations of both were blamed for the strong retrospective patterns observed in halibut assessments. During the 1980s, the assessments consistently overestimated biomass (Parma 1993), but the direction of the errors reversed in the early 1990s. A strong pattern of initial underestimation of abundance was observed, affecting most strongly the stocks at the center of the range in the Gulf of Alaska (Figure 1). For a number of years the assessments indicated abrupt downward trends in biomass, but the final-year estimates used to determine catch quotas remained relatively stable. While the period of overestimation in the 1980s was attributed to likely increases in catchability (Parma 1993), the reversal of the retrospective pattern was linked to a remarkable decrease in halibut size-at-age that took place in the 1980s and into the 1990s (Clark et al. 1999; Sullivan et al. 1999) reducing the selectivity of the young age classes. This decrease was most marked in Alaska, where fish of age 11, a modal age in the catch, are now about 25% smaller in length than in the1970s (Figure 2), and

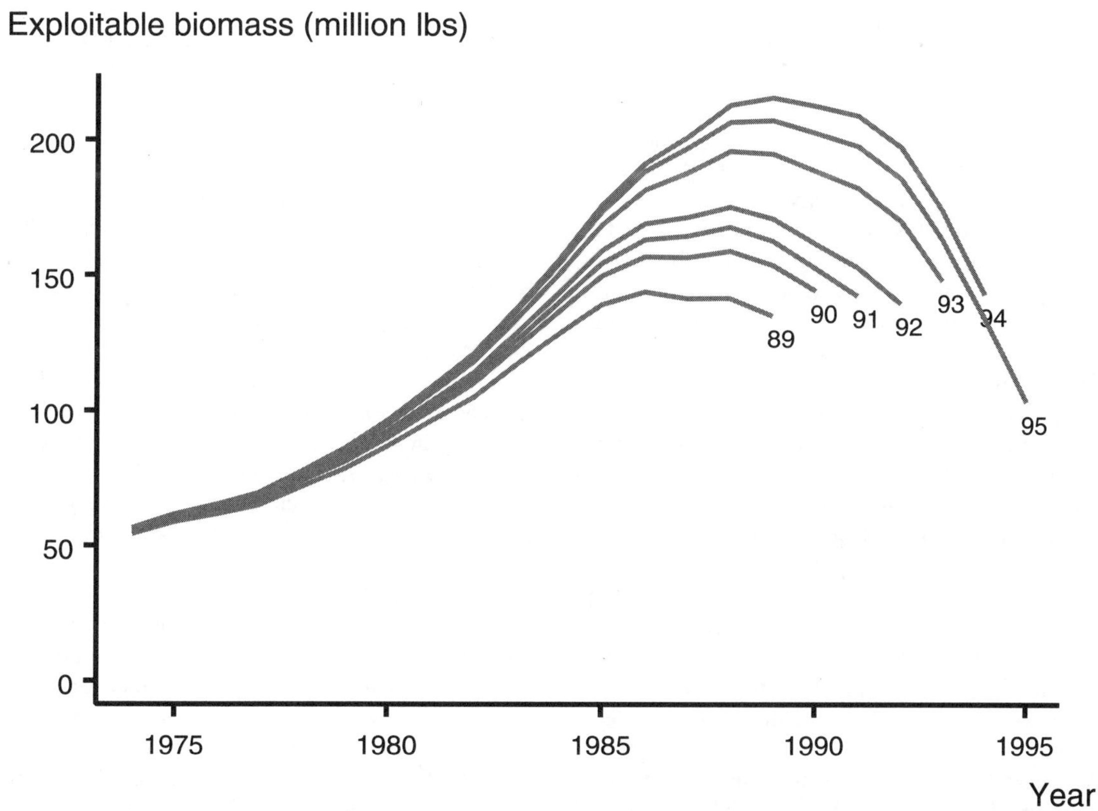

FIGURE 1. Retrospective pattern of Pacific halibut estimates of exploitable biomass in the Gulf of Alaska (IPHC Area 3A) obtained using the old assessment model (CAGEAN). Each line represents a retrospective assessment; noted how initial biomass estimates increase with the addition of new data.

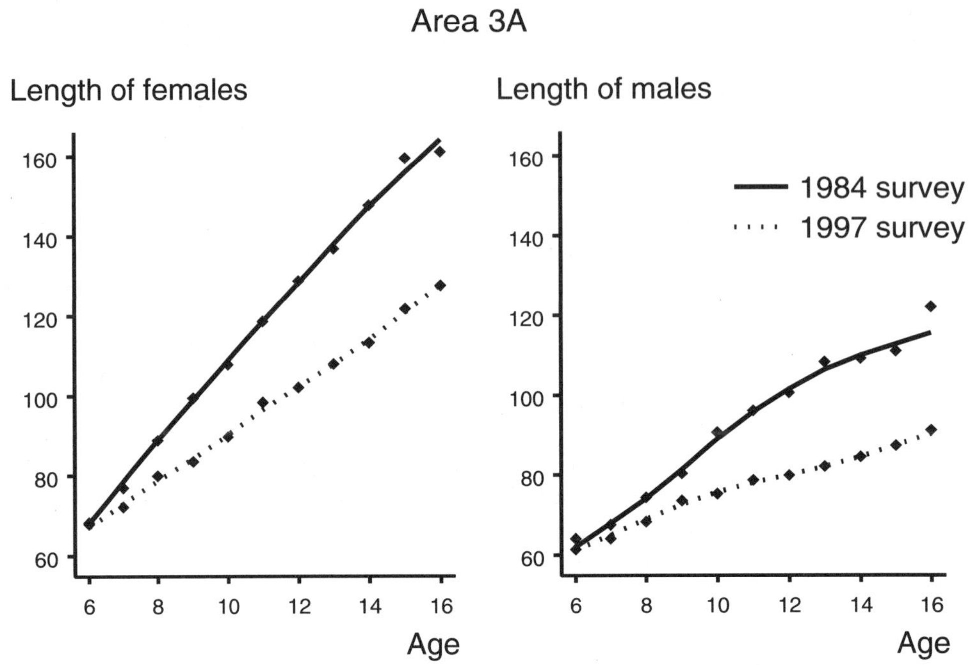

FIGURE 2. Mean length at age of male and female Pacific halibut caught in setline surveys in the Gulf of Alaska in two different periods.

about 50% lighter. Changes in size-at-age can be expected to affect the age composition of the catch because setline gear is selective for large fish sizes, and, in addition, there is a minimum size limit of 32 inches (81 cm) so that smaller fish are discarded at sea. Survey samples show that the fraction of fish of age eight (approximately the age at recruitment) that are of legal size dropped from 67% in 1976 to 20% in 1994 (Figure 3). As a consequence, smaller size-at-age in the early 1990s resulted in poor representation of the young age classes in the landings (Figure 4). Under the assumption of constant selectivity, this translated into a substantial drop in recruitment estimates. In subsequent years, as year classes gradually became selected and showed their actual strengths, recruitment estimates were corrected upwards, producing the strong retrospective pattern in Figure 1. The consistency of the negative bias was such that it prompted suggestions to correct estimates in an *ad hoc* manner (i.e., by inflating biomass estimates by some factor derived from previous errors). While *ad hoc* corrections would have been prudent in the case of overestimation of abundance, the staff discarded them as being too risky for halibut, particularly when surveys were not available to corroborate population trends. A new model was developed instead to try to correct the problem.

The New Assessment Models

The two fundamental assumptions made in the old assessment model, namely constant catchability and constant selectivity at age, were relaxed in the new models developed by the IPHC staff (Sullivan et al. 1999). To account for the effects of the reduction in size-at-age, commercial selectivity was modeled as a (time-varying) function of size (instead of age), so that changes over time in the growth schedule resulted in changes over time in the implied age-specific selectivity. The model is thus age- and size-structured. To determine growth trends, size-at-age data from the commercial catch and the surveys were used as input, in addition to the usual catch and age composition data. The series of surveys resumed in 1993 was used together with the old series discontinued in 1986 to provide relative abundance indices, while commercial catchability was allowed to change over time. The new assessment methods are described in detail in Sullivan et al. (1999), and only slight modifications have been introduced for this analysis. I therefore concentrate on the fundamental model assumptions that are relevant for this study and refer the reader to Sullivan et al. (1999) for further discussion of specific model assumptions and input data.

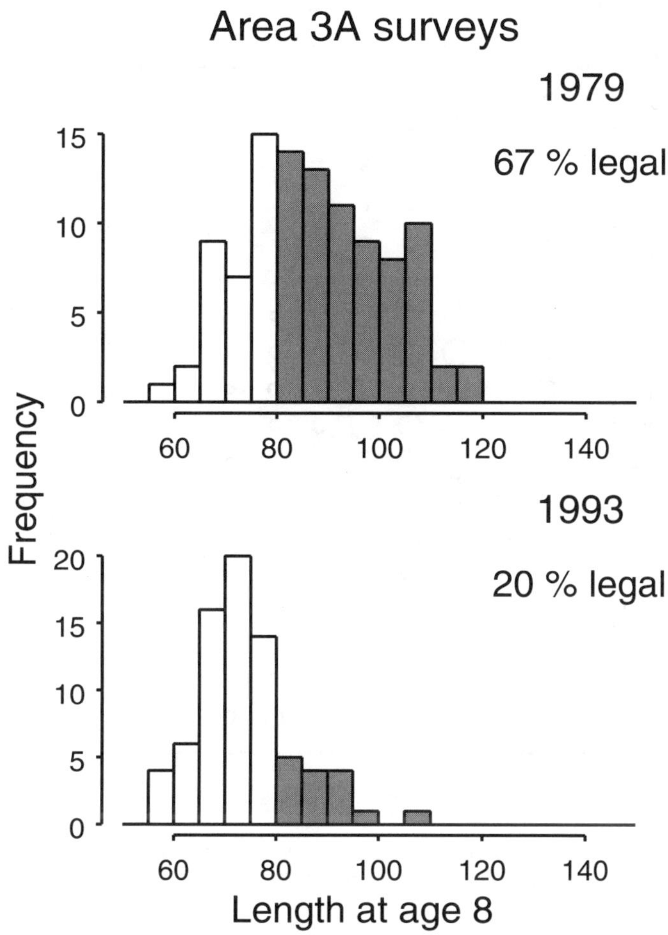

FIGURE 3. Size frequency distribution of eight-year-old halibut caught in setline surveys in 1979 and 1994 in the Gulf of Alaska. Note the drop in the fraction of legal-size fish (area shaded).

Fundamental assumptions about process variability

Trends in commercial catchability ${}_{c}q_t$ were modeled as a random walk process of the form

$$\begin{aligned} \ln({}_{c}q_{t+1}) &= \ln({}_{c}q_t) + {}_{cq}\varepsilon_t \\ \text{where} \quad {}_{cq}\varepsilon_t &\sim N\big(0, {}_{cq}\sigma^2\big), \end{aligned} \quad (3)$$

and ${}_{cq}\sigma^2$, the variance of the ${}_{cq}\varepsilon_t$ random variables, was specified *a priori* so that catchability could change gradually from year to year (see Tables A1–A3 in the Appendix). A break in the series was incorporated in 1995 to allow more flexibility during the transition to ITQ management, which affected significantly the conduct of the fishery. The use of time-series models to represent stochastic trends follows Gudmundsson (1994) and Fournier et al. (1998). An advantage of time-series models over parametric functions of time is their greater flexibility and the fact that the influence of the data on determining temporal trends is locally restricted. Similar formulations were used to represent trends in growth and selectivity parameters.

The selectivities of the commercial fishery and the surveys were modeled as explicit functions of fish size (respectively ${}_{c}s_t(X)$ and ${}_{s}s_t(X)$), and so abundance at size, in addition to age, had to be predicted. The distribution of log-length at age was approximated using Gaussian distributions with mean $\mu_{a,t}$ and variance σ^2. In the absence of size-selective mortality, mean log-length at age was assumed to propagate according to

$$\exp(\mu_{a+1,t+1}) = \alpha_t + \beta \exp(\mu_{a,t}),$$

with mean log-length at age six (first age-class in

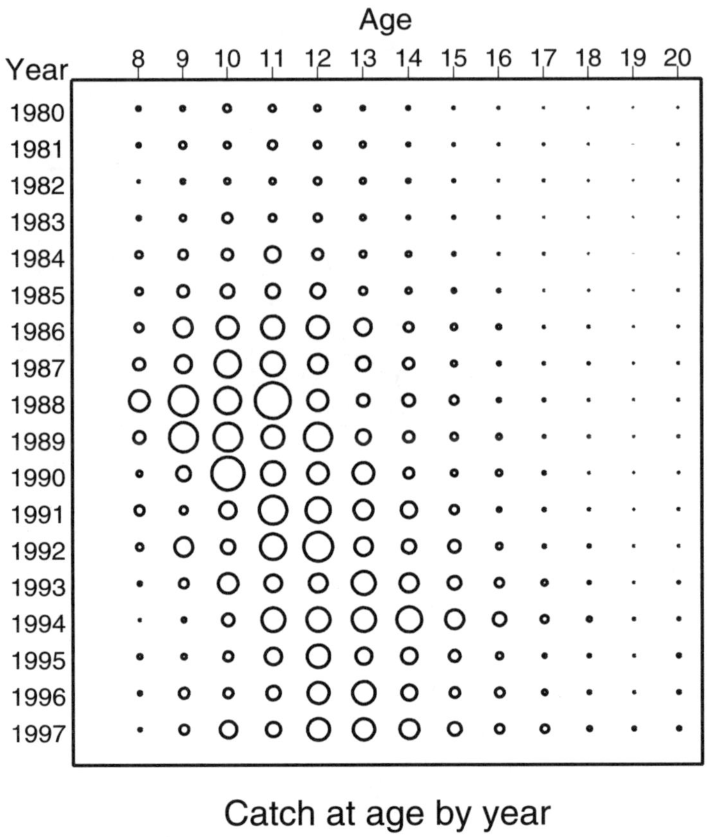

FIGURE 4. Bubble plot of catch at age and year in IPHC Area 3A (circle radii proportional to catch numbers). Note the poor representation of young age groups in the early 1990s.

the model) $\mu_{6,t}$, and growth intercept α_t modeled as random times series:

$$\begin{aligned} \mu_{6,t+1} &= \mu_{6,t} + {}_{\mu}\varepsilon_t \qquad {}_{\mu}\varepsilon_t \sim N\big(0, {}_{\mu}\sigma^2\big) \\ \ln(\alpha_{6,t+1}) &= \ln(\alpha_{6,t}) + {}_{\alpha}\varepsilon_t \quad {}_{\alpha}\varepsilon_t \sim N\big(0, {}_{\alpha}\sigma^2\big) \end{aligned} \tag{4}$$

As before, variances of the time series were specified *a priori*. By selectively removing the largest fish, the fishery is expected to affect the size-distribution of survivors (Deriso and Parma 1988). The mean log-length after the harvest was thus corrected for this effect, as explained in Sullivan et al. (1999) (see equation for $\mu^+_{a,t}$ in Table A1). The variances of the growth parameters were set large to allow sufficient flexibility to capture the dramatic decrease in size at age observed in the data, especially in the survey catches. Surveys are not affected by the minimum size limit, so survey observed and predicted sizes at age decreased more than commercial sizes at age (Figure 5).

The size-based selectivity functions were parameterized as half-normal curves and, in the case of the commercial fishery, the effect of the minimum size limit was incorporated by setting ${}_{c}s_t(X) = 0$ for $X < \ln(81)$ (see Table A1). If $p(X_{a,t})$ denotes the distribution of size (log-length X) at age a and year t, and $s_t(X_{a,t})$ is the size-based selectivity function (for either the commercial fishery or the surveys), then the implied selectivity at age $S_{a,t}$ results from integrating $s_t(X_{a,t})$ over $X_{a,t}$ as

$$S_{a,t} = \int_{-\infty}^{\infty} s_t(X_{a,t})\, p(X_{a,t})\, \mathrm{d}X_{a,t}.$$

A decrease in size at age of the magnitude estimated for Area 3A (Figure 5) resulted in a substantial reduction in commercial age-specific selectivity (Figure 6), even though the commercial selectivity at length ${}_{c}s_t(X)$ shifted somewhat towards smaller fish sizes (see Table A1).

A critical assumption in the Pacific halibut assessment relates to the treatment of survey selectivity. Surveys are usually used in catch-at-age models

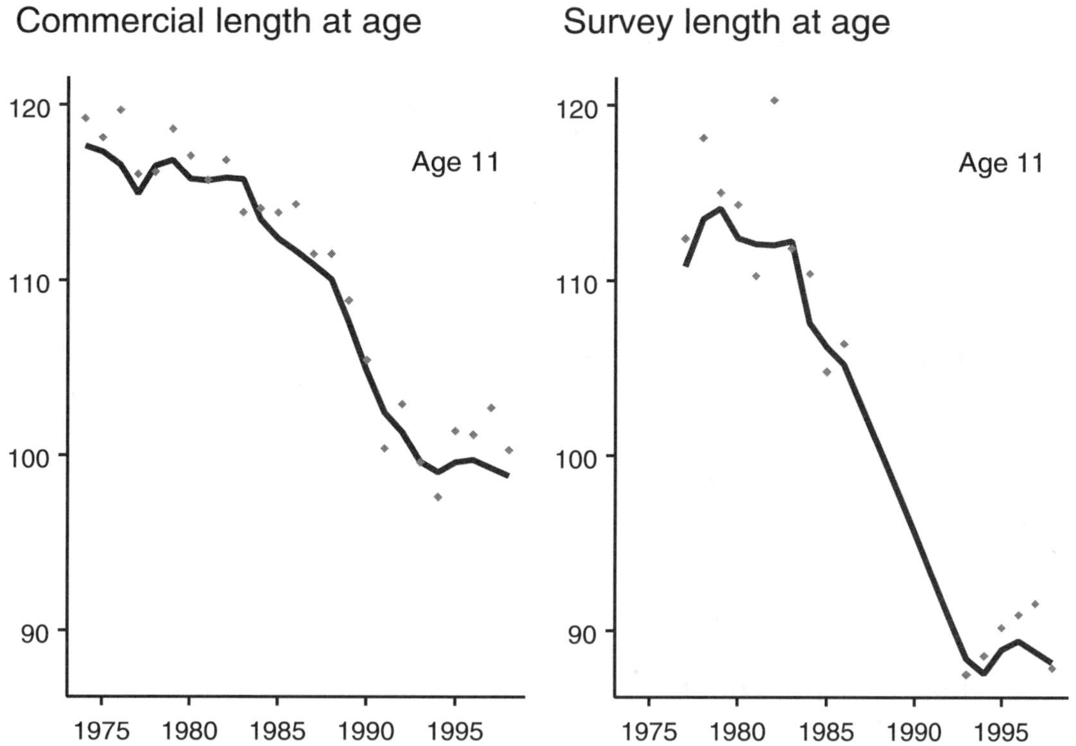

FIGURE 5. Observed and predicted median length at age 11 in the commercial and survey catches of Pacific halibut in the central Gulf of Alaska.

to provide age-specific indices of abundance. This implies that survey catchability *at age* is assumed constant, at least in expectation. In the case of Pacific halibut, however, the substantial change in size-at-age led us to wonder about the relative importance of *age* versus *size* in determining survey catchability. While hooking efficiency certainly increases with fish size (Kaimmer 1999), the fact that estimated size-based selectivity schedules differed markedly among management regions (Sullivan et al. 1999) is inconsistent with size being the dominant factor. Age effects, related for example to availability of fish on the grounds (ontogenetic migrations from nursery grounds) are likely. The two effects are confounded in the catch at age observations and it would be hopeless to try to estimate them both without auxiliary information. Thus, two versions of the assessment model were constructed based on rather extreme possibilities: (1) survey selectivity at length is constant over time (${}_{s}s_t(X) = {}_{s}s(X)$), or (2) survey selectivity at length varies but the implied selectivities at age are constant (${}_{s}S_{a,t} = {}_{s}S_a$). The problem is further complicated by the fact that the J-type of hook used in the early surveys was replaced in 1984 by the much more efficient Circle hook, following the practice in the commercial fleet. Both models had to accommodate this break, as the size-specific selectivities of the two hooks differ (Table A3; Sullivan et al. 1999).

The remaining model equations used to reconstruct stock abundances at age are summarized in Tables A1–A2 in the Appendix.

Parameter estimation

Unknown parameters in these models comprise a series of coefficients (recruitment and F_t for $t = 1974, \ldots, 1998$; initial abundance, selectivity, commercial catchability, and size-at-age parameters for $t = 1974$, survey catchability and selectivity parameters, and growth coefficients α_{1974} and β) and a series of random variables representing stochastic time-series trends (e.g., those denoted as ${}_{x}\varepsilon_t$ in equations (3) and (4), where x is the affected model parameter). Parameters are estimated by maximizing the posterior probability density

$$P(\theta \mid \text{data}) = \frac{P(\text{data} \mid \theta)P(\theta)}{K},$$

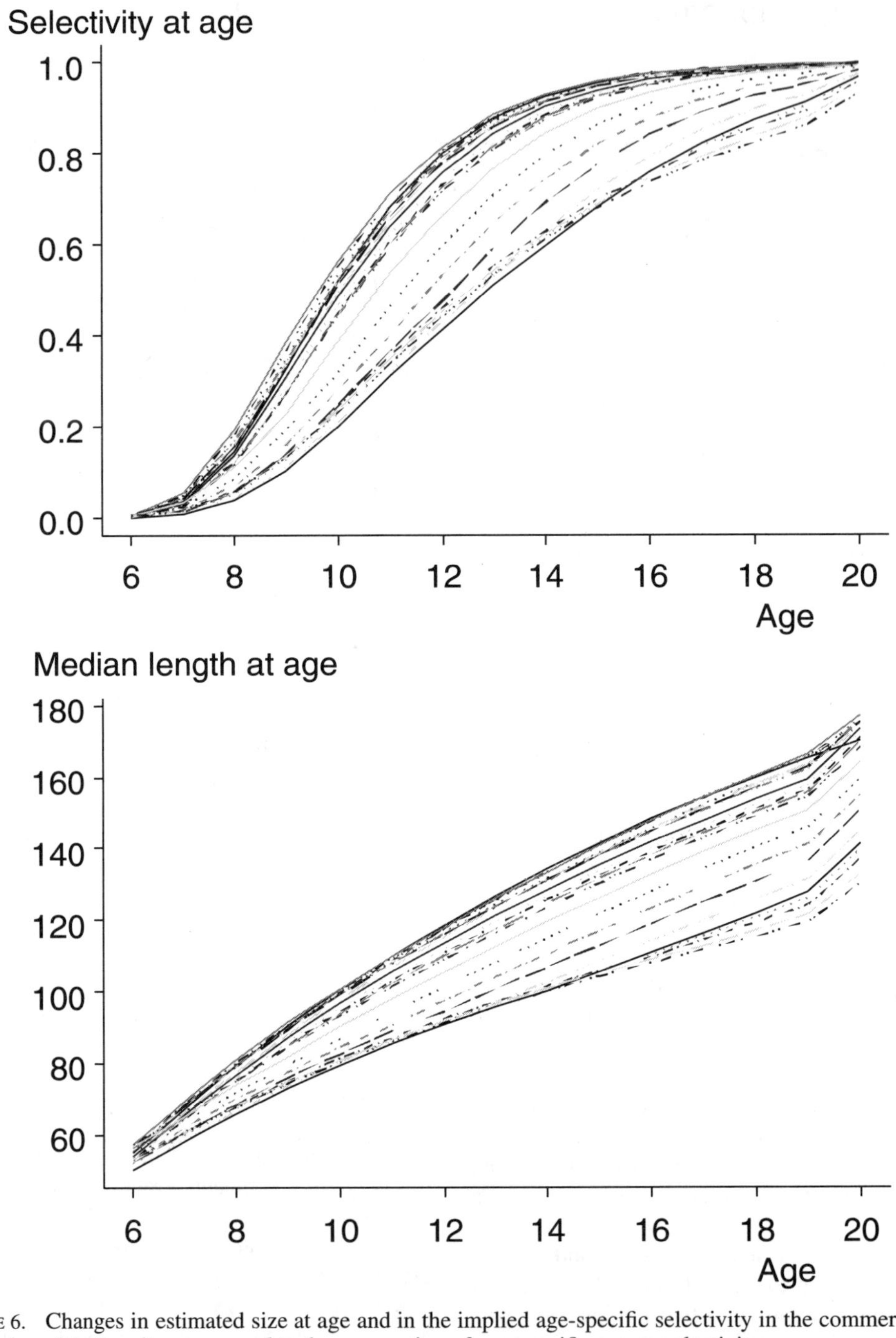

FIGURE 6. Changes in estimated size at age and in the implied age-specific selectivity in the commercial catch of IPHC Area 3A. Results correspond to the assumption of age-specific survey selectivity.

where $\theta_i, i = 1, \ldots, n_{\text{par}}$, denote the actual estimated parameters, as described in Table A3 together with their assumed prior distributions $P(\theta_i)$, and $P(\text{data} \mid \theta)$ is the likelihood of the observations. The prior distributions constrain the process variability represented by the ${}_x\varepsilon_t$ parameters. These are assumed to be Gaussian with zero mean and variance ${}_x\sigma^2$ specified *a priori*. Estimation of this large number of parameters is accomplished using the software AD Model Builder (Otter Research Ltd. 1999) developed by D. A. Fournier. This software uses automatic differentiation to compute the analytical derivatives for input to a quasi-Newton optimization algorithm. The optimization is conducted in steps or phases of increasing complexity as specified by the user, which greatly improves convergence.

Data used for the regular assessments comprise commercial fishery data (effort and catch-at-age since 1974, and mean and variance of log-length at age in the catch). There are also data collected during setline grid surveys, which comprise catch per unit of effort (CPUE), age composition of the catch, and mean and variance of log-length at age. For Area 3A, survey data are available for 1977–1986 and 1993–1998. With the exception of the effort data, the coefficient of variation (CV) of each of the observations is also available, as estimated by conventional sampling equations or bootstrap (details in Sullivan et al. 1999).

These different data types are fitted by assuming lognormal errors throughout, with variances approximately proportional to the estimated sampling variances. The overall residual variance for all observations is assumed unknown and estimated as

$$\hat{\sigma}^2 = \frac{RSS}{n_{\text{obs}}},$$

where n_{obs} is the total number of observations (=1655), and RSS is the weighted sum of squared residuals for the eight different data components: commercial catch-at-age, commercial effort, survey *CPUE* and age proportions, and mean and variance at age in the commercial and survey catches. If $Y_{i,d}$ denotes the i-th observation of type d and $\hat{Y}_{i,d}$ its corresponding model prediction (see equations in Table A3),

$$RSS = \sum_{d=1}^{8} \lambda_d \sum_{i=1}^{n_{\text{obs}_d}} \left[\frac{\ln(\hat{Y}_{i,d}) - \ln(Y_{i,d})}{\text{CV}(Y_{i,d})} \right]^2,$$

where λ_d are relative weights, especified as discussed in Sullivan et al (1999). Replacing the variance of the normal likelihood function by its maximum likelihood estimator (MLE) $\hat{\sigma}^2$ results in the *condensed* log-likelihood function, which has the form

$$\ln L = -0.5 n_{\text{obs}} \ln \left(\frac{RSS}{n_{\text{obs}}} \right) + \text{constant}$$

The log of the survey catchability $\ln({}_s q)$, used to predict $\ln(CPUE)$, was also replaced by its analytical MLE $\ln({}_s\hat{q})$. Due to the large number of observations, replacing $\ln({}_s\hat{q})$ for $\ln({}_s q)$ and $\hat{\sigma}^2$ for σ^2 is almost identical to integrating the posterior density over the unknown parameters $\ln({}_s q)$ and σ^2 (see Walters and Ludwig 1994), assuming uninformative priors on both (Jeffreys' priors: $P(\ln({}_s q))$ is uniform and $P(\sigma^2) \propto 1/\sigma^2$). Thus, maximizing the joint posterior distribution $P(\theta \mid \text{data})$ (integrated over $\ln({}_s q)$ and σ^2) is equivalent to minimizing the objective function

$$f = -\ln L + \text{penalties},$$

where the term "penalties" corresponds to the normal prior distributions on the ${}_x\varepsilon_t$ parameters,

$$\text{penalties} = 0.5 \sum_x \sum_t \frac{{}_x\varepsilon_t^2}{{}_x\sigma_t^2}.$$

This procedure for obtaining point estimates differs from other state-space models that allow for both stochasticity in the dynamics and measurement errors (e.g., Mendelssohn 1988; Sullivan 1992; Schnute 1994). In maximum likelihood (ML) estimation, unobserved random variables which amount to process error (the ε parameters in our model) need to be integrated out to compute a likelihood function (using for example the Kalman filter) before the latter can be maximized to estimate the fundamental model parameters in the usual way. Here, instead, we maximize the *penalized* likelihood, which, in the Bayesian paradigm, corresponds to the joint posterior density of all model parameters including the ε parameters. The maximum of the penalized likelihood is the mode of $P(\theta \mid \text{data})$. A problem with these so-called MAP (maximum *a posteriori*) estimators is that they depend on the parametrization of the model; that is, they are not *equivariant* to 1-1 transformations (Ripley 1996, p. 334), in the sense that ML estimators are. This means that if, say, $\hat{\theta}$ is the MLE of θ, then $f(\hat{\theta})$ is the MLE of $f(\theta)$. This does not hold for a MAP estimator because, when parameters are transformed, the prior density needs to be transformed as well using the *change of variable formula* (Bickel and Doksum 1977, p. 448). For nonlinear transformations, this changes the form of the objective function and likely the location of the maximum. As a consequence, if $\tilde{\theta}$ denotes a vector of MAP estimates, the use of $f(\tilde{\theta})$ to estimate derived quantities of management interest (such as current exploitable biomass) is ambiguous. To fully describe all parameters of interest we need to compute their marginal posterior densities, which involves integrating $P(\theta \mid \text{data})$. We deal with that in section 4, but first explore the results of the two models in terms of the MAP point estimators. While other point estimators, such as the posterior means, are sometimes used in Bayesian estimation, the strength of Bayesian analysis is not on the point estimates, but on the use of the full posterior distribution to evaluate consequences of management decisions.

Results of assessments based on age- and length-specific survey selectivity

The change in model structure resulted in coast-wide estimates of present biomass more than double the previous estimates (Clark et al. 1999). The rather grim picture of drastically declining recruitment coast-wide was replaced by a much more optimistic forecast once the effects of the reduction in size at age were taken into account. Results for Area 3A, where the decrease in growth rates was most dramatic, were very sensitive to the different assumptions about survey selectivity. Estimated recruitment and total biomass are substantially higher under the assumption of length-specific selectivity than under age-specific selectivity (Figure 7).

Retrospective analyses were conducted to evaluate consistency among successive estimates under the two models. Because surveys were suspended between 1987 and 1992, the analysis could not be carried all the way back into the past without substantially affecting the database used for the assessments. Therefore, only retrospective assessments for the periods 1974–1993, 1974–1994, and so forth up to 1974–1998 were conducted.

Results indicate that neither of the models shows a consistent and strong retrospective pattern such as was observed under CAGEAN. The age-specific version does show a slight tendency to initially underestimate year-class strength, as would be expected if the vulnerability of the young age classes to the survey was at least partially a function of fish length (Figure 8). The assumption of length-specific selectivity results in somewhat more consistent estimates of exploitable biomass among successive assessments (Figure 8).

A Bayesian Analysis of Assessment Uncertainty

When several models are considered, such as in the assessment of Pacific halibut stocks, statistical uncertainty can first be evaluated conditional on the structure of each of the models being correct. For simplicity, I have dropped from the notation the conditioning on model structure, and will only add it later when results from the two models are combined (section 6).

The discrete probability distribution of any quantity of management interest Ψ, that is a function $f(\theta)$ of the vector of model parameters, can

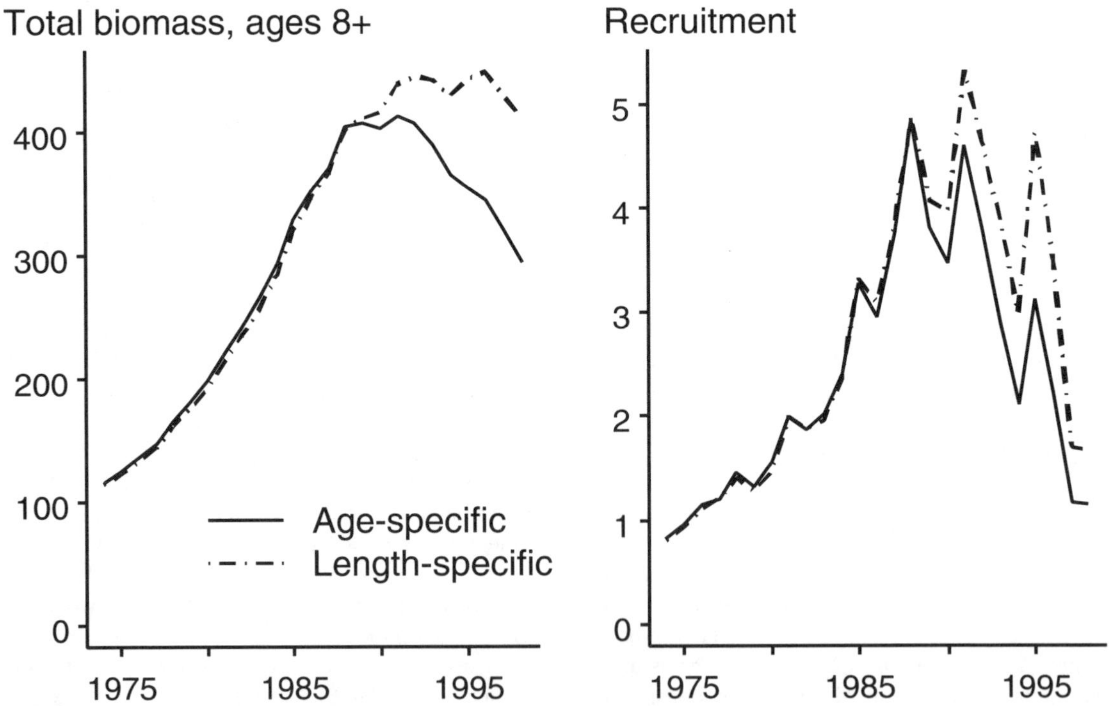

FIGURE 7. Estimates of total adult biomass and recruitment of Pacific halibut in IPHC Area 3A obtained under two different assumptions about survey selectivity: (a) selectivity is age-specific, and (b) selectivity is length-specific.

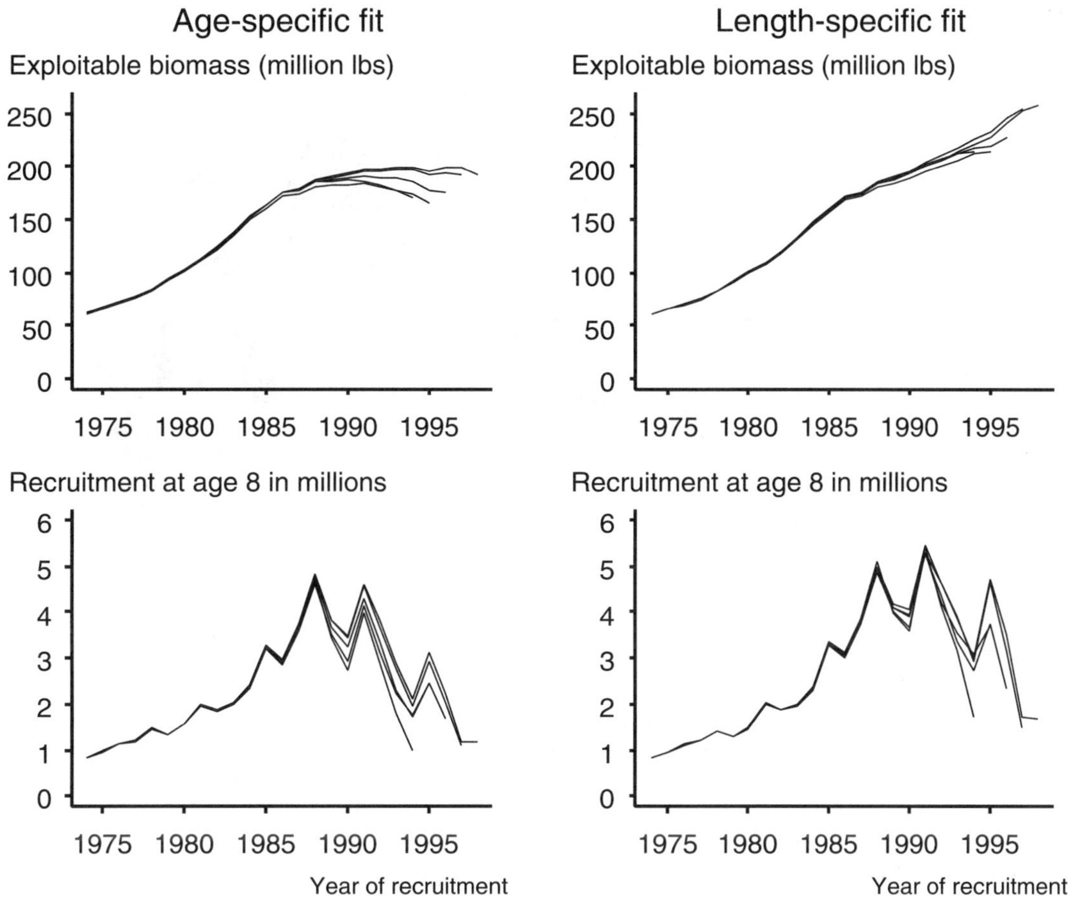

FIGURE 8. Retrospective assessments for Pacific halibut in IPHC Area 3A under the new assessment models based on two different assumptions about survey selectivity: (a) selectivity is age-specific, and (b) selectivity is length-specific.

be obtained by integrating the posterior density $P(\theta \mid \text{data})$ over θ, as

$$P(\Psi_i < f(\theta) \leq \Psi_{i+1} \mid \text{data}) = \int_{\Psi_i < f(\theta) \leq \Psi_{i+1}} P(\theta \mid \text{data})\, d\theta,$$

where Ψ_i are discrete values of Ψ. In complex models such as ours, the large dimensions of θ make it computationally intractable to integrate the posterior density by conventional numerical methods. MCMC can be used instead to generate a sample from $P(\theta \mid \text{data})$, from which the distribution of any derived parameters of interest can be approximated. AD Model Builder (Otter 1999) has an efficient implementation of MCMC, which uses a multivariate-normal distribution (truncated due to parameter bounds) to "jump" over the parameter space. Because this *jumping distribution* is not symmetric, the Metropolis-Hastings rule (Gelman et al. 1995) is used to accept/reject the sampled candidate points. The covariance of the multivariate-normal jumping distribution is proportional to the covariance matrix of the parameters, estimated from the Hessian matrix at the mode of $P(\theta \mid \text{data})$; its scaling is adjusted based on the first 1500 MCMC draws to achieve an acceptance rate of sampled points between 0.15 and 0.4. One million MCMC draws were done and one out of every 200 sets of θ parameters in the series was stored for later use (in order to have approximately independent draws), after discarding the first ten thousand. The algorithm converged before the million samples, as indicated by comparison of successive segments in the series. The acceptance rate remained at about 0.30, which implies that the fixed covariance used in the jumping distribution performed adequately.

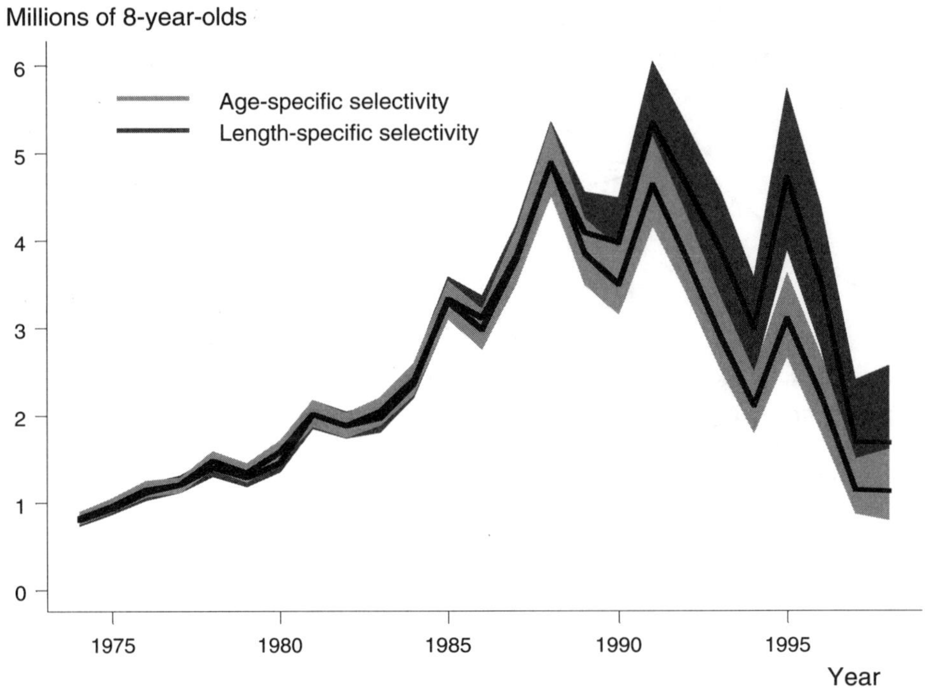

FIGURE 9. Estimates of recruitment trends under two different assumptions about survey selectivity. Shaded areas cover 95% intervals centered at the medians of the MCMC draws indicated by the solid lines.

MCMC simulations were conducted separately for the two models and a sample $\{\theta^1, \ldots, \theta^{n_{\text{mcmc}}}\}$ from the multivariate posterior distribution was obtained in each case. Marginal posterior distributions of historical recruitment trends for the two models (Figure 9) overlap in the early years but differ substantially in recent years, after the decrease in size at age in the mid-1980s. Recent recruitments are significantly higher under the length-specific fit. As a result, the marginal distributions of predicted exploitable biomass conditioned on the two models (Figure 10) are also significantly different. The modes of the two distributions are separated by more than two standard deviations (of the first distribution). Under the age-specific fit, the probability that biomass is as high or higher than the value that is most likely under the length-specific fit is less than 0.05. Most of the uncertainty about the size of the exploitable biomass thus results from uncertainty in the structure of the model, here restricted to two rather extreme assumptions about survey selectivity. The uncertainty conditional on model structure is relatively small; the posterior distributions of the current exploitable biomass have CVs close to 12% in both models.

Posterior distributions of derived quantities $f(\theta)$ can also be approximated by a modified likelihood profile. The following steps are completed automatically by the AD Model Builder software by simply specifying a likelihood-profile option. First, a series of conditional maximizations are conducted while forcing $f(\theta)$ to take a series of fixed values Ψ_i on each side of its value at the mode of $P(\theta \mid \text{data})$. The series of conditional maxima is then adjusted for the gradient of $f(\theta)$ at each point, as in the *change in variables formula* (Bickel and Doksum 1977). The resulting profile

$$L_\Psi = \max_{\theta: f(\theta)=\Psi} \{P(\theta \mid \text{data})\} \left| \frac{\partial f(\theta)}{\partial(\theta)} \right|^{-1}_{\theta=\hat{\theta}(\Psi)},$$

where $\hat{\theta}(\Psi)$ denotes the value of θ at the conditional maximum, approaches the true posterior after normalization. The approximation improves when $P(\theta \mid \text{data})$ approaches a multivariate normal distribution.

In the case of Pacific halibut, the posterior marginal distribution of exploitable biomass was well-approximated by the likelihood profile, as indicated by the similarity between the MCMC and likelihood-profile distributions (Figure 10). Also, the mode of the MCMC posterior coincided with the mode of $P(\theta \mid \text{data})$ (i.e., the MAP point estimate), which is convenient because the latter is normally

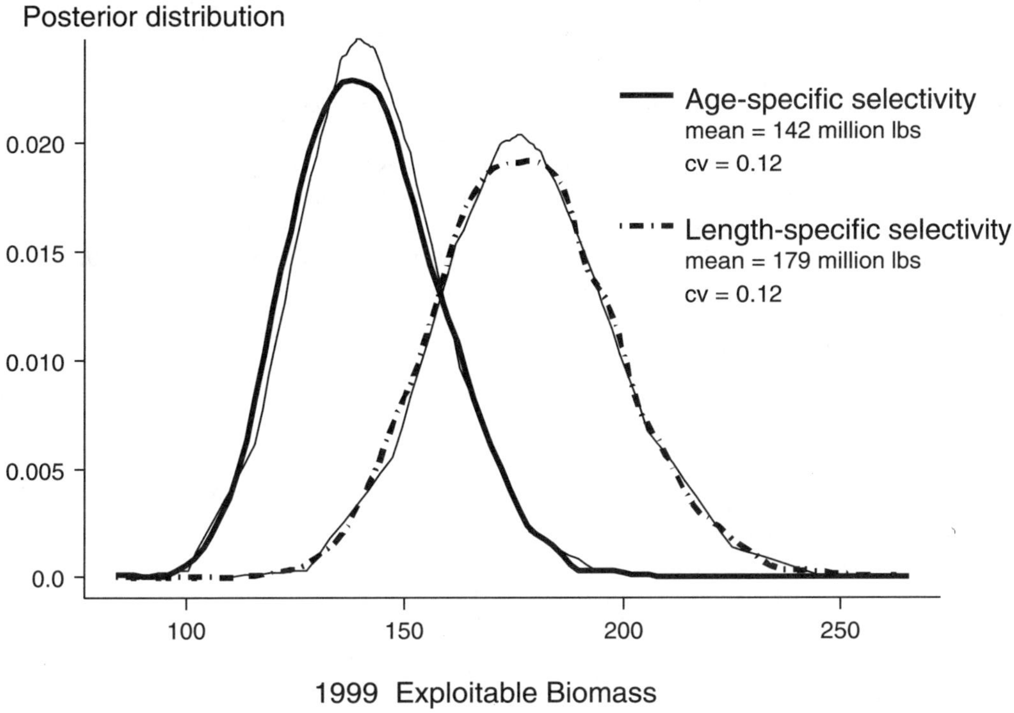

FIGURE 10. Posterior distributions of predicted exploitable biomass conditioned on two different assumptions about survey selectivity (constant at age or at length), computed with MCMC (thick lines) and approximated by the modified likelihood profiles (thin lines).

used to set catch quotas. As discussed earlier, these two modes do not have to coincide so this was a positive finding of the analysis.

Projections and Risk Analysis

While approximations such as the one based on the likelihood profile may be adequate to assess uncertainty in estimated parameters, projections used for policy evaluation and risk analysis require drawing samples from the posterior distribution, which makes algorithms such as MCMC very suitable. Evaluating future performance of any given series of management actions or control rule involve (1) simulating future stock trajectories under the evaluated control rule, and (2) estimating the distribution of some defined indicators of performance (e.g., average yield, probability that biomass falls below some threshold, etc.). Let Z denote any such performance indicator, the posterior distribution of Z for a given control rule C_j is

$$P(Z \mid C_j, \text{data}) = \int_\theta P(Z \mid \theta, C_j) P(\theta \mid \text{data})\, d\theta \tag{5}$$

This integral can be approximated from the empirical distribution of Z^i generated in the projections. Each set of parameters θ^i stored from the MCMC draws (or a subsample of them) is used to reconstruct historical abundance and compute other parameters needed to do the forward simulations. Stochastic projections are then done implementing C_j, which amounts to generating samples from $P(Z \mid \theta, C_j)$.

I illustrate the method by conducting short-term (10-year) projections for Pacific halibut abundance in Area 3A, using MCMC results separately for the two models. A full policy evaluation is beyond the scope of this paper, as it would require combining results from all IPHC areas to do long-term projections under different catch control rules. For this illustration, only the harvest policy currently followed by IPHC is implemented in the simulations, although any other policy could be similarly implemented instead. The current IPHC policy consists of limiting total removals to 20% of the estimated exploitable biomass. The control rule thus involves (1) a constant harvest rate target $H = 0.20$, and (2) a procedure to come up with a point estimate $\hat{B}$ of the actual exploitable biomass B to be used for setting catch quotas. While IPHC staff presents results obtained with the two alternative selectivity assumptions as part of the annual assessment (see for example Clark and Parma, unpublished[2]), quota recommendations are based on the more conservative estimates (i.e., those based on age-specific survey selectivity).

The simulation model is identical in structure to the assessment model, but additional assumptions were needed about process variability (e.g., future recruitment, growth rates etc.) and future assessment uncertainty (i.e. the relationship between $\hat{B}$ and B in the future) or other sources of implementation uncertainty. I describe those assumptions below.

Process variability

For short-term projections, a simple autoregressive model was considered adequate to simulate future recruitments. The mean log-recruitment $\bar{r}^i$, serial autocorrelation ρ^i, and residual variance ${}_\eta\sigma^{2^i}$, were estimated for each of the stored MCMC draws ($i = 1, \ldots, n_{\mathrm{mcmc}}$), and used to simulate future recruitment as

$$R_t^i = \exp\left(\bar{r}^i + \varepsilon_t^i\right)$$
$$\varepsilon_{t+1}^i = \rho^i \varepsilon_t^i + \eta_{t+1}^i$$

The series of η_t^i were generated from a normal distribution with zero mean and variance ${}_\eta\sigma^{2^i}$, and ε_o^i was set equal to the terminal $\ln(R)$ deviation in the corresponding i-th MCMC draw.

Future sizes at age in the population and in the catch, and future commercial selectivity were simulated using the same assessment model explained in section 3. Growth and selectivity parameters were set at the values estimated for the final period of the assessment so that future trends in growth rates were ignored. Changes in size at age, however, are still likely because the predicted distributions of size at age had yet to stabilize. In particular, projected sizes for the older age classes are smaller than those observed in the catches of 1998 because of the cumulative effect of reduced growth rates in the past.

More complex recruitment models involving the effect of total spawning biomass in the Gulf (see, for example, those in Clark et al. 1999) would be required for long-term projections. Also, other assumptions about future trends in growth, including the possibility of density-dependence should be considered. A policy evaluation analysis that incorporates these aspects is presented in Parma (in press).

Implementation uncertainty

Departures between the target harvest rate H and the implemented one were attributed exclusively to errors in the estimates of stock size. Thus, implemented yields were set as a fixed fraction of the estimated exploitable biomass

$$Y_t^i = H\hat{B}_t^i,$$

which was simulated as

$$\hat{B}_t^i = B_t^i \exp(\delta_t^i) \tag{6}$$

with log-normal, auto-correlated errors

$$\delta_{t+1}^i = \rho\delta_t^i + \upsilon_{t+1}^i, \qquad \upsilon_t^i \sim N\left(0, {}_\upsilon\sigma^2\right).$$

A high value of $\rho = 0.8$ was used to reflect the fact that, halibut being relatively long-lived, errors in successive estimates tend to be highly correlated. Simulations done with artificial data and standard catch-at-age models (NRC 1998) indicate that such values are reasonable. The standard deviation of the independent component of the error υ_t was set at 0.15.

An important point is that the first biomass estimate $\hat{B}_1^i$ in all the n_{mcmc} simulations was set at the point biomass predicted by the *actual* assessment from the mode of the posterior distribution. In other words, for each of the i simulations, the first assessment error was set at the difference between the log of the "true" biomass (i.e., the biomass simulated from the i-th MCMC draw) and the log of the *actual* estimated biomass (i.e., the MAP estimate). While this is simple when there is only one assessment model, we need to deal with the fact that we have two assessments, each of which gives rise to n_{mcmc} simulations. The case of the simulations based on the age-specific selectivity is conceptually straightforward because the simulations and point estimates used for quota setting are based on the same model (remember the IPHC control rule involves using the age-specific results for setting quotas). For the simulations based on the length-based selectivity, on the other hand, the choice of an adequate model for future assessment errors is not as clear-cut. The error in the first year has to be the difference between the true value (i.e., the B_1^i simulated from the length-based MCMC results) and the point estimate $\hat{B}_1$ *derived from the age-based model*. This makes sense because it is the latter that is used in the control rule. But what will happen in the

[2]Clark, W. and A. M. Parma. Assessment of the Pacific halibut stock in 1998 (unpublished) http://www.iphc.washington.edu/halcom/research/stock/sa98.all.pdf

future? If the "wrong" model continued to be used to set quotas no matter what, then a persistent bias could result and the assumptions in equation (6) would be inappropriate. In reality, however, we expect to be able to model selectivity better once the recent series of surveys expands to cover more completely the lifespan of a few cohorts. Also, if growth rates stabilize, the difference between results from the two models may diminish with more years of data, as results will become more robust to assumptions about survey selectivities during the 1980s. So instead of assuming a fixed bias, I assumed that if, indeed, the population behaved as the length-based model predicts, future biomass estimates would gradually change to finally oscillate around the true values, as described by the measurement error model above. The effect of the wrong choice of model would persist for some time, to the extent that the initial error propagates due to the assumed correlation between successive errors. Other different assumptions could of course be made as discussed in the last section.

Performance indicators

Two indicators were used for this illustration, average yield and probability that the exploitable biomass dropped below the historical minimum attained in the mid-1930s and again in the mid-1970s. The latter was estimated as

$$P(B_{1999:2008} < B_{\min}) = \frac{1}{n_{\text{mcmc}}} \sum_{i=1}^{n_{\text{mcmc}}} I_{[B_t^i < \min(B_{1974}^i, \ldots, B_{1998}^i), t=1999, \ldots, 2008]},$$

where $I_{[(.)]}$ denotes an indicator variable which takes a value of one if condition (.) is met and is zero otherwise. Exploitable biomass is used here as a proxy for spawning biomass, which has been used in past evaluations of harvesting strategies (Clark et al. 1999), although the latter used short and long-term projections reflecting the emphasis on sustainability in the choice of the harvest rule. It is important to remark that the minimum biomass record is not equated here with an overfishing threshold. In fact, estimates of recruitment corresponding to periods of minimum spawning biomass are above average, and there is no indication the halibut stocks were ever recruitment-overfished. When the stocks reached record low levels in the 1970s, and of course the effect of this on future recruitment was unknown, tight catch limits were imposed on the catches to promote stock rebuilding. Managers still prefer to try to maintain the stocks within their historical range, which I consider to be a sensible goal, given the long history of sustained exploitation of the Pacific halibut fishery.

Results

Projections of exploitable biomass based on the length-specific selectivity assumptions are substantially more optimistic (ca. 40% higher) than those based on the age-specific fits (Figure 11). The mean biomass and 95% intervals projected for year 2002, for example, are 201 million lbs (95% I: 147,264) for the length-specific fit, and 138 million lbs (95% I: 98, 186) for the age-based fit. Such differences could be expected considering that recruitment estimates (average of the MCMC results) for the last five years are about 50% higher under the length-specific selectivity than under the assumption of constant selectivity at age. In addition to higher recruitments, projected biomasses are higher for the length-based model because harvest rates implemented in the forward simulations are lower, at least initially. This is because catch quotas in the first year are determined using the most conservative model irrespective of which model is "true". Average yields are 25 million pounds and 31 million pounds, respectively, for the projections based on the length-specific and age-specific fits. Higher yields would be possible under the length-based model if estimates from that model were used to implement the 20% harvest rate instead of the more conservative ones. Risks, however, would increase. Under the current harvest rule, there is a substantial probability ($P(B_t < B_{\min}) = 0.27$) that that exploitable biomass will drop below the historical minimum in the early 1970s, if the population behaves consistently with the age-specific fits and growth rates do not increase in the future. That risk is very small ($P(B_t < B_{\min}) = 0.04$) under the alternative, more optimistic model. The recent decline in recruitment was more pronounced in Area 3A than in other areas (Clark and Parma, unpublished[2]), so the risk for the stock as a whole of dropping below its minimum record is likely to be lower than the risk computed here for Area 3A.

Combining Results from Different Models

So far we have assessed uncertainty and estimated average outcomes of a given control rule conditioned on model structure. In order to evaluate the

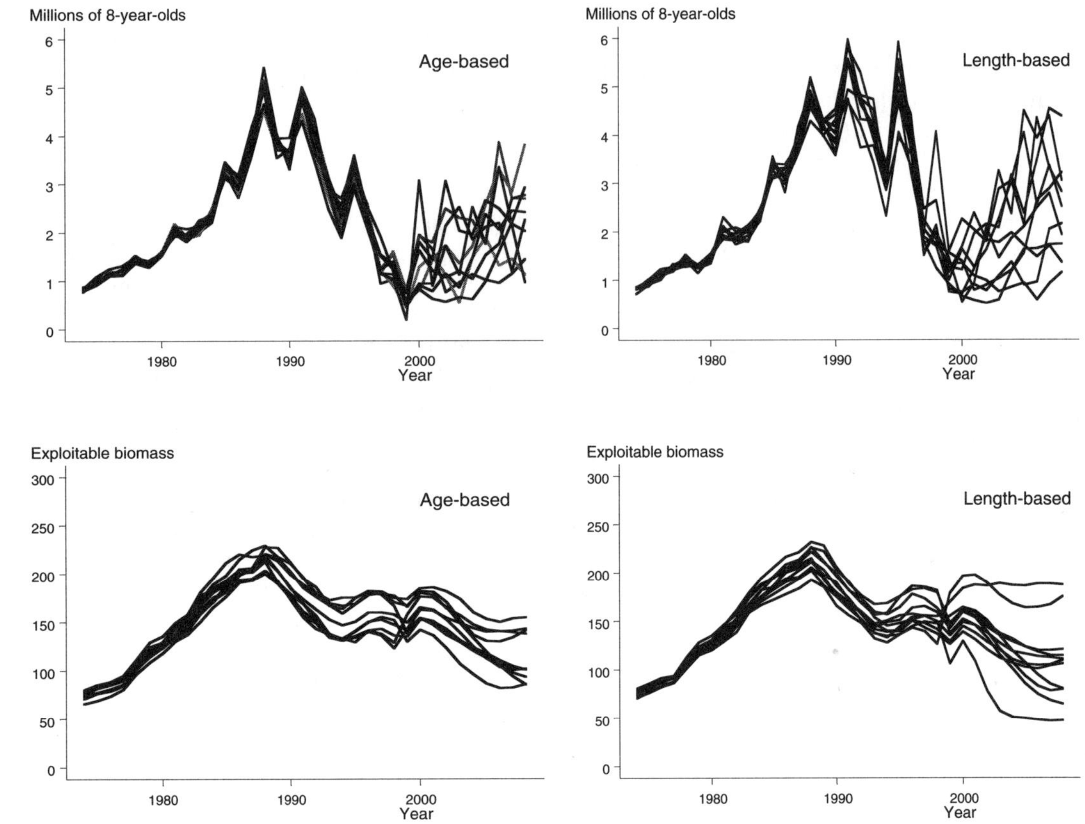

FIGURE 11. Sample of historical trends in recruitment and exploitable biomass and 10-year forward projections obtained using MCMC and the two different model structures used for halibut assessments.

overall uncertainty, or to compute expected outcomes of management decisions over all models considered, we need to assign probabilities to the alternative models. To emphasize conditioning on model structure, we rewrite equation (5.1) for the posterior distribution of a given indicator Z as

$$P(Z \mid C_j, \text{data}, M_k) = \int_{\theta_k} P(Z \mid \theta_k, C_j, M_k) \times P(\theta_k \mid \text{data}, M_k)\, d\theta_k,$$

where M_k denotes model k and C_j is the control rule. The unconditional posterior distribution of Z is a sum of conditional distributions, each weighted by the probability of its underlying model:

$$P(Z \mid C_j, \text{data}) = \sum_k P(Z \mid C_j, \text{data}, M_k) \times P(M_k \mid \text{data}).$$

Thus, this posterior distribution includes all sources of uncertainty admitted in the analysis, including model uncertainty. Below, I discuss some technical issues regarding the meaning and computation of $P(M_k \mid \text{data})$.

Note that in order to combine results from different models the probabilities that we need are not the prior probabilities $P(M_k)$, assigned to the models prior to analyzing the data, but the *posterior* probabilities $P(M_k \mid \text{data})$. From Bayes' theorem,

$$P(M_k \mid \text{data}) = \frac{P(\text{data} \mid M_k)P(M_k)}{\sum_l P(\text{data} \mid M_l)P(M_l)}, \tag{7}$$

where $P(\text{data} \mid M_k)$ measures the support given to model M_k by the data. The latter needs to be integrated (not maximized) over the parameter space so that

$$P(\text{data} \mid M_k) = \int_{\theta_k} P(\text{data} \mid \theta_k, M_k)P(\theta_k \mid M_k)\, d\theta_k, \tag{8}$$

where $P(\theta_k \mid M_k)$ is the prior density of θ_k and $P(\text{data} \mid \theta_k, M_k)$ its likelihood function. $P(\text{data} \mid M_k)$, which is sometimes called an *integrated likelihood*, is a *predictive* probability of the data, that is the probability of obtaining the data values actually observed, calculated *before* the data became available (Kass and Raftery 1995). In the case of two models, the ratio $P(\text{data} \mid M_1)/P(\text{data} \mid M_2)$, known as Bayes factor (Kass and Raftery 1995), measures the evidence provided by the data in favor of M_1 as opposed to M_2.

Several approximations exist to estimate the integral (6.2) (Kass and Raftery 1995; Hoeting et al. 1999). In the halibut case, because a sample from the posterior distribution $P(\theta_k \mid \text{data}, M_k)$ for each of the two models was available from the MCMC results, it was convenient to use the method proposed by Kass and Raftery (1995) based on *Importance Sampling*, where

$$P(\text{data} \mid M_k) = \frac{n_{\text{mcmc}}}{\sum_{i=1}^{n_{\text{mcmc}}} \frac{1}{P(\text{data}|\theta_k^i, M_k)}}.$$

This corresponds to the harmonic mean of the likelihood values computed for the n_{mcmc} parameter vectors θ_k^i stored from the MCMC trials. For the specific likelihood function used in the halibut assessment, the equation above equals

$$P(\text{data} \mid M_k) = \kappa \frac{n_{\text{mcmc}}}{\sum_{i=1}^{n_{\text{mcmc}}} \left[\frac{RSS^i}{n_{\text{obs}}}\right]^{n_{\text{obs}}/2}}, \tag{9}$$

where n_{obs} is the number of observations in the likelihood (=1655), and κ is a multiplicative constant which is common to both models and so can be ignored.

The posterior distributions of log-likelihood values obtained for the two models favor the age-based selectivity model (Figure 12). Not only is the value of the log-likelihood at the mode of the posterior higher ($\ln L = -383.65$ compared with $\ln L = -386.74$), but also the posterior distribution of log-likelihood values is shifted to the right

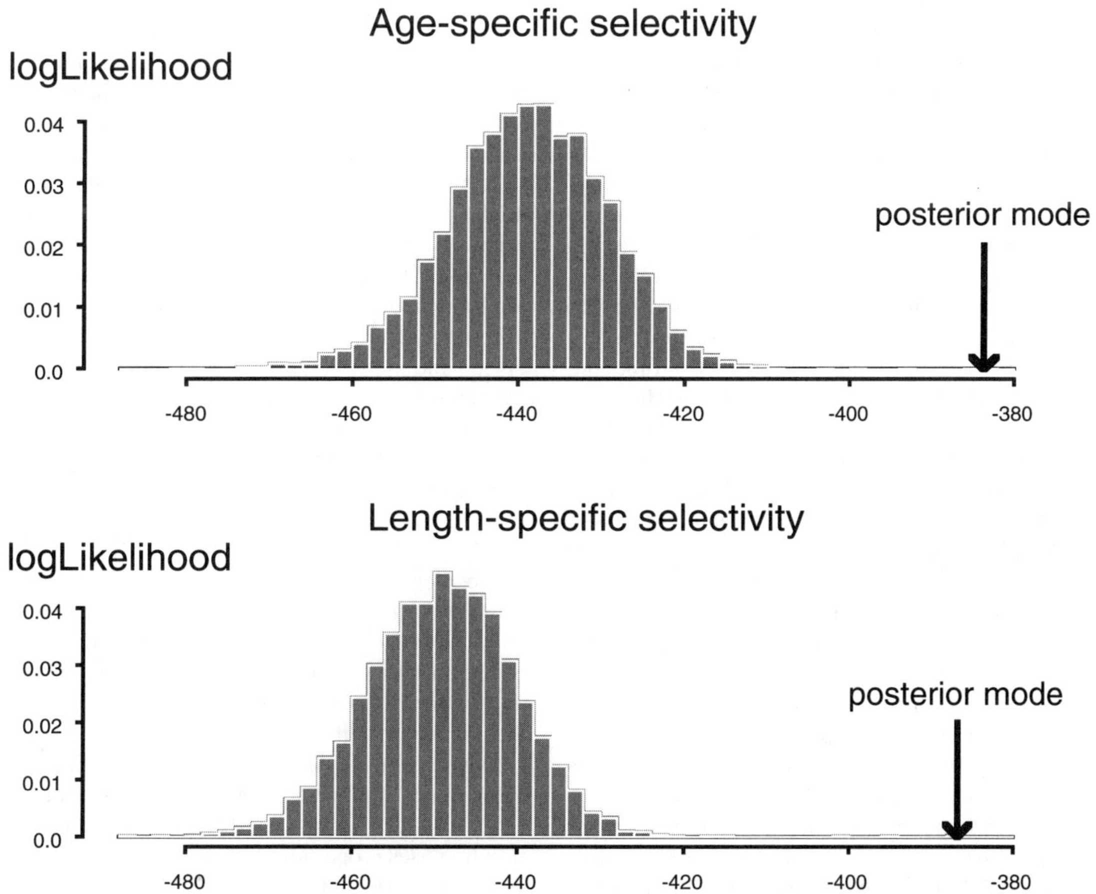

FIGURE 12. Histograms of log-likelihood values for samples from the posterior distributions $P(\theta_k \mid \text{data}, M_k)$ for the two models used in the halibut assessments, obtained using MCMC. Arrows indicate the log-likelihood values at the mode of each $P(\theta_k \mid \text{data}, M_k)$.

with respect to that of the length-specific selectivity model. Note that due to the high dimensions of θ, the most frequent log-likelihood values are much smaller than the log-likelihood value at the mode of $P(\theta_k \mid \text{data}, M_k)$. If the two models are assigned equal probabilities *a priori*, the difference in distributions of likelihood values shown in Figure 12 results in posterior probabilities equal to 0.99 and 0.01, respectively, for the age-specific and length-specific selectivity models. Thus, the data overwhelmingly favor the age-specific model.

Given the large number of observations used in the halibut assessment, small differences in the *RSS*, which may be hard to distinguish in a graph of residuals, will have a strong effect on the integrated likelihood (equation 6.3) and, in turn, on the posterior probabilities of the alternative models.

Discussion

In the new halibut assessment models, temporal trends in growth, selectivity and catchability are represented as stochastic processes with variances specified *a priori*. This allows incorporation of measurement and process error, thus providing an intermediate alternative between the rigid assumptions of separable catch-at-age models, and the unconstrained flexibility of VPA models. The approach is Bayesian, in the sense that catchability, selectivity, and the growth parameters are treated as random variables with prior distributions specified by the time-series models. Measurement errors are specified in the likelihood components of the objective function, while process errors correspond to the penalties.

Although the models have a large number of parameters, their values appear to be well-determined conditional on each of the two model structures considered. The CV of present exploitable biomass is low (ca. 12%) for both, which results from the rather rigid assumptions made about survey selectivity and the fact that natural mortality M is assumed known. The choice of a working value for M is however uncertain. The value of 0.2 used in the past was recently reduced to 0.15 based on reanalysis of age composition data from lightly exploited areas. Also, recent work by Clark (1999) suggests that values towards the lower end of the range of estimates are safer for the stock without much yield being forfeited, especially when target harvest rates are derived from analysis of stock-productivity conditioned on the same value of M. With regard to survey selectivity, the two models considered are extremes in how the process might be represented. Size and age effects are both likely in reality, but would be difficult to separate with the data at hand. More flexible assumptions regarding selectivity and natural mortality may be possible in the future, once the new series of surveys resumed in 1993 expands to cover more completely the life span of a few cohorts. Also, the strength of the Bayesian framework could be further explored by incorporating information on hook selectivity obtained from underwater observations (Kaimmer 1999), and specifying prior distributions on other model parameters (e.g., M.)

The advantages of a Bayesian approach are clear when several alternative models are considered, whether they involve different model structures or just different parameter values of a given model structure. Instead of choosing one "best" model according to some criteria, results from different models are combined by weighting model-specific outcomes by the posterior probabilities of the models. When a few alternative models are considered, such as in the halibut assessments, two approaches are possible to quantify and present uncertainties and risks. In the approach favored here, the identity of the models is maintained so that each M_k represents a composite hypothesis (with free parameters θ_k). Results are presented conditional on model structure first, including computation of the posterior distributions $P(\theta_k \mid \text{data}, M_k)$, and the posterior probabilities of the composite hypotheses $P(M_k \mid \text{data})$ are computed in a second step. Alternatively, model uncertainty and parameter uncertainty can be combined, and posterior probabilities for the simple hypotheses $P(\theta_k, M_k \mid \text{data})$ can be approximated directly. The latter is the approach followed by Patterson (1999) in his implementation of MCMC. Both approaches are of course equivalent, as

$$P(\theta_k, M_k \mid \text{data}) = P(\theta_k \mid \text{data}, M_k)P(M_k \mid \text{data}).$$

The choice is a matter of computational efficiency and personal preference. The first approach may be more stable numerically when $P(\theta_k, M_k \mid \text{data})$ is multimodal due to the mixture of the different M_k s; estimating the probabilities $P(M_k \mid \text{data})$ (i.e., the mixture parameters) by MCMC can be difficult when the modes are too separate. The first approach is suitable for implementation with AD Model Builder because the MCMC's jumping distribution uses properties of the conditional posterior $P(\theta_k \mid \text{data}, M_k)$ at the mode.

With either approach, the use of methods such as MCMC or Sampling Importance Resampling (e.g., Kinas 1996) make it very easy to conduct simulations of future stock trajectories, as they readily provide samples from the joint posterior distribution of all fundamental and derived model parameters needed (e.g., abundance at age in the final year). In addition, parameters of stock-recruitment models (required for long-term projections) can be estimated jointly with the rest, so that their uncertainty and correlation with other model parameters are directly carried into the policy evaluation.

One of the most difficult aspects in the evaluation of feedback control rules by simulations (even a control rule as simple as a constant harvest rate) is that we need to simulate the implementation process. When the control rule uses some point estimate, such as the estimated exploitable biomass, the simulation involves making assumptions about future assessment errors. When simple assessment models are used, it is possible to simulate future observations and apply the actual assessment model very much like in reality (e.g., as in McAllister and Kirkwood 1998; Walters 1998). Unfortunately, this would be difficult computationally with very complex models, and so simple representations of future assessment errors are commonly used instead. The choice of assumptions is, however, nontrivial, especially when alternative model structures are admitted. While the MCMC (or alternative methods) provide a probability distribution for the assessment errors at the start of the simulations, the performance of any given control rule will depend on how it is implemented not just the first year but in the long run. So, for example, if the wrong model is used to set catch quotas, how biased will future estimates of abundance be? How long will it take for the problem to become apparent? Some uncertainties are unlikely to be resolved ever, others may be reduced with more years of data. Yet, in some cases the structural uncertainty may persist but become less of a problem in the future. Answers to these questions will be case-specific. But as the best approach to characterize future implementation uncertainty is not clear-cut, sensitivity of the results with respect to some of the key parameters should be analyzed.

The most difficult problem in dealing with structural uncertainty is that of assigning probabilities to the different models considered plausible. Ideally, we want to construct decision tables that link actions (control rules, not just single-year actions) and outcomes for different hypotheses about the managed system (NRC 1998), and we also want to evaluate their likelihood. This is a critical step, especially when the alternative hypotheses lead to contrasting "best" decisions. Different future scenarios may be viewed as more or less likely by different groups depending on interests, philosophies and levels of risk-aversion. In some cases, the analyst may want to go an extra step in the decision analysis and compute average outcomes of alternative management decisions over all possible states of nature and future dynamics. This, however, may be dangerous, as averaging may cover some serious risks. But even if computing average outcomes is not the goal, it is critical to try to assign probabilities based on scientific criteria to prevent unduly favoring certain actions by overemphasizing ever more optimistic or pessimistic scenarios. In highly contested cases, panels of independent experts have been appointed to approach the problem with as much objectivity as possible (Deriso 2001, this volume). But even in less controversial situations, assessment scientists, normally perceived as objective players in this game, have different levels of risk aversion, which even inadvertently may affect model choice.

In this context, the evaluation of posterior probabilities and Bayes factors can be very helpful by providing an objective measure of the evidence the data give to the alternative models. In the case of halibut, the computed Bayes factor overwhelmingly favored the age-specific selectivity model over the length-specific one. Some questions, however, need to be explored further before this result can be used to discount the length-specific model. As discussed earlier, when the number of observations is very large, small differences in goodness of fit may heavily favor one of the models over the others. Bayes factors may thus be very sensitive to the addition of new data (unstable over time) or to minor changes in the weights given to the different data components. This problem is not peculiar to Bayes factors, but would also affect likelihood ratio tests based on comparing maximum likelihood values, or any other non Bayesian method. While it is standard practice to evaluate the sensitivity of point estimates with respect to some of these choices (e.g., the λ values), evaluating the sensitivity of results derived by MCMC would be very demanding computationally, at least for complex models such as the ones used here. Some of the available approximations (see, for example, Kass and Raftery 1995) may facilitate this task. At the minimum, we need to see how Bayes factors respond to the inclusion of more years of data.

In addition to these caveats, posterior probabilities are not the only criterion that can be used to evaluate alternative models. Other diagnostics, such autocorrelation in residuals and performance in retrospective analyses, are most useful to detect model misspecification and should be considered. Retrospective analyses have been used extensively to guide model development in Pacific halibut assessments, and we expect their diagnostic value to increase in a few years when more survey data become available.

Acknowledgements

I thank Dave Fournier for support during different stages of implementation of MCMC in AD Model Builder. Discussions with Jim Ianelli, Vivian Haist, Pat Sullivan, and D. Fournier during a workshop funded by IPHC were an excellent start for this project. I thank Pat Sullivan, Bruce Leaman, Bill Clark, and two anonymous reviewers for helpful comments on this manuscript. The Atlantic States Marine Fisheries Commission provided financial support for presentation and publication of this research.

References

Bickel, P. J. and K. A. Doksum, 1977. Mathematical statistics. Holden-Day Inc., Oakland, California.

Clark, W. G. 1999. The effects of an erroneous natural mortality rate on a simple age-structured stock assessment. Canadian Journal of Fisheries and Aquatic Sciences 56:1721–1731.

Clark, W. G., S. R. Hare, A. M. Parma, P. J. Sullivan, and R. J. Trumble, 1999. Decadal changes in halibut growth and recruitment. Canadian Journal of Fisheries and Aquatic Sciences 56:1–11.

Cordue, P. L. 1998. Designing optimal estimators for fish stock assessment. Canadian Journal of Fisheries and Aquatic Sciences 55:376–386.

Deriso, R. B. and A. M. Parma. 1988. Dynamics of age and size for a stochastic population model. Canadian Journal of Fisheries and Aquatic Sciences 45:1054–1068.

Deriso, R. B., T. J. Quinn, II, and P. R. Neal, 1985. Catch-age analysis with auxiliary information. Canadian Journal of Fisheries and Aquatic Sciences 42:815–824.

FAO. 1995. Precautionary approach to fisheries. Part 1: guidelines on the precautionary approach to capture fisheries and species introductions. FAO Fish Technical Paper No. 350/1. Rome, Italy.

Fournier, D. A., J. Hampton, and J. R. Sibert, 1998. MULTIFAN-CL: a length-based, age-structured model for fisheries stock assessment, with application to South Pacific albacore, *Thunnus alalunga*. Canadian Journal of Fisheries and Aquatic Sciences 55:1–12.

Gelman, A., J. B. Carlin, H. S. Stern, and D. B. Rubin, 1995. Bayesian data analysis. Chapman and Hall, London.

Gudmundsson, G. 1994. Time-series analysis of catch-at-age observations. Appl. Statist. 43:117–126.

Hilborn, R. and C. Walters, 1992. Quantitative fisheries stock assessment: choice, dynamics and uncertainty. Chapman and Hall, New York.

Hoeting, J. A., D. Madigan, and A. Raftery, and C. T. Volinsky. 1999. Bayesian model averaging: a tutorial. Statistical Science 14:382–417.

Kaimmer, S. M. 1999. Direct observations on the hooking behavior of Pacific halibut. Fish. Bull. 97:873–883.

Kass, R. E. and A. E. Raftery, 1995. Bayes factors. J. Am. Stat. Assoc. 90:773–795.

Kinas, P. G. 1996. Bayesian fishery stock assessment and decision making using adaptive importance sampling. Canadian Journal of Fisheries and Aquatic Sciences 53:414–423.

Otter Research Ltd. 1999. AD Model Builder documentation on line. http://otter-rsch.com/admodel.htm.

McAllister, M. K. and G. P. Kirkwood, 1998. Using Bayesian decision analysis to help achieve a precautionary approach for managing developing fisheries. Canadian Journal of Fisheries and Aquatic Sciences 55:2642–2661.

Mendelssohn, R. 1988. Some problems in estimating population sizes from catch-at-age data. Fishery Bulletin 86:617–630.

National Research Council (NRC). 1998. Improving fish stock assessments. National Academy Press, Washington D.C.

Parma, A. M. 1993. Retrospective catch-at-age analysis of Pacific halibut: implications on assessment of harvesting policies. Pages 247–265 *in* G. Kruse, D. M. Eggers, R. J. Marasco, C. Pautzke, and T. J. Quinn II, editors. Proceedings of the International Symposium on Management Strategies for Exploited Fish Populations. Alaska Sea Grant, Fairbanks, Alaska (AK-SG-93-02).

Parma, A. M. In press. In search of robust harvest rules for Pacific halibut in the face of uncertain assessments and decadal changes in productivity. Mote Symposium on Targets, Thresholds, and the Burden of proof in Fisheries Management. Bulletin of Marine Science.

Patterson, K. R. 1999. Evaluating uncertainty in harvest control law catches using Bayesian Markov Chain Monte Carlo virtual population analysis with adaptive rejection sampling and including structural uncertainty. Canadian Journal of Fisheries and Aquatic Sciences 56: 208–221.

Punt, A. E. and R. Hilborn, 1997. Fisheries stock assessment and Bayesian analysis: the Bayesian approach. Rev. Fish. Biol. Fish. 7:35–63.

Restrepo, V. R., G. G. Thompson, P. M. Mace, W. L. Gabriel, L. L. Low, A. D. MacCall, R. D. Methot, J. E. Powers, B. L. Taylor, P. R. Wade, and J. F. Witzig, 1998. Technical guidance on the use of precautionary approaches to implementing national standard 1 of the Magnuson-Stevens Fishery Conservation and Management Act. NOAA Tech. Memorandum NMFS-F/SPO-##.

Richards, L. J. and J.-J. Maguire, 1998. Recent International agreements and the precautionary approach: new directions for fisheries management science. Canadian Journal of Fisheries and Aquatic Sciences 55:1545–1552.

Ripley, B. D. 1996. Pattern recognition and neural networks. Cambridge University Press, Cambridge.

Schnute, J. T. 1994. A general framework for developing sequential fisheries models. Canadian Journal of Fisheries and Aquatic Sciences 51:1676–1688.

Sullivan, P. J. 1992. A Kalman filter approach to catch-at-length analysis. Biometrics 48:237–257.

Sullivan, P. J. and D. A. McCaughran, 1995. Pacific halibut assessment: changing strategies for changing times. Pages 129–144 *in* G. T. Sakagawa, editor. Assessment methodologies and management. Proceedings of the World Fisheries Congress, Theme 5. Oxford & IBH Publishing Co., New Delhi.

Sullivan, P. J., A. M. Parma, and W. G. Clark, 1999. The Pacific halibut stock assessment of 1997. International Pacific Halibut Commission Scientific Report 79. Seattle, Washington.

Walters, C. 1998. Evaluation of quota management policies for developing fisheries. Canadian Journal of Fisheries and Aquatic Sciences 55:2691–2705.

Walters, C. and D. Ludwig, 1994. Calculation of Bayes posterior probability distributions for key population parameters: a simplified approach. Canadian Journal of Fisheries and Aquatic Sciences 51:713–722.

Walters, C. J. and J.-J. Maguire, 1996. Lessons for stock assessment from the Northern cod collapse. Reviews in Fish Biology and Fisheries 6:125–137.

Walters, C. and A. M. Parma, 1996. Fixed exploitation rate strategies for coping with effects of climate change. Canadian Journal of Fisheries and Aquatic Sciences 53:148–158.

Appendix

TABLE A1. Equations describing the age- and size-structured model used for Pacific halibut assessments.

Process	Notation
Abundance dynamics	a: age 6, ..., 19; plus group $a = 20^+$
	t: 1974, ..., 1997
$N_{a+1,t+1} = N_{a,t}\mathrm{e}^{-M}(1 - {}_\mathrm{b}H_{a,t})(1 - {}_\mathrm{c}S_{a,t}(1 - \mathrm{e}^{-F_t}))$	$N_{a,t}$: abundance at age a and year t
$N_{20^+,t+1} = N_{20,t+1} + N_{20^+,t}\mathrm{e}^{-M}(1 - {}_\mathrm{b}H_{20^+,t}) \times (1 - {}_\mathrm{c}S_{20^+,t}(1 - \mathrm{e}^{-F_t}))$	${}_\mathrm{c}S_{a,t}$: age-specific commercial selectivity
	${}_\mathrm{b}H_{a,t}$: bycatch mortality rate[3]
	F_t: fishing mortality
	M: natural mortality
Commercial catchability	
$\ln({}_\mathrm{c}q_{t+1}) = \ln({}_\mathrm{c}q_t) + {}_q\varepsilon_t$ where ${}_q\varepsilon_t \sim N(0, {}_q\sigma^2)$	${}_\mathrm{c}q_t$: commercial catchability
Commercial selectivity	
${}_\mathrm{c}s_t(X) = \begin{cases} 0 & \text{for } X < \ln(81) \\ \mathrm{e}^{\frac{-(X - {}_\mathrm{c}X_t^\mathrm{full})^2}{2{}_\mathrm{c}v_t}} & \text{for } \ln(81) \le X \le {}_\mathrm{c}X_t^\mathrm{full} \\ 1 & \text{for } X > {}_\mathrm{c}X_t^\mathrm{full} \end{cases}$	$X_{a,t}$: log-length at age a and year t; ${}_\mathrm{c}s_t(X)$: length-specific commercial selectivity
${}_\mathrm{c}X_{t+1}^\mathrm{full} = {}_\mathrm{c}X_t^\mathrm{full} + {}_{{}_\mathrm{c}X^\mathrm{full}}\varepsilon_t$ where ${}_{{}_\mathrm{c}X^\mathrm{full}}\varepsilon_t \sim N\left(0, {}_{X^\mathrm{full}}\sigma_t^2\right)$	${}_\mathrm{c}X_t^\mathrm{full}$: log-length at full selectivity
$\ln({}_\mathrm{c}v_{t+1}) = \ln({}_\mathrm{c}v_t) + {}_{\mathrm{c}v}\varepsilon_t$ where ${}_{\mathrm{c}v}\varepsilon_t \sim N\left(0, {}_v\sigma_t^2\right)$	${}_\mathrm{c}v_t$: variance-like parameter of half-normal selectivity function
${}_\mathrm{c}S_{a,t} = \int_{-\infty}^{\infty} {}_\mathrm{c}s_t(X_{a,t})\, p(X_{a,t})\, \mathrm{d}X_{a,t}$	$p(X)$: probability density of X
Growth	
$X_{a,t} \sim N(\mu_{a,t}, \sigma^2)$	$\mu_{a,t}$: mean of $X_{a,t}$ before the harvest
$\mu_{6,t+1} = \mu_{6,t} + {}_\mu\varepsilon_t$ where ${}_\mu\varepsilon_t \sim N(0, {}_{\mu_6}\sigma^2)$	σ^2: variance of $X_{a,t}$
$\exp(\mu_{a+1,t+1}) = \alpha_t + \beta\exp(\mu_{a,t}^+)$	
$\ln(\alpha_{6,t+1}) = \ln(\alpha_{6,t}) + {}_\alpha\varepsilon_t$ where ${}_\alpha\varepsilon_t \sim N(0, {}_\alpha\sigma^2)$	$\mu_{a,t}^+$: mean of $X_{a,t}$ after the harvest
$\mu_{a,t}^+ = \dfrac{\int_{-\infty}^{\infty} X_{a,t}(1 - {}_\mathrm{c}s_t(X_{a,t})(1 - \mathrm{e}^{-F_t}))p(X_{a,t})\,dX_{a,t}}{\int_{-\infty}^{\infty} (1 - {}_\mathrm{c}s_t(X_{a,t})(1 - \mathrm{e}^{-F_t}))p(X_{a,t})\,dX_{a,t}}$	
Survey selectivity	${}_\mathrm{s}s_t(X)$: length-specific survey selectivity
${}_\mathrm{s}s_t(X) = \begin{cases} \mathrm{e}^{\frac{-\left(X - {}_\mathrm{s}X_t^\mathrm{full}\right)^2}{2{}_\mathrm{s}v_t}} & \text{for } X \le {}_\mathrm{s}X_t^\mathrm{full} \\ 1 & \text{for } X > {}_\mathrm{s}X_t^\mathrm{full} \end{cases}$	${}_\mathrm{s}S_{a,t}$: age-specific survey selectivity; ${}_\mathrm{s}X_t^\mathrm{full}$: log-length at full selectivity
${}_\mathrm{s}S_{a,t} = \int_{-\infty}^{\infty} {}_\mathrm{s}s_t(X_{a,t})p(X_{a,t})\,\mathrm{d}X_{a,t}$	${}_\mathrm{s}v_t$: variance-like parameter of half-normal selectivity function
(a) Length-specific:	${}_\mathrm{sJ}X^\mathrm{full}$, ${}_\mathrm{sC}X^\mathrm{full}$: log-length at full selectivity for J- and Circle hooks
${}_\mathrm{s}X_t^\mathrm{full} = {}_\mathrm{sJ}X^\mathrm{full}$ or ${}_\mathrm{sC}X^\mathrm{full}$	
${}_\mathrm{s}v_t = {}_\mathrm{sJ}v$ or ${}_\mathrm{sJ}v$	${}_\mathrm{sJ}v$, ${}_\mathrm{sC}v$: variance-like parameters of selectivity functions of J- and Circle hooks
(b) Age-specific:	
${}_\mathrm{s}X_{t+1}^\mathrm{full} = {}_\mathrm{s}X_t^\mathrm{full} + {}_{{}_\mathrm{s}X^\mathrm{full}}\varepsilon_t$ where ${}_{{}_\mathrm{s}X^\mathrm{full}}\varepsilon_t \sim N(0, 1)$	
$\ln({}_\mathrm{s}v_{t+1}) = \ln({}_\mathrm{s}v_t) + {}_{\mathrm{s}v}\varepsilon_t$ where ${}_{\mathrm{s}v}\varepsilon_t \sim N(0, 1)$	${}_\mathrm{sJ}S_a$, ${}_\mathrm{sC}S_a$: (constant) age-specific selectivity for J- and Circle-hooks
constrained so that ${}_\mathrm{s}S_{a,t} = {}_\mathrm{sJ}S_a$ or ${}_\mathrm{sC}S_a$	

[3]Estimated internally by the model assuming numbers killed by size categories are known without error.

TABLE A2. Equations describing model predictions of observed variables.

Predicted variable	Notation
Commercial catch at age $C_{a,t} = N_{a,t} e^{-\frac{M}{2}} (1 - {}_{\mathrm{b}}H_{a,t})\, {}_{\mathrm{c}}S_{a,t} (1 - e^{-F_t})$	$C_{a,t}$: catch at age a and year t
Fishing effort $E_t = \dfrac{F_t}{{}_{\mathrm{c}}q_t}$	E_t: predicted effort ${}_{\mathrm{c}}q_t$: commercial catchability in year t
Survey CPUE $CPUE_t = {}_{\mathrm{sH}}q_t \sum_{a=6}^{20^+} {}_{\mathrm{sH}}S_{a,t} N_{a,t}$	${}_{\mathrm{sH}}q_t$: survey catchability for hook type H in year t ${}_{\mathrm{sH}}S_{a,t}$: survey selectivity for hook type H in year t
Survey age composition ${}_{\mathrm{sH}}P_{a,t} = \dfrac{{}_{\mathrm{sH}}S_{a,t} N_{a,t}}{\sum_{i=6}^{20^+} {}_{\mathrm{sH}}S_{i,t} N_{i,t}}$	${}_{\mathrm{sH}}P_{a,t}$: survey age proportions for hook type H in year t
Mean and variance of log-length at age ${}_{\mathrm{c}}\mu_{a,t} = \dfrac{\int_{-\infty}^{\infty} X_{a,t}\, {}_{\mathrm{c}}s_t(X_{a,t}) p(X_{a,t})\, dX_{a,t}}{{}_{\mathrm{c}}S_{a,t}}$ ${}_{\mathrm{c}}\sigma^2_{a,t} = \dfrac{\int_{-\infty}^{\infty} X^2_{a,t}\, {}_{\mathrm{c}}s_t(X_{a,t}) p(X_{a,t})\, dX_{a,t}}{{}_{\mathrm{c}}S_{a,t}} - ({}_{\mathrm{c}}\mu_{a,t})^2$	${}_{\mathrm{c}}\mu_{a,t}$: mean log-length at age and year in the commercial catch ${}_{\mathrm{c}}\sigma^2_{a,t}$: variance of log-length at age and year in the commercial catch Note: for survey equations replace subscript c for s

TABLE A3. Estimated parameters and prior distributions.

Process	**Estimated parameters and constraints**	**Value or Prior distribution**
Natural mortality	$\boldsymbol{M}$	0.15
Fishing mortality	$\theta_1 = \overline{\ln(F_t)}$ $\theta_{2:26} = \ln(F_t) - \overline{\ln(F_t)},\ \sum_{i=2}^{26}\theta_i = 0$	$U(-\infty, \infty)$ $U(-5, 5)$
Recruitment	$\theta_{27} = \overline{\ln(N_{6,t})}$ $\theta_{28:52} = \ln(N_{6,t}) - \overline{\ln(N_{6,t})},\ \sum_{i=28}^{52}\theta_i = 0$	$U(-\infty, \infty)$ $U(-15, 15)$
Initial abundance	$\theta_{53} = \overline{\ln(N_{a,1974}),}\ a = 7, \ldots, 20^+$ $\theta_{54:67} = \ln(N_{a,1974}) - \overline{\ln(N_{a,1974})},\ \sum_{i=54}^{67}\theta_i = 0$	$U(-\infty, \infty)$ $U(-15, 15)$
Commercial selectivity	$\theta_{68} = {}_{\mathrm{c}}X_{1974}^{\mathrm{full}}$ $\theta_{69:92} = {}_{X^{\mathrm{full}}}\varepsilon_t,\ {}_{\mathrm{c}}X_t^{\mathrm{full}} \leq \ln(130), t = 1974, \ldots, 1997$ $\theta_{93} = \ln({}_{\mathrm{c}}v_{1974})$ $\theta_{94:117} = {}_{v}\varepsilon_t,\ t = 1974, \ldots, 1997$	$U(\ln(81), \ln(130)$ $N(0, 0.05^2)$ $U(-5, -2.5)$ $N(0, 0.05^2)$
Commercial catchability	$\theta_{118} = \ln({}_{\mathrm{c}}q_{1974})$ $\theta_{119:142} = {}_{q}\varepsilon_t, \quad t = 1974, \ldots, 1997$	$U(-15, -10)$ $N(0, 0.03^2)$; ${}_{q}\varepsilon_{1994}$ unconstrained[4]
Growth	$\theta_{143} = \alpha_{1974}$ $\theta_{144:165} = {}_{\alpha}\varepsilon_t, \quad t = 1974, \ldots, 1995$ $\theta_{166} = \beta$ $\theta_{167} = \sigma$	$U(4, 20)$ $N(0, 0.1^2)$ $U(0.6, 1)$ $U(-\infty, \infty)$
Size at recruitment	$\theta_{168} = \mu_{1974}$ $\theta_{169:192} = {}_{\mu}\varepsilon_t, \quad t = 1974, \ldots, 1997$	$U(-\infty, \infty)$ $N(0, 0.1^2), t < 1996; N(0, 0.02^2), t \leq 1996$
Initial size at age	$\theta_{193} = \alpha_0$ $\theta_{194} = \beta_0$ used to compute $e^{\mu_{a+1,1974}} = \alpha_0 + \beta_0 e^{\mu_{a,1974}}$	$U(-\infty, \infty)$ $U(0.6, 1)$
Survey age catchability	$\theta_{195} = \ln({}_{\mathrm{s}}q)$	$U(-\infty, \infty)$
Survey selectivity, two hook types used (H=J and C)[5]	(a) length-specific: $\theta_{196:197} = {}_{\mathrm{sJ}}X^{\mathrm{full}},\ {}_{\mathrm{sC}}X^{\mathrm{full}}$ $\theta_{198:199} = \ln({}_{\mathrm{sJ}}v), \ln({}_{\mathrm{sC}}v)$ (b) age-specific: constrained so ${}_{\mathrm{sH}}S_{a,t} = {}_{\mathrm{sH}}S_a$ $\theta_{196:197} = {}_{\mathrm{sJ}}X_{1997}^{\mathrm{full}},\ {}_{\mathrm{sC}}X_{1984}^{\mathrm{full}}$ $\theta_{198:199} = \ln({}_{\mathrm{sJ}}v_{1997}), \ln({}_{\mathrm{sc}}v_{1984})$ $\theta_{200:228} = {}_{{}_{\mathrm{s}}X^{\mathrm{full}}}\varepsilon_t$ and $\ln({}_{{}_{\mathrm{s}}v}\varepsilon_t)$ $t = 1997, \ldots, 1982; 1984, \ldots, 1986; 1993, \ldots, 1997$	 $U(\ln(66), \ln(130))$ $U(-\infty, \infty)$ $U(\ln(66), \ln(130)$ $U(-\infty, \infty)$ $N(0, 1)$

[4]To allow more variability during transition to ITQs.

[5]J-hook used during 1977–1984 and Circle-hook during 1984–1986 and 1993–1998

American Fisheries Society Symposium 27:137–155, 2002

Bayesian Analysis of Stock Survival and Recovery of Spring and Summer Chinook of the Snake River Basin

RICHARD B. DERISO
Inter-American Tropical Tuna Commission, Scripps Institution of Oceanography, La Jolla, California 92093, USA

Abstract.—Spring and summer chinook salmon *Oncorhynchus tshawytscha* populations of the Snake River basin provide the setting for an application of Bayesian analysis to derive risks of population survival and recovery for these endangered populations. The Bayesian approach is appealing because it provides a theoretical framework within which uncertainty about population dynamics is directly translated into measures of probability of achieving various population abundance targets, given certain types of actions in the future. Uncertainty about parameters governing the population dynamics is based on an application of the Bayes Theorem to the likelihood of observations about past recruitment, as viewed in the context of a generalized Ricker spawner and recruitment model. Uncertainty about future dynamics is based on simulations of population abundance over the next 100 years, and they contain both model parameter uncertainty and annual stochastic elements affecting survival. Results show substantial reductions in mortality rate (on the order of 0.5–0.7 per year, as compared with rates in recent years) are required in order for the populations to meet recovery and survival standards set for the next 48–100 years. The level of mortality reduction needed to achieve these standards can assist in guiding potential hydropower system management options.

Introduction

The recent low numbers of spring and summer chinook salmon *Oncorhynchus tshawytscha* in the Snake River basin are a source of concern to many. They were listed as threatened under the Endangered Species Act in 1992 and reclassified as endangered in August 1994. Their fate depends on their ability to withstand many sources of mortality in the future. In particular, chinook salmon can experience both downriver passage mortality and upriver return passage mortality, through the dam systems of the Snake River and Columbia River. Over the past 50 years, eight main-stem dams have been constructed in the migratory path of these salmon populations. The decline in productivity and survival of these populations has been coincident in time with the construction and operation of the four lower Snake River dams. In the current study, my focus is on the degree to which passage mortality affects the ability of Snake River spring and summer chinook salmon to achieve survival and recovery standards, which are essentially specified probabilities of reaching given population abundance targets by specified times in the future. The level of passage mortality reduction needed to achieve these standards can assist in guiding potential hydropower system management options. To address this issue, I developed a prospective analysis of chinook abundance for seven populations of the Snake River basin. The prospective analysis is the result of several workshops and extensive interaction by the author with various members of the project PATH (Plan for Analyzing and Testing Hypotheses; Marmorek et al. 1998).

This evaluation of the fate of chinook salmon takes into account two major kinds of uncertainties. First, there are many alternative hypotheses about the dynamics of chinook and about their past levels of mortality. Second, the future of chinook will be affected by many types of variability, some arising from basic stochasticity about the mean response of the populations to changes in spawner levels but others because of variability in downriver passage mortality as they migrate from nursery areas to the ocean as age two year-olds, variability in harvest rates in the river during their return from the ocean to spawn, and variability in other components of survival in the river during their return. Both types of uncertainties are addressed by employing a Bayesian statistical approach to the problem. The Bayesian approach has been found to be useful for many fisheries problems (Quinn and Deriso 1999).

The Bayesian approach is appealing because it provides a theoretical framework within which uncertainty about population dynamics is directly translated into measures of probability of achieving various population abundance targets, given certain types of actions in the future. Indeed, it makes it possible for one to define fishery management goals in terms of probabilities, as, for example, a goal could be an 80% probability that the population has doubled over the next 20 years.

The admission of uncertainty in the Bayesian model extends over all 88 parameters of the population dynamics model employed in this study, which include parameters describing common year-effects (such as some portion of ocean mortality) that affect all analyzed chinook populations, parameters quantifying annual in-river passage mortality, parameters affecting the shape of a generalized three-parameter spawner-recruit model, and the overall process variance about the model. The use of annual passage mortality parameters directly considers nonstationary processes in their population dynamics due to physical changes in the passage process (such as construction of fish by-pass technologies). The use of annual year-effects directly considers nonstationary processes in their population dynamics due to physical changes in their environment (such as large-scale oceanographic cycles).

Uncertainties about future events affecting chinook are addressed by development of a population projection model that contains many sources of stochasticity. Stochasiticity is included foremost by the admission of a stochastic relationship between spawners and resultant recruits and by the admission of uncertainty about fundamental parameters governing their dynamics. Additional uncertainty is included, by varying the time sequence of "year-effects," the initial-condition spawning stock abundance, and the time sequence of water transit times through the Columbia River and consequent relation to in-river passage mortality. A number of sensitivity analyses are also presented where structural changes are made to either the input data or to the basic model framework. The purpose for evaluating these uncertainties is to determine the degree to which they affect the level of passage mortality reduction needed to achieve survival and recovery standards.

Prospective analysis builds upon one of the best models deduced for chinook salmon from a model selection process described in Deriso et al. (in press). In particular, the maximum likelihood model (MLM) developed in that paper (labeled model 1 in that paper) provides a means to set up a Bayesian simulation model (BSM) for the current problem. As in the MLM model, the current BSM model is based on analysis of spawner and recruitment data for thirteen spring/summer chinook stocks of the Columbia River. The model framework relies on a generalized Ricker spawner and recruit model in which the generalization is the incorporation of an additional parameter to allow for depensatory mortality at low spawning stock levels and specific terms describing passage mortality and common-effects. Probability calculations are made by combining the generalized MLM model with a Monte Carlo Markov Chain (MCMC) algorithm (Gelman et al. 1995) and a population projection model. The resultant model framework produces probability estimates of stock recovery and survival as prescribed by the jeopardy standard criteria listed in Appendix I.

Spawner and Recruitment Data

Time series estimates of spawner and recruitment abundance, for thirteen populations of chinook salmon, were analyzed in this study (Table 1), and are described in detail in Beamesderfer et al. (1997). They represent populations in three downriver subbasins in the lower Columbia River—those of the Wind River, Klickitat River, and Warm Springs River; three populations in the John Day subbasin system in the lower Columbia River—the John Day Main-Stem, John Day Middle Fork, and John Day North Fork; and seven populations in upriver subbasins all branching from the Snake River—those of the Imnaha, Minam, Bear Valley, Marsh Creek, Sulfur Creek, Poverty Flats, and Johnson Creek Rivers. Lower Columbia River stocks are included in the analysis because the population model described below relies on the assumption that Snake River and lower Columbia River spring/summer chinook salmon populations share a "common" mortality factor that can be used to partition total mortalities. Recruitment is the abundance of year-classes, at the lowermost dam (Bonneville Dam), which are estimated by making "back calculations," based on expansion of spawner estimates to account for losses that occur during the migration of salmon from Bonneville Dam to their spawning ground. Spawner estimates are estimates of annual abundance of sexually mature salmon, at their respective spawning grounds.

Table 1 summarizes the number of main-river dams located below each river subbasin, along with the number of years of spawner and recruitment information available. Further descriptions of these subbasins, including a map, are given in Schaller et al. (1999).

Methods

Bayesian simulation model

The Bayesian Simulation Model (BSM) is a method in which a generalized Ricker Spawner and Recruitment (S-R) Model is fitted to observations; then population projections are made. The fitting of the

TABLE 1. Summary information on thirteen spring and summer chinook populations (adapted from Deriso et al., in press).

Subbasin	Brood years of paired spawner-recruit data	Number of main-stem dams below subbasin
1. Wind	1973–1990	1
2. Klickitat	1966–1990	1
3. Springs	1969–1990	2
4. John Day Mainstem	1959–1990	3
5. John Day Mid Fork	1959–1990	3
6. John Day North Fork	1959–1990	3
7. Imnaha	1952–1990	8
8. Minam	1954–1990	8
9. Bear Valley	1957–1990	8
10. Marsh Creek	1957–1990	8
11. Sulphur Creek	1957–1990	8
12. Poverty Flat	1957–1990	8
13. Johnson Creek	1957–1990	8

model and the projections of the model are linked together by probability distributions for the parameters of the model, in which the distributions are based on the likelihood of the observations. The S-R model is described, followed by a description of the method for calculation of probability distributions, and then followed by a description of the population projection model.

Generalized Ricker Spawner-Recruit model

The population model is based on the Ricker-type Spawner-Recruitment Model, similar in structure to the Ricker models used in Deriso et al. (in press), except generalized to allow for depensatory mortality at low spawner levels, which can be written in generic form:

$$R = \alpha\beta S^{(1+p)}e^{-\beta S}.$$

Depensatory mortality occurs for low spawner values when p is greater than zero. The β and α parameters represent benchmark spawner-inverse and recruitment values: when $S = 1/\beta$ and $p = 0$, then $R = \alpha e^{-1}$, the maximum recruitment possible with the standard Ricker model. Therefore, α is e^1 times the maximum recruitment of a Ricker model. This parametric representation was chosen for the Ricker-type parameters (β and α) because it offers reduced parameter correlation over more common representations (such as given in Quinn and Deriso 1999), which in turn improves the efficiency of numerical Bayesian integration schemes.

The year and area-specific terms of the S-R model are better seen in the logarithmic form of the model:

$$\ln(R_{t,i}) = a_i + \delta_t - m_{t,i} + \ln(\beta_t) + (1+p)\ln(S_{t,i}) - \beta_i S_{t,i} + \varepsilon_{t,i}, \qquad (1)$$

where

$R_{t,i}$ = observed returns (recruitment) originating from spawning in year t and river subbasin i,

$S_{t,i}$ = observed spawning in year t and river subbasin i,

a_i = transformed Ricker α parameter, which depends on river subbasin i,

β_i = Ricker β parameter, which depends on river subbasin i,

$m_{t,i}$ = in-river passage mortality which depends on year t and river i,

$\varepsilon_{t,i}$ = normally distributed mixed process error and recruitment measurement error term N $(0, V_\varepsilon)$

p = depensation parameter ($p > 0$), and

δ_t = year-effect parameter in year t.

In-river passage mortality $m_{t,i}$ is described by a two-level parameterization scheme, as in the MLM model:

$$m_{t,i} = X^*n + \mu_t,$$

where n is the number of "first level" dams, X is the dam passage mortality per first level dam, and μ_t is the net dam passage mortality from the Snake River subbasins to John Day Dam, expressed as an instantaneous mortality rate for brood years $t \geq 1970$. The "first level" (number n of X's in each row of Table 2) treats mortality as a process proportional to the number of dams passed by a salmon during its transit to the ocean, excluding those dams and/or populations treated in the second level. At the second level (Y in Table 2), the incremental mortality experienced by upstream stocks relative to

TABLE 2. Types of passage mortality estimates. (X = fixed estimate of mortality per dam; Y = annually varying estimates of mortality due to passage through 4–5 dams; BON = Bonneville; TDD = Dalles; JDA = John Day; McN = McNary; IHR = Ice Harbor; LOMO = Lower Monumental; LGS = Little Goose; LGR = Lower Granite) Adapted from Deriso et al., in press.

	Dams							
Brood year	BON	TDD	JDA	McN	IHR	LOMO	LGS	LGR
1952–1954	X			X				
1955–1958	X	X		X				
1959–1965	X	X		X	X			
1966	X	X	X	X	X			
1967	X	X	X	X	X	X		
1968–1969	X	X	X	X	X	X	X	
1970–1972	X	X	X	Y	Y	Y	Y	
1973–present	X	X	X	Y	Y	Y	Y	Y

downriver stocks is estimated by μ_t. The μ term is a "net" effect mortality estimate because it reflects the overall impacts of dam passage over the complete life cycle of upstream stocks, including the effect of transporting some of the Snake River smolts around the dams to below the most downriver dam in barges and trucks. The μ term is usually estimated as a positive value, but it can sometimes be a negative value, as, for example, when the net benefit of barge transportation of upstream smolts induces a total passage mortality for upstream stocks below that experienced by downriver populations. Mortality in any given year for any given population is obtained in Table 2, by adding the number of X's (number of "first level" dams passed) plus a second annual term μ_t, provided that at least one Y is listed for a set of years. A symbol is first listed on the diagram for a given dam for the year of initial service (lagged two years to standardize to brood year).

The population model (1) contains the potential for depensatory mortality at low spawner levels, whenever p is greater than zero. In applications presented later, the likelihood of such mortality is quite low. A weakness of this approach is that the likelihood only evaluates the potential for depensatory mortality within the range of observed data. It is possible that depensation could occur at spawning levels below those observed in the past. To address that possibility, a further level of depensation was added to the model (1), for a sensitivity analysis presented later. This additional depensation is modeled by assuming that recruitment survival (as given by R/S) declines below minimum observed spawning levels, at a rate different than that given in (1). Specifically, this possibility is written as the equation

$$R = R(\text{as given in equation 1})^*(S/S_{\min})^d \quad \textit{for} \quad S < S_{\min}.$$

Figure 1(A, B) show the different types of spawner-recruitment curves one obtains as the parameter d is varied. Figure 1 only shows the portion of the curve pertaining to low values of spawners. When d is set to zero, then we obtain the original model (1). Minimum observed spawning levels, during years in which the likelihood function is calculated, are given later in Table 5, along with other quantities of interest.

Bayesian Structure

Bayesian statistical conclusions about the probability of future events rely on two components: (1) probability statements about the parameters governing the population dynamics process, and (2) probability statements about the population projections into the future conditional on a given set of parameter values. The first component, probability statements about parameters, is given by the posterior probability $p(\theta \mid \text{data})$ of the parameters, which is proportional to the likelihood $p(\text{data} \mid \theta)$ of the observed data and the prior distribution $p(\theta)$ for the parameters. In particular,

$$p(\theta \mid \text{data}) \propto p(\theta)p(\text{data} \mid \theta),$$

where the parameter vector θ contains 88 parameters, for the application in this paper, which are δ's (year-effects) for brood years 1952–1989, μ's (Snake River to John Day passage mortality) for brood years 1970–1990, river-specific S-R parameters a, β, plus a first-level dam mortality parameter, the depensation parameter p, and a process error variance V_ε. Those parameters are all implemented in the Bayesian analysis, as written in equation (1), except for the S-R parameters (β, p) and process error variance V_ε, which are written on logarithmic scale. Note that the

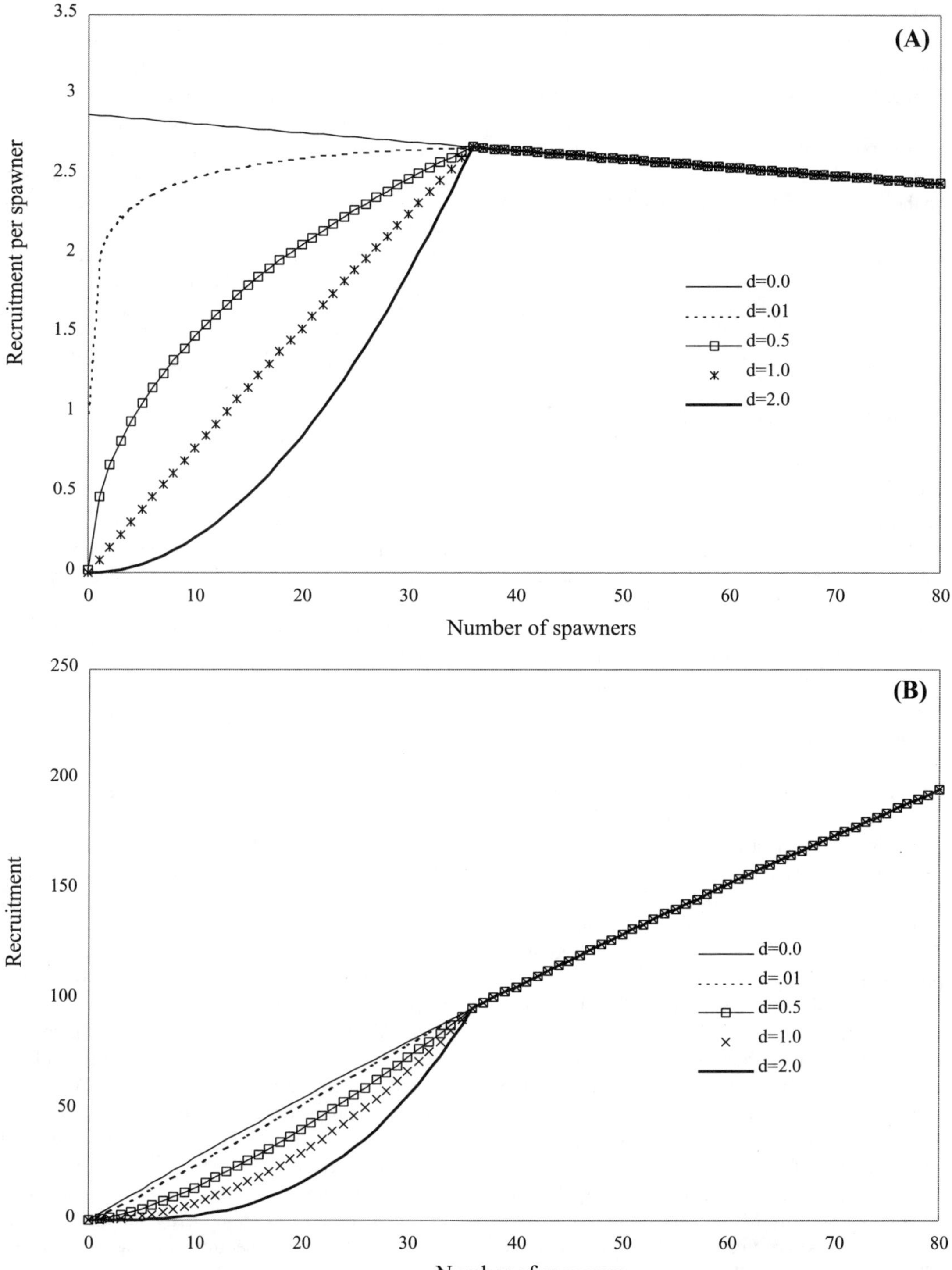

FIGURE 1(A). Recruitment survival (R/S) at low numbers of spawners for Johnson Creek. Spawner-recruit parameters are MLE estimates with parameter μ set to 2.0. The curves only show the low end of the range of spawners. (B). Recruitment at low numbers of spawners for Johnson Creek. Spawner-recruit parameters are MLE estimates with parameter μ set to 2.0. The curves only show the low end of the range of spawners.

arithmetic scale chosen for the passage mortalities allows those parameters to take on either positive or negative values, so that it is feasible (although it seldom occurred in application) that, in some projected years, "net" passage mortality actually benefits upriver populations (relative to the downriver populations). The likelihood function $p(\text{data} \mid \theta)$ is based on the assumption of log-normally distributed errors for recruitment in equation (1), which can be written in generic form as

$$p(\text{ data} \mid \theta) \propto \exp\Big[-N/2\ln(V_\varepsilon) - 1/2\sum\{\ln(R) - \ln(\hat{R})\}^2/V_\varepsilon\Big],$$

for which N is the number of observed recruitment data for all 13 populations, and the caret symbol denotes predicted value. The uniform probability distribution $U(-\infty, \infty)$ is used for the prior distribution, for all linear parameters (linear as described below); uniform $U(-12, 12)$ were used for prior distributions for the other parameters, where the number "12" is sufficiently large to allow parameters to span a range of meaningful values. The second component to the Bayesian approach, regarding population projections, is described in detail below.

The two components interact in the following manner. First, a random sample is drawn from the posterior probability of the parameters. A 100-year population projection is made with those parameters. Several annual sources of variation are introduced in that projection, as described later in the paper. This two-step process (select parameter from posterior, then project) is repeated several thousands of times, to ensure good coverage of the posterior distribution and random distributions used in the projections.

The MCMC algorithm designed for this analysis takes advantage of the partial linear structure of the model. In particular, the year-effects parameters, the passage mortality rate parameters, and the S-R model a parameters occur linearly in the model. A two-step MCMC algorithm was constructed so that the Gibbs steps are taken for the linear parameters (a conditional, multivariate normal posterior distribution is calculated) and the Metropolis steps (with normal jump functions) are taken for the other parameters (Gelman et al. 1995). The MCMC approach works by forming a Markov chain of parameter values selected from the posterior distribution. The chain is calculated by stepping through the parameter space to obtain updates of parameters, which form the next step of the chain. The chain is updated 41,000 times in our application, to ensure good coverage of the posterior distribution. The first 1,000 chain steps are not used to calculate posterior probabilities (the so-called "burn in" initialization period), and every fifth step thereafter is used as a sample from the cumulative distribution functions of the posterior distributions. Thus, 8,000 samples are utilized. Several tests were made to ensure convergence of the chain, with respect to calculation of jeopardy standards; good convergence is expected because model (1) could be written as a linear model (conditional on V_ε) if not for the log transformation applied to keep β and p positive definite. In a linear model, the Gibbs algorithm is an exact solution.

Population Projections Model

Population projections are made for the seven Snake River chinook populations listed in Table 1 (stocks indexed 7–13). The projection model simulates the life cycle of chinook salmon. These stream-type chinook populations migrate down the Columbia River past all the dams to the ocean in their second year of life. These populations return from the ocean and migrate up the Columbia River past all the dams to their natal streams, one, two, and three years later, as three, four, and five year olds. The spawning population for a given year is composed of four and five year-old fish from two parental broods. The three year-old fish are not used in the spawning population projections because three year-old stream-type chinook salmon are characterized as jacks (which are almost exclusively composed of precocious males). Dam passage can induce mortality during their downriver migration and later as they return to their natal streams. This paper will refer both to water year and brood year, which are alternative ways of characterizing a cohort: the year in which a particular cohort is spawned is its "brood year" and then the year in which they migrate down the river is "water year." Water year equals brood year plus two years.

An overview of the general structure of the model is provided here. The projection model calculates posterior probabilities of survival and recovery in accordance with jeopardy standards listed in Appendix I. Each projection is made for 100 years, beginning with 1996 as the first year of projection of spawner abundance. The initial values required to initiate each projection are based on projected age-specific returns for brood years beginning in 1991, except where actual estimates of returns are available. In each projected year, spawners $S_{t,i}$ are calculated as

$$S_{t,i} = \sum\nolimits_a f_{t,a,i} s_{t,i} R_{t-a,i}, \qquad (2)$$

in which a fraction $f_{t,a,i}$ of $R_{t-a,i}$ (total recruitment produced in brood year $t - a$) returns back from the ocean to the lowermost dam in year t. Recruits then experience an upriver survival $s_{t,i}$ to the spawning ground. The a subscript denotes age and the i subscript denotes river subbasin. The projected spawners in equation (2) are then applied to equation (1) in order to project the next generation of recruits. Repeated cycling through equations (1) and (2) generates a 100-year simulation of the salmon populations. Important details left to be discussed include (a) how common year-effects in equation (1) are selected in the projections, (2) how downriver survival is projected, (3) how upriver survival is projected, (4) how maturity proportions are projected, and (5) how stochasticity in the spawner and recruitment relationship is projected. A description of each component of the population projection model follows below.

Projecting Upriver Survival of Recruits: The $s_{t,i}$ Term

Many types of losses affect returning adults to the lowermost dam on the Columbia system. The losses can be placed in three categories: (1) a conversion factor from Bonneville dam (the lowermost dam) through Lower Granite dam (the uppermost dam), which accounts for all nonfishery-related losses of Snake River chinook during their upriver passage; (2) an exploitation fraction, which is the total loss of chinook due to in-river fisheries; and (3) prespawn mortality, which accounts for nonfishery losses between Lower Granite Dam and spawning grounds. The product of those three components equals upriver survival:

$$s_{t,i} = (\text{conversion factor})(1 - \text{exploitation fraction}) \times (1 - \text{prespawn mortality}).$$

Estimates of annual conversion rates for each population were provided to me (H. Schaller, U.S. Fish and Wildlife Service, Vancouver, Washington, personal communication). By inspection of loess regression plots, I decided to use the conversion estimates from the years 1985–1995 for the base case, as representative of values in recent years (Table 3). Within that 11-year time frame, there is no significant autocorrelation in conversion estimates for Middle Fork or South Fork chinook salmon of the Snake River, and so a random selection process is used in the projections.

TABLE 3. Conversion rates for Snake River spring and summer chinook salmon. (1) Middle Fork applies to Bear Valley, Marsh Creek, and Sulphur Creek; (2) South Fork applies to Poverty Flat and Johnson Creek.

Year	Minam River	Imnaha River	Middle Fork(1)	South Fork(2)
1985	0.74	0.75	0.74	0.77
1986	0.66	0.72	0.66	0.79
1987	0.75	0.65	0.75	0.56
1988	0.69	0.60	0.69	0.52
1989	0.48	0.58	0.48	0.69
1990	0.63	0.67	0.63	0.70
1991	0.49	0.59	0.49	0.69
1992	0.75	0.66	0.75	0.56
1993	0.74	0.81	0.74	0.88
1994	0.92	0.78	0.92	0.64
1995	0.54	0.64	0.54	0.74

Columbia River and tributary harvest rates were applied to population projections, in accordance with the harvest rate policy followed by fishery management on the system. Harvest rules are established for main-stem and tributary harvest rates, as a function of projected spawner abundance relative to MSP (maximum sustainable production) levels. The rules were provided to me (H. Schaller, personal communication), along with MSP levels currently in use by management, as listed in Tables 4(A, B) and 5, respectively. In the current version of the model, only Snake River populations are projected into the future, and thus, main-stem harvest rules were applied to their aggregate abundance. The rules differ by spring runs and summer runs. The Imnaha stock is a mixed spring/summer run; in the BSM model, half of the population is treated as a spring run and half the population is treated as a summer run. Tributary harvest rules are based on projected escapement from the main-stem harvest, aggregated into spring chinook runs and summer chinook runs. Prespawning mortality is assumed to equal 10%, based on discussions of evidence about the issue within PATH.

Projecting common year-effects

There is a significant first-order auto-correlation present in the maximum likelihood (ML) estimates of δ (the year-effects parameters in equation (1).

TABLE 4A. Upriver spring chinook Columbia River Fisheries Management Plan harvest rate schedule. Run size is given as a percentage (%) of maximum sustainable production abundance (MSP) for combined abundance of spring chinook stocks. Run size has been adjusted for 1977–1990 average conversion rate and 90% prespawning survival.

Run size % of MSP	Mainstem harvest rate	Tributary harvest rate
<22%	0.03	0
22%–44%	0.082	0
45%–112%	0.14	0
113%–125%	0.25	0.05
126%–175%	0.3	0.15
176%–200%	0.35	0.2
>200%	0.4	0.25

Regression of $\delta(t+1)$ versus δ(t) has an $R^2 = 0.271$ (significant at $p = 0.0008$). The autocorrelation apparent in the year-effects parameters was captured in the population projections, by a type of Markov process with empirical probability densities. The method implemented in the model consists of the following steps:

a. For each 100-year population projection, a sample of the posterior density of the year-effect parameters is obtained.
b. From that sample, ML estimates are calculated for the multinomial probabilities that characterize the sign of a δ, such as $\mathrm{P}(\delta(t) > 0 \mid \delta(t-1) > 0)$. This produces the odds that a positive year-effect is followed by another positive year-effect.
c. The sequence of positive and negative year-effects in the projections are simulated to follow a multinomial distribution, with probabilities set equal to the P matrix developed in step (b).
d. Once step (c) determines that, say, a positive δ is to be simulated for a given year, then the particular positive value simulated is chosen at random from the whole set of positive values obtained in step (a).

Steps (a) and (b) are computed once for each 100-year simulation. Steps (c) and (d) are computed for each simulated year.

Projecting survival downriver

The key issue, in the projections of survival downriver, pertains to treatment of passage mortality. In Deriso et al. (in press), we showed that there is a weak correlation between water travel time (WTT), a measure of the time it takes water to travel down the Columbia River, and passage mortality. The velocity of water, as measured by the WTT, may be correlated with passage mortality because the longer fish are in the dam reservoirs the longer they are exposed to predators, and thus, under conditions of higher velocity water, the fish experience less delay when passing dams.

As a first step towards modeling the relationship between those variables, I investigated properties of the time series of WTT itself. The historical

TABLE 4B. Upriver summer chinook Columbia River Fisheries Management Plan harvest rate schedule. Run size is given as a percentage (%) of maximum sustainable production abundance (MSP) for combined abundance of summer chinook stocks. Run size has been adjusted for 1977–1990 average conversion rate and 90% pre-spawning survival.

Run size % of MSP	Mainstem harvest rate	Tributary harvest rate
<25%	0.02	0
25%–49%	0.05	0
50%–99%	0.1	0
100%–129%	0.15	0
130%–149%	0.2	0.05
150%–169%	0.25	0.1
170%–200%	0.3	0.2
>200%	0.35	0.25

TABLE 5. Maximum sustainable production abundance (MSP) and minimum observed spawners (through brood year 1990) in numbers of spawners for spring and summer chinook of the Snake River. MSP values are those assumed by the Columbia River Fisheries Management Plan.

Subbasin in Snake River	MSP	Minimum spawners
Bear Valley	1244	42
Marsh Creek	456	16
Sulphur Creek	333	12
Poverty Flat	3497	76
Johnson Creek	553	36
Imnaha River	1538	170
Minam River	810	41

record of unregulated water transit times shows little autocorrelative relationship; there is no significant first-order autocorrelation of WTT($t+1$) versus WTT(t) ($R^2 = 0.011$ for water years 1929–1990). There is little structure revealed from a non-parametric view either: let w-bar be the average WTT for all years; then, P(WTT($t+1$) > w-bar | WTT(t) > w-bar) = 0.467, and P(WTT($t+1$) > w-bar | WTT < w-bar) = 0.50. On the other hand, there have been relatively few above average WTT years recently. For example, only 38% of the WTT values for 1972–1992 exceed w-bar.

The relationship between passage mortalities and WTT was modeled by a two-step procedure. Population projections affected by this procedure are those projections made for the "recent condition" scenario, as defined in detail later in the paper, but, in overview, it is one in which passage mortalities from the 1978–1992 water years are utilized. For any given simulated year of a "recent condition" scenario, the first step of the procedure is random selection of a WTT from the 1978–1992 water years, in proportion to a smoothed version of the empirical cumulative distribution function (CDF) of long-term WTTs (water years 1929–1992). Suppose that the WTT for 1984 is one such random sample; then, the second step of the procedure is that passage mortality selected for the given simulated year is a sample of the posterior density of in-river passage mortality for 1984. The smoothing of the CDF is accomplished also in two steps: (1) randomly choose a 0.33 probability interval with ending point 0.33–1.0 from the empirical CDF of the long-term WTTs; (2) from the chosen interval, randomly select one of the water years 1978–1992, whose empirical CDF value falls within the chosen interval.

Projecting maturity proportions

The proportion of recruits returning to any given population can vary by subbasin and age of recruit, according to historical data. For each projected year, randomly selected proportions were chosen from those estimated for historical data (brood years 1963–1993) on each population. Those 31 historical years are the maximum number of historical data years in common for all Snake River populations, except Poverty Flats, which has 30 years (no samples in 1984).

Projecting stochasticity in recruitment

Stochasticity is simulated in the spawner and recruitment model (1) by random selection of a normally distributed random variable, with variance equal to 61% of V_ε. This stochasticity is intended to represent process error in recruitment survival, but V_ε includes both recruitment measurement error and process error. The 61% estimate is based on a partitioning of the residual sum of squares (RSS) from the ML fitting of model (2) in Deriso et al. (in press), in which spawning measurement error is modeled. From that analysis, the RSS of total recruitment error is 76.4, and the RSS of spawning measurement error is 8.04. By assuming the RSS of spawning measurement error equals recruitment measurement error, then the estimate 68.36 (= 76.4 – 8.04) is the RSS for recruitment process error. The ratio of this last RSS to the RSS for equation (1) is 0.61 (= 68.36/112.4). For each population and each projection year, a random process error is added to the logarithm of the recruitment model (1). The error is chosen from a normal distribution with mean zero, and variance is selected from the posterior distribution of process error variance.

Jeopardy Standards

The jeopardy standard summary in Appendix I is used to guide development of appropriate output calculations. Some interpretation is needed to place the standards on a sound probability basis. With regards to the survival standard (item 1 of Appendix I), it is straightforward to estimate the probability that a given population is above a given threshold in any given year; one simply calculates the frequency of monte carlo projections, for which the population is above the threshold in that given year. By averaging those probabilities over all projection years, one finds the average probability a given population is above a given threshold over all projection years. This "average probability" is the item (1c) for the 100-year projection time period; the 24-year time period is handled similarly. Threshold values for each of the Snake River populations are listed in Table 6.

The jeopardy standards require calculation of the "average probability" for both recent and historical conditions in the Columbia River. PATH members chose the passage mortalities for brood years 1976–1990 to represent recent conditions, and we chose the passage mortalities for brood years 1952–1969 to represent historical conditions. Other parameters are not assumed to differ between scenarios; in particular, I assume that the δ year-effects or upriver survivals will not depend on whether a scenario is for "recent conditions" or "historical conditions."

Initial population conditions differ between scenarios. Projections begin with the following initial conditions for the "recent condition" scenario: estimates of spawners for 1991–1995 and recruits produced by those spawners, where available. The estimates of spawners are multiplied by a lognormal random variable, with standard deviation of 24% to account for measurement error. The 24% standard deviation is based on limited, within stream, comparisons of redd counts (C. Petrosky, Idaho Department of Fish and Game, 600 S. Walnut, Boise, Idaho 83707, personal communication). The "historical conditions" scenario was obtained by random selection from observed values for the 1957–1969 brood years.

Item 1 d of Appendix I is the ratio of "average probability" for the recent conditions scenario relative to the historical conditions scenario. For example, a 90% ratio of average probability for a given population and given threshold means that the average probability for a "recent condition" scenario is 90% of the average probability for a "historical condition" scenario.

TABLE 6. Survival and recovery thresholds for the jeopardy standards for Snake River spring and summer chinook index populations. Average spawners listed are averages for all years of data prior to 1971 brood year.

Population	Survival standard threshold	Recovery standard threshold	Historical average spawners
Bear Valley	300	950	1611
Imnaha	300	900	1521
Marsh Creek	150	450	735
Minam River	150	500	823
Poverty Flats	300	1150	1884
Sulphur Creek	150	350	523
Johnson Creek	150	350	589

The recovery standard (item 2 of Appendix I) is a standard based on the probability that an eight-year geometric mean of spawners is above a given threshold in a given projection year; year-24 and year-48 probabilities are calculated. I calculated the average number of spawners for each population, for all years of data on pre-1971 brood years, as specified in Appendix I. The threshold values for the recovery standard are 60% of those historical averages rounded to the nearest 50 spawners. The thresholds for both criteria are given in the table below. I went beyond the requirements of the jeopardy standard and calculated the eight-year geometric mean spawners for both "recent condition" scenario and "historical condition" scenario. I also provided the ratio of probability for the "recent condition" scenario to probability for the "historical condition" scenario, so as to make the results of item 2 more comparable to those in item 1.

Projections with Change in Future In-River Passage Mortality

Management actions to increase the chances of Snake River population survival and recovery can involve changes in future in-river passage mortality. A number of calculations were made to investigate consequences of changes in the in-river passage mortality, by either fixed incremental units or by fixed units subject to a constraint. The results pertain to any management action that results in an incremental decrease in the density-independent component of recruitment mortality. In the projections, a fixed incremental change was implemented by altering the "recent condition" scenario, so that, in each simulated year, passage mortality m is replaced by m', where

$$m' = m - \Delta,$$

in which Δ is the fixed increment change. In some simulated years, the fixed increment change can be so large as to change the sign of m' to a negative number. Such enhanced production by salmon is considered unlikely by some PATH participants, and thus, a sensitivity analysis was made in which the increments were constrained, such that m' always remains a nonnegative number. In both sets of calculations, there is a ten-year linear "ramp-up" of Δ, beginning in projection year 1, with an increment of 10% of Δ, and linearly increasing to 100% in year 10 and

TABLE 7A. Average probabilities of spawners being above survival threshold using 24-year simulation period.

Snake River subbasin	Survival threshold	Average probability of being above threshold: Historical	Average probability of being above threshold: Recent	Recent probability/ historical probability
Imnaha	300	0.984	0.706	0.718
Minam	150	0.967	0.609	0.630
Bear Valley	300	0.993	0.576	0.581
Marsh Creek	150	0.980	0.478	0.488
Sulphur Creek	150	0.900	0.462	0.513
Poverty Flat	300	0.974	0.628	0.644
Johnson Creek	150	0.946	0.525	0.555

TABLE 7B. Average probabilities of spawners being above survival threshold using 100-year simulation period.

Snake River subbasin	Survival threshold	Average probability of being above threshold: Historical	Average probability of being above threshold: Recent	Recent probability/ historical probability
Imnaha	300	0.981	0.747	0.761
Minam	150	0.964	0.692	0.717
Bear Valley	300	0.992	0.718	0.724
Marsh Creek	150	0.979	0.637	0.651
Sulphur Creek	150	0.895	0.573	0.640
Poverty Flat	300	0.972	0.627	0.645
Johnson Creek	150	0.943	0.555	0.588

TABLE 7C. Average probabilities of spawners being above recovery threshold using 24-year simulation period.

Snake River subbasin	Recovery threshold	Average probability of being above threshold: Historical	Average probability of being above threshold: Recent	Recent probability/ historical probability
Imnaha	850	0.945	0.280	0.297
Minam	450	0.884	0.196	0.222
Bear Valley	900	0.961	0.202	0.211
Marsh Creek	450	0.933	0.127	0.136
Sulphur Creek	300	0.833	0.166	0.200
Poverty Flat	850	0.904	0.147	0.163
Johnson Creek	300	0.926	0.202	0.218

TABLE 7D. Average probabilities of spawners being above recovery threshold using 48-year simulation period.

Snake River subbasin	Recovery threshold	Average probability of being above threshold: Historical	Average probability of being above threshold: Recent	Recent probability/ historical probability
Imnaha	850	0.945	0.292	0.309
Minam	450	0.880	0.211	0.239
Bear Valley	900	0.961	0.284	0.295
Marsh Creek	450	0.936	0.238	0.254
Sulphur Creek	300	0.822	0.220	0.268
Poverty Flat	850	0.906	0.154	0.170
Johnson Creek	300	0.925	0.215	0.232

thereafter. The ramp-up is designed to incorporate delays and the incremental nature of management changes feasible for the system (such as breaching of four of the Snake River dams or spawning habitat improvements).

Sensitivity Analyses

Five different variations to the modeling were developed. The first version is my "base case," followed by four others designed to examine sensitivity of the results:

1. The base case, which has been described in previous sections. This method does not contain the additional depensation term illustrated in Figure 1, and it assumes that "observed" spawners correctly measure true spawners in the likelihood function $p(\text{data} \mid \theta)$.
2. Depensation is assumed to occur below minimum observed spawner levels. The model described in Figure 1, with depensation parameter $d = 1.0$, is applied.
3. A consequence of spawner measurement error was modeled by application of results from model number 2 in Deriso et al. (in press). That model provides ML estimates of spawners. The ML estimates of spawners were used, instead of "observed" spawners—the data used in the base case. Recruitment process error was rescaled in the run to produce the same residual sum of squares as decided for the base case.
4. Age composition error was accounted for by application of a missing data algorithm I developed in an unpublished report. Estimated recruitment values from that analysis were used instead of "observed" recruitment—the data used in the base case. Recruitment process error was rescaled in the run to produce the same residual sum of squares as decided for the base case.
5. Year-effect parameters δ's were set to zero. This model had a favorable BIC (Bayesian Information Criterion) score (see Quinn and Deriso 1999 for a definition of BIC; see Deriso et al. (in press), model 29 in that paper).

Results of the Base Case

Tables 7(A–D) summarize jeopardy standard probability results for the base case of the BSM. The jeopardy standards in Appendix I are vague regarding specific quantification of the standards. If we take the least stringent of the ones listed in Appendix I, then the survival standard is satisfied if 80% of the populations have a probability ratio of at least 70% in the right-hand column of Tables 7(A, B). In Table 7(A), 14% (one of seven) of the populations exceed 70% probability ratio, and therefore, the 24-year survival standard is not satisfied. In Table 7(B), 43% (three of seven) of the populations exceed 70% probability ratio, and therefore, the 48-year survival standard is not satisfied. The least stringent recovery standard is satisfied if 80% of the populations have a probability of at least 50% in the fourth column of Tables 7(C, D). None of the populations have a 50% probability rate in Tables 7(C, D), and therefore, the recovery standard is not satisfied.

Jeopardy standards for survival and recovery can be met in principle by making changes to passage mortality. In this paper, the analysis of potential changes pertains to any management action that results in an incremental decrease in the density-independent component of recruitment mortality. Figures 2(A–D) show how the probabilities are affected by fixed incremental changes in passage mortality for each of the survival and recovery standards. Reference lines are drawn on the graphs at roughly the midpoint of the range of options listed in Appendix I, for each standard. All standards require at least 80% of the populations (which is six of the seven) to lie above the reference lines. The shorter time horizon survival standard (Figure 2A) probably cannot be achieved realistically. The other three standards would require reductions in mortality rate on the order of 0.5–0.7 per year. Similar results were obtained for model simulations with constrained incremental changes.

Estimated posterior probability distributions for the 88 parameters are given in Appendix II. The 90% posterior quartiles are quite large for most of the parameters, which shows that the population projections are made over a wide range of parameter values. Of particular interest is the negligible amount of depensation, listed in logarithms as $\ln(p)$, estimated from the observed data. The wide range of parameter values used in the projections lessens the sensitivity of jeopardy standard calculations to modest shifts in ML estimates of the parameters, as shown by the sensitivity results given below.

Results of the sensitivity analysis are summarized in Table 8. Results in the table show that the base case results are quite similar to results obtained from models 2–4. Model 5, the no year-effect model, gives a substantially more pessimistic view of the future of Snake River chinook salmon.

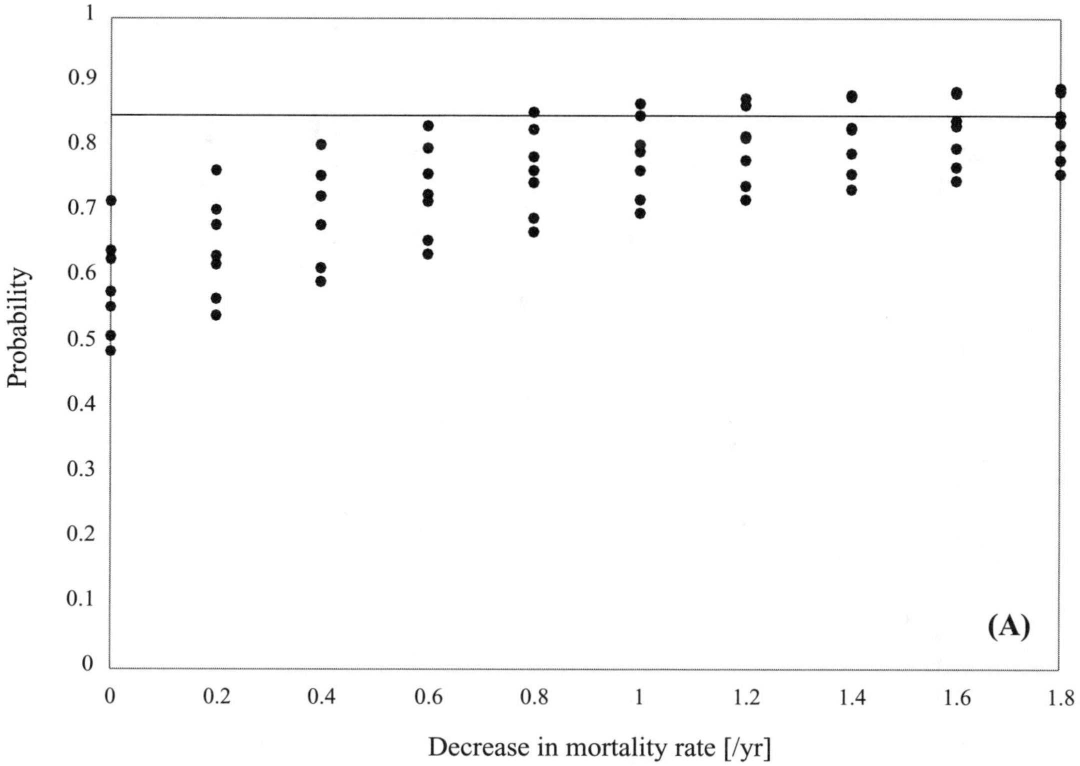

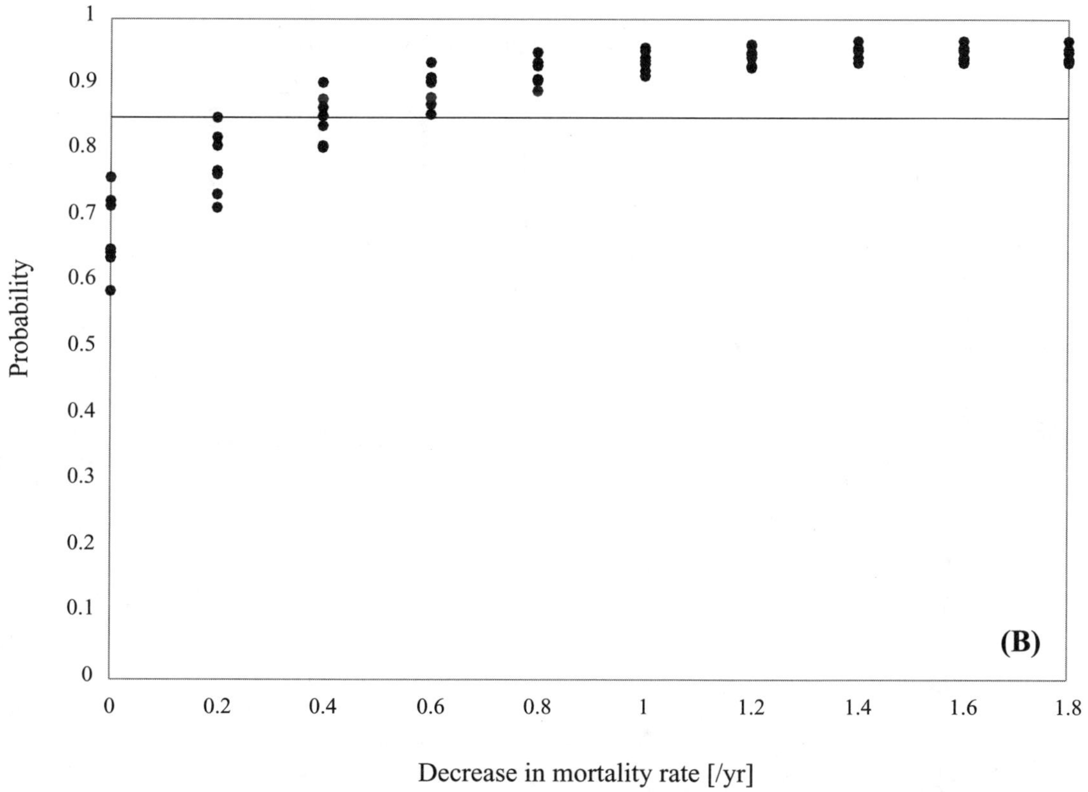

FIGURE 2 (A–D). Probability of achieving various threshold standards are graphed versus specified decreases in annual mortality rate. Figure 2(A) pertains to 24-year survival threshold; 2(B) pertains to 100-year survival threshold; 2(C) pertains to 24-year recovery threshold; 2(D) pertains to 48-year recovery threshold. Each dot represents a separate population of spring/summer chinook salmon of the Snake River basin. Line drawn at reference level needed to satisfy jeopardy standard.

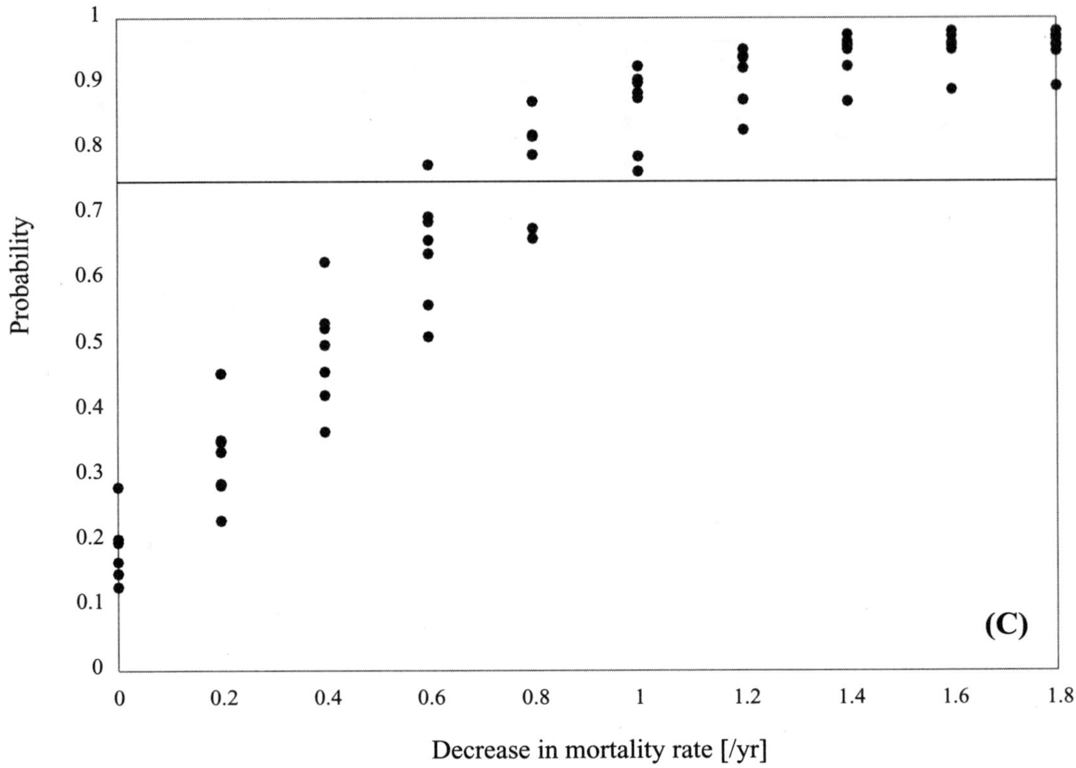

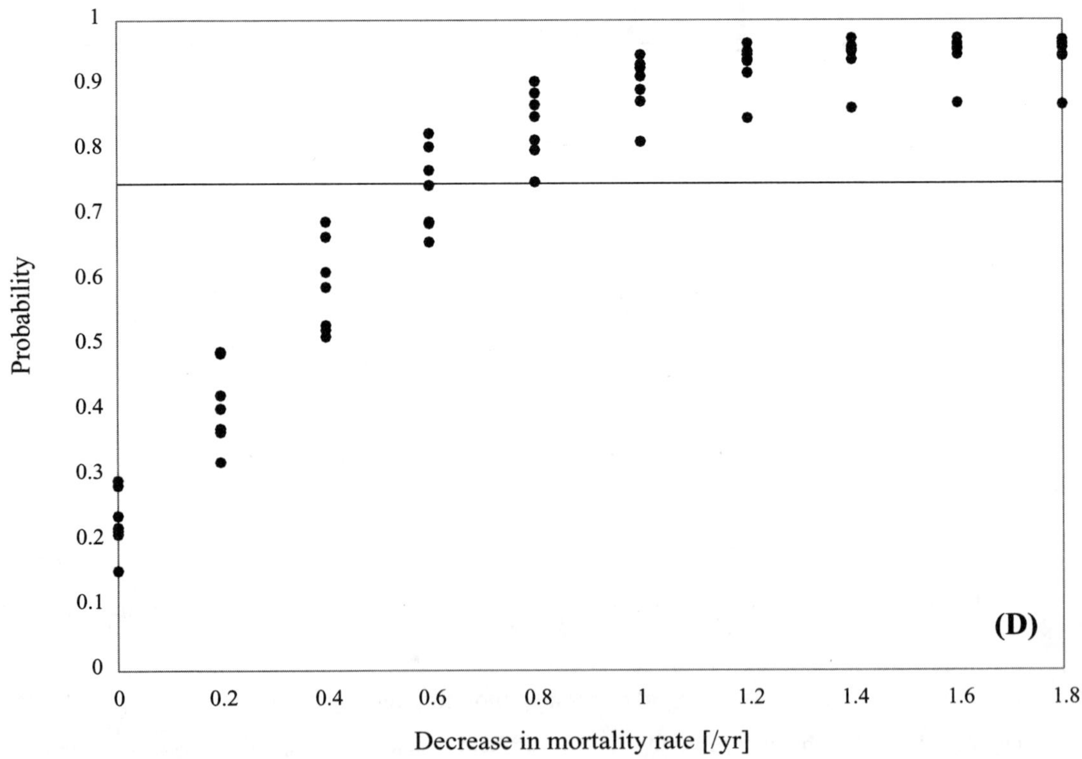

FIGURE 2. (*Continued*)

TABLE 8. Results of sensitivity analysis of jeopardy standard probability calculations to alternative models. See text for description of the models implemented.

	Model				
Probability average across stocks	1 Base case	2 Depensation $d = 1.0$	3 Spawner error	4 Age error	5 Year-effect $\delta = 0$
24-year survival threshold	0.59	0.57	0.61	0.60	0.45
with $\Delta = 0.4$	0.68	0.66	0.70	0.69	0.56
with $\Delta = 0.8$	0.76	0.74	0.78	0.78	0.67
100-year survival threshold	0.68	0.65	0.71	0.69	0.50
with $\Delta = 0.4$	0.85	0.84	0.88	0.86	0.76
with $\Delta = 0.8$	0.92	0.91	0.94	0.92	0.87
24-year recovery threshold	0.19	0.17	0.23	0.19	0.05
with $\Delta = 0.4$	0.49	0.46	0.56	0.49	0.26
with $\Delta = 0.8$	0.78	0.75	0.84	0.79	0.62
48-year recovery threshold	0.23	0.22	0.29	0.23	0.07
with $\Delta = 0.4$	0.59	0.57	0.68	0.59	0.38
with $\Delta = 0.8$	0.84	0.83	0.89	0.84	0.75

Discussion

A number of hydropower system management actions are contemplated to decrease future in-river passage mortality and to increase the chances of Snake River population survival and recovery. Without substantial action, results indicate that the chances are remote for recovery of the Snake River populations. The results on incremental changes in mortality can be carried directly to other analyses that measure the size of Δ one obtains by taking a specified management action. Thus, this paper serves as a kind of mapping system, in which alternative management actions, which have been quantified in terms of the Δ they induce, are mapped back to this paper and evaluated as to whether they achieve desired management goals.

A number of alternative actions and hypotheses about what needs to be fixed in the Columbia River have been suggested (Marmorek et al. 1998), such as breaching of four Snake River dams, improvements to spawning habitat, improvements to in-river passage technology, and reductions in potential disease-bearing hatchery fish. Those alternative hypotheses, along with others concerning the structure of the population model itself, required PATH to place the entire modeling concept developed in this paper within a broad context of a decision analysis (Marmorek and Peters 2001; Peters and Marmorek 2001). One feature of decision analysis is that a composite probability statement can be made by weighting results in accordance with probabilities associated with alternative hypotheses. Suppose for a given management action, action "A," that there are N alternative hypotheses about the benefit of that action and we label them as $\Delta_1, \ldots, \Delta_N$, and further suppose that we are provided with the probabilities of the validity of those hypotheses as $p_1, \ldots, p_N$. Denote by $r(\Delta_i)$, the probability that a given population, say the Imnaha stock, will achieve the 24 year survival standard given the ith hypothesis, is true. The probability that management action "A" will achieve the 24-year survival standard for the Imnaha stock equals $\Sigma p_i r(\Delta_i)$.

The method described in this paper can be generalized to deal with incorporation of more information into probability statements. The "prior" probabilities $p(\theta)$ specify information about parameters different from the actual data used in the likelihood function. For example, several studies recently have addressed the general issue of depensation in salmon (Myers et al. 1995; Hilborn and Liermann 1998; Liermann and Hilborn 1997). In principle, probability statements from those studies could replace my assumption of prior uniform distribution of log-transformed depensation.

Acknowledgments

I am grateful for the constructive suggestions made by several scientists involved in the PATH project but especially to those individuals involved in a series of working group meetings on modeling the life cycle of spring/summer chinook, including D. Marmorek, C. Peters, H. Schaller (who also provided a thorough review of this manuscript),

C. Paulsen, C. Petrosky, R. Hinrichsen, and C. Toole. I appreciate the invitation by J. Berkson to present this paper in the workshop. This paper is based on a report that was developed by the author as part of the PATH project.

Literature Cited

Beamesderfer, R. C. P., H. A. Schaller, M. P. Zimmerman, C. E. Petrosky, O. P. Langness, and L. LaVoy. 1997. Spawner-recruit data for spring and summer chinook salmon populations in Idaho, Oregon, and Washington. Section 2.2. Pages 1–78 *in* D. M. Marmorek and C. N. Peters, editors. Plan for analyzing and testing hypotheses (PATH): retrospective and prospective analyses of spring/summer chinook reviewed in FY1997. Vancouver, BC.

Deriso, R., D. Marmorek, and I. Parnell. In press. Retrospective patterns of differential mortality and common year effects experienced by spring chinook of the Columbia River. Canadian Journal of Fisheries and Aquatic Science.

Gelman, A., J. B. Carlin, H. S. Stern, and D. B. Rubin. 1995. Bayesian data analysis. Chapman and Hall, London.

Hilborn, R., and M. Liermann. 1998. Standing on the shoulders of giants: learning from experience. Reviews in Fish Biology and Fisheries 8:273–283.

Liermann, M., and R. Hilborn. 1997. Depensation in fish stocks: a hierarchic Bayesian meta-analysis. Canadian Journal of Fisheries and Aquatic Science 54:1976–1984.

Marmorek, D. R., C. N. Peters, J. J. Anderson, L. Barnthouse, L. Botsford, N. Bouwes, P. Budy, T. Cooney, R. Deriso, J. Geiselman, A. Giorgi, R. Hinrichsen, J. Hayes, M. Jones, L. Krasnow, O. P. Langness, D. Lee, C. McConnaha, C. M. Paulsen, R. Peterman, C. E. Petrosky, C. Pinney, H. Schaller, S. Smith, C. Toole, E. Weber, J. Williams, P. Wilson, and R. W. Zabel. 1998. Plan for analyzing and testing hypotheses (PATH): weight of evidence report. Compiled and edited by ESSA Technologies Ltd., Vancouver, BC.

Marmorek, D., and C. Peters. In press. Finding a path towards scientific collaboration: insights from the Columbia River Basin. Conservation Ecology.

Myers, R. A., N. J. Barrowman, J. A. Hutchings, and A. A. Rosenberg. 1995. Population dynamics of exploited fish stocks at low population levels. Science 269:1106–1108.

Peters, C., and D. Marmorek. 2001. Application of decision analysis to evaluate recovery actions for threatened Snake River spring and summer chinook salmon (*Oncorhynchus tshawytscha*). Canadian Journal of Fisheries and Aquatic Science in review. Manuscript available from C. Peters (4612 Briggs Rd., Vernon, British Columbia V1B 3J4).

Quinn, T. J. II., and R. B. Deriso. 1999. Quantitative fish dynamics. Oxford University Press, New York.

Schaller, H. A., C. E. Petrosky, and O. P. Langness, 1999. Contrasting patterns of productivity and survival rates for stream-type chinook salmon (*Oncorhynchus tshawytscha*) populations of the Snake and Columbia Rivers. Canadian Journal of Fisheries and Aquatic Science 56:1031–1045.

Appendix I. Spring/summer chinook "Jeopardy Standard" summary

C. Toole (National Marine Fisheries Service [NMFS], Portland, Oregon) provided a numerical interpretation of the jeopardy standards proposed by NMFS in recent years, through personal communication. I edited his interpretation and provide the edited version below. The standards have not been endorsed by NMFS or any other public agency to the author's knowledge.

1. Survival Standard
   a. Set threshold levels for each population. Estimates below were used by NMFS for the following stocks:

Population	Number of spawners annually
Bear Valley/Elk	300
Imnaha	300
Marsh Creek	150
Minam River	150
Poverty Flats	300
Sulphur Creek	150
Johnson Creek	150

   b. Using simulation models, project population levels over 24 years into the future and 100 years into the future.
   c. Determine likelihood that each population will be above its threshold level over each of the two time periods. This is determined from the cumulative distribution of all simulations encompassing the time period. For example, if 500 simulations each projected population levels for a 100-year period, the resulting distribution would consist of 50,000 values.
   d. Express probability in step c. as a proportion of probability of being above threshold during a historical period in which stocks were believed to be relatively healthy. NMFS did not define this historical period, but accepted model results based on all available years prior to 1976. Estimation of the historical probability follows the same process described in step c., except the simulation model is calibrated only to observations during the historical period.
   e. NMFS' jeopardy standard is that a "high percentage" of available populations must have a "high likelihood," relative to the historical

probability, of being above the threshold level over each time period. NMFS defined "high percentage" as 80% of available populations. NMFS did not define "high likelihood." C. Toole suggests that 70% be considered an approximation of this standard. Other PATH members have suggested reporting results for a range of probabilities between 70% and 95%.

2. Recovery Standard
   a. Set recovery population level. Relevant population recovery goal in a NMFS "Proposed Recovery Plan" is eight-year geometric mean of annual redd counts equivalent to 60% of the pre-1971 brood-year average redd counts. Although the NMFS draft recovery level is expressed as redd counts, analyses for the biological opinion converted these to estimates of number of spawners.
   b. Using simulation models, project population levels 24 and 48 years into the future.
   c. Determine likelihood that the 8-year geometric mean of each population will be above its recovery level in the 48th year of a simulation (i.e., geometric mean of years 41–48). This is determined from the cumulative distribution of all simulations. For example, if 500 simulations each project an 8-year geometric mean population level for the 48th year of the simulation, the resulting distribution would consist of 500 values. Also repeat (c) for the 24th year of the simulation.
   d. NMFS' jeopardy standard is that a "high percentage" of available populations just have a "moderate to high likelihood" of being above the recovery level within 48 years. NMFS defined "high percentage" as 80% of available populations. NMFS did not define "moderate to high likelihood." C. Toole suggests that 50% be considered an approximation of this standard and that results reported for a range of probabilities from 50% to 95%.

Appendix II

Posterior probability distributions were calculated for the 88 estimated parameters in the BSM model. They are listed below.

Highest posterior densities (analogous to quartiles)

Parameter		5%	25%	Median	75%	95%
a for area	1	6.40	6.71	7.09	7.41	7.66
a for area	2	6.99	7.39	7.81	8.13	8.38
a for area	3	8.80	9.32	9.84	10.24	10.56
a for area	4	7.54	7.90	8.35	8.75	9.07
a for area	5	7.95	8.36	8.84	9.24	9.56
a for area	6	9.02	9.52	10.06	10.50	10.85
a for area	7	8.90	9.74	10.34	10.97	11.78
a for area	8	8.39	9.24	9.83	10.44	11.26
a for area	9	9.18	9.95	10.61	11.22	12.20
a for area	10	8.59	9.39	10.06	10.64	11.67
a for area	11	8.07	8.92	9.52	10.12	11.01
a for area	12	8.78	9.63	10.22	10.85	11.62
a for area	13	7.82	8.68	9.27	9.86	10.75
δ for year	1952	−0.72	−0.01	0.49	1.00	1.73
δ for year	1953	−1.30	−0.59	−0.08	0.42	1.14
δ for year	1954	−1.42	−0.83	−0.43	−0.02	0.55
δ for year	1955	−0.74	−0.24	0.12	0.47	0.99
δ for year	1956	−1.24	−0.73	−0.39	−0.05	0.46

(*Continued*)

(*Continued*)

Parameter		Highest posterior densities (analogous to quartiles) 5%	25%	Median	75%	95%
δ for year	1957	−0.48	−0.10	0.16	0.42	0.81
δ for year	1958	0.07	0.43	0.68	0.92	1.28
δ for year	1959	0.17	0.38	0.53	0.69	0.91
δ for year	1960	0.12	0.34	0.50	0.65	0.87
δ for year	1961	0.07	0.29	0.44	0.59	0.81
δ for year	1962	−0.27	−0.05	0.10	0.25	0.47
δ for year	1963	−0.40	−0.18	−0.03	0.13	0.34
δ for year	1964	−0.69	−0.47	−0.32	−0.16	0.06
δ for year	1965	−0.01	0.22	0.37	0.52	0.74
δ for year	1966	−0.18	0.02	0.15	0.29	0.48
δ for year	1967	0.23	0.46	0.63	0.80	1.04
δ for year	1968	0.75	1.05	1.26	1.47	1.78
δ for year	1969	−0.22	0.07	0.27	0.47	0.76
δ for year	1970	−0.68	−0.39	−0.18	0.03	0.32
δ for year	1971	−0.59	−0.28	−0.08	0.12	0.40
δ for year	1972	−0.56	−0.27	−0.06	0.15	0.43
δ for year	1973	−0.38	−0.10	0.09	0.27	0.54
δ for year	1974	−0.70	−0.43	−0.24	−0.06	0.20
δ for year	1975	−0.88	−0.61	−0.42	−0.23	0.05
δ for year	1976	−1.06	−0.78	−0.58	−0.40	−0.12
δ for year	1977	−1.30	−1.02	−0.84	−0.65	−0.39
δ for year	1978	−0.94	−0.66	−0.48	−0.29	−0.02
δ for year	1979	−0.89	−0.63	−0.44	−0.26	0.01
δ for year	1980	−0.49	−0.22	−0.03	0.16	0.42
δ for year	1981	−0.83	−0.56	−0.37	−0.19	0.08
δ for year	1982	−0.45	−0.16	0.02	0.21	0.47
δ for year	1983	0.18	0.45	0.64	0.83	1.11
δ for year	1984	−0.17	0.10	0.29	0.47	0.74
δ for year	1985	−0.18	0.11	0.30	0.49	0.78
δ for year	1986	−0.71	−0.44	−0.25	−0.06	0.20
δ for year	1987	−0.95	−0.66	−0.46	−0.26	0.04
δ for year	1988	−0.23	0.07	0.28	0.48	0.78
δ for year	1989	−0.93	−0.65	−0.47	−0.28	−0.02
X dam effect		0.00	0.15	0.25	0.35	0.52
μ for year	1970	−0.39	0.14	0.51	0.87	1.40
μ for year	1971	0.75	1.27	1.65	2.02	2.54
μ for year	1972	1.32	1.85	2.23	2.60	3.12
μ for year	1973	−0.27	0.24	0.61	0.96	1.50
μ for year	1974	0.88	1.38	1.75	2.10	2.62
μ for year	1975	1.83	2.36	2.73	3.10	3.64
μ for year	1976	0.38	0.89	1.26	1.62	2.13
μ for year	1977	−0.09	0.42	0.77	1.13	1.63
μ for year	1978	0.62	1.12	1.48	1.84	2.34
μ for year	1979	0.35	0.87	1.23	1.60	2.11

(*Continued*)

(*Continued*)

Parameter		Highest posterior densities (analogous to quartiles)				
		5%	25%	Median	75%	95%
μ for year	1980	−0.70	−0.17	0.19	0.55	1.07
μ for year	1981	−0.71	−0.19	0.18	0.53	1.03
μ for year	1982	−0.28	0.26	0.61	0.97	1.49
μ for year	1983	−0.43	0.10	0.46	0.82	1.34
μ for year	1984	0.71	1.24	1.61	1.97	2.52
μ for year	1985	1.21	1.73	2.09	2.45	2.99
μ for year	1986	0.13	0.65	1.02	1.38	1.88
μ for year	1987	0.92	1.46	1.84	2.21	2.76
μ for year	1988	0.98	1.51	1.87	2.25	2.77
μ for year	1989	0.77	1.30	1.67	2.03	2.54
μ for year	1990	1.85	2.38	2.75	3.11	3.65
$ln\ \beta$ area	1	−6.10	−5.86	−5.56	−5.29	−5.08
$ln\ \beta$ area	2	−6.77	−6.29	−6.01	−5.68	−5.38
$ln\ \beta$ area	3	−8.01	−7.28	−6.95	−6.57	−6.15
$ln\ \beta$ area	4	−6.36	−6.17	−5.93	−5.62	−5.29
$ln\ \beta$ area	5	−7.23	−6.98	−6.72	−6.41	−5.98
$ln\ \beta$ area	6	−9.00	−8.23	−7.88	−7.44	−6.99
$ln\ \beta$ area	7	−7.95	−7.74	−7.47	−7.14	−6.65
$ln\ \beta$ area	8	−7.31	−7.11	−6.86	−6.54	−6.11
$ln\ \beta$ area	9	−8.69	−7.99	−7.68	−7.32	−6.79
$ln\ \beta$ area	10	−8.47	−7.46	−7.15	−6.82	−6.30
$ln\ \beta$ area	11	−7.12	−6.64	−6.40	−6.13	−5.67
$ln\ \beta$ area	12	−7.89	−7.66	−7.37	−6.98	−6.59
$ln\ \beta$ area	13	−7.21	−6.62	−6.35	−6.02	−5.65
$ln\ p$		−11.67	−10.29	−8.49	−6.67	−4.83
$ln\ V_{\varepsilon}$		−1.18	−1.10	−1.04	−0.96	−0.89

American Fisheries Society Symposium 27:157–169, 2002

Incorporating Uncertainty into Marine Mammal Management

DEBRA PALKA
National Marine Fisheries Service, 166 Water Street, Woods Hole, Massachusetts 02543, USA

Abstract.—Management schemes for marine mammals developed by the United States and International Whaling Commission (IWC) have sought to achieve their management objectives by developing control laws designed to calculate acceptable levels of human-caused mortality, while explicitly incorporating some types of uncertainty, and while being robust to other types of uncertainty. The United States developed the "potential biological removal" control law in managing commercial fisheries to reduce incidental catches of marine mammals. The IWC developed, but has not yet used the "catch limit algorithm" control law in managing commercial harvests of baleen whales. In both cases, to develop and test the control law, quantitative management objectives were specified, and only reliably and easily collected data were required. Then, given these specifications and requirements, simulations were used to define the control law and test its robustness to uncertainties in assumptions and data. Finally, to identify unforeseen uncertainties, the management schemes include rules and guidelines on reviews, monitoring programs, and data collection and analyses.

Introduction

Previous marine mammal stock assessments commonly generated a wide range of interpretations about stock status and needed management actions because of limited biological data or, when data were available, large uncertainties in those data. When there were conflicting scientific opinions on stock status, managers often had difficulty in deciding on appropriate management actions, thus, frequently resulting in no management action. The problem was not so much that the scientific process failed, but that the management system had not developed approaches to deal with uncertainties in marine mammal population dynamics data and stock assessment methods.

In this paper, the term "population" refers to a biologically separate group of animals, and "stock" refers to a management unit that may be part of, or all of a population. In theory, the goals of management schemes are to manage populations, so populations were used to develop the control laws. However, in practice, it is only possible to delineate stocks, so stocks are the actual management unit. The term "unknown stock identity" refers to a group of animals that could be part of, one complete, or more than one, population.

In an ideal world, scientists and managers would have information on the following: marine mammal stock structure, distribution, and abundance; natural and human-caused mortality rates; rates, sensitivities, and elasticities of increase; and extinction probabilities. This information would be available for circumstances as they are currently, as they were before any exploitation, and as they would be at maximum net productivity levels. In addition, the scientists and managers would have information on the short- and long-term population and stock effects of present and proposed management actions. For most marine mammals, all or part of this information is missing. In general, available biological data are sparse, and where they do exist, time series are usually short and/or incomplete; consequently, the effects of present and proposed management actions are uncertain.

Because of this background, two management schemes have been developed to manage and conserve marine mammals while explicitly accounting for some uncertainties. The International Whaling Commission (IWC) developed a management scheme for commercial harvesting of baleen whales, with the goal to conserve populations and obtain the highest possible continuing yield. The United States developed a scheme for managing incidental takes of marine mammals in commercial fisheries, with the goal to conserve marine mammal populations and to minimize changes in the fishery. Though the goals differ, a similar approach has been taken toward management. In each management scheme, a management decision rule, or control law, was adopted to determine the allowable harvest or take. Each control law was developed and tested using simulations. Each management scheme includes the chosen control law along with a monitoring program and guidelines on data collection and analyses.

In both cases, risks and benefits of potential control laws were investigated through simulations. The simulations involved the following: 1) specifying quantitative objectives and performance statistics; 2) generating a marine mammal population; 3) removing animals from the simulated population following the control law; 4) using predetermined life history parameters to regulate population growth; 5) sampling the simulated population to estimate abundance and removals; 6) cycling through steps 3–5 over many years; and 7) compiling performance statistics. To determine the effectiveness of a control law through simulation, objectives were compared with performance statistics from many scenarios, each of which differed in its model assumptions and/or in the types and magnitudes of uncertainty typically encountered during assessments. This process resulted in the development of control laws that incorporated some types of uncertainty, were robust toward other types of uncertainty, met management objectives, and used only data that could realistically be collected.

This paper reviews the background, objectives, and development of the IWC and U.S. marine mammal management schemes, with particular emphasis on how uncertainties were incorporated.

IWC Management Scheme

Background and objectives

The IWC is an intergovernmental body established in 1946 under the International Convention for the Regulation of Whaling to provide for the proper conservation of whale populations and make possible the orderly development of the whaling industry. In1975, the IWC adopted the "New Management Procedure" (NMP), the first attempt to manage commercial whaling of baleen whales using a scientific basis, with the goal to ensure sustainability (IWC 1977). The NMP specified that whale populations below their maximum sustainable yield level (MSYL) were to be protected, and catches from other populations would not cause the population to go below their MSYL. Thus, all populations would be managed to be at or above their MSYL. Two supplementary specifications were as follows: 1) populations subjected to stable harvest levels for a considerable period in the past would be allowed to be harvested at those levels in the absence of any definite evidence of decline; and 2) populations not previously subjected to significant exploitation would be limited to a harvest of 5% of the estimated population size.

The NMP ran into difficulties because necessary data were not available. For most stocks, there were no reliable estimates of MSYL relative to current or past abundance, and the relationship between current abundance and MSYL was unknown. When MSYL was estimated, annual updates often led to widely fluctuating catch limits, especially for stocks close to MSYL. In addition, the NMP had no guidelines on how, when, and what kinds of data should be collected, or on how to cope with uncertainties in stock identity, population size, and life history parameters. Moreover, the long-term consequences of the NMP were unknown.

Because of these difficulties, in 1986, the IWC implemented a moratorium on commercial whaling until knowledge of whale populations was improved and a more satisfactory approach to management was developed. As a result, a new management approach, called the "Revised Management Procedure" (RMP), was developed. An essential element of the RMP was the explicit specification of three quantitative objectives, which were agreed upon by all member nations: 1) catch limits should be as stable as possible; 2) catches should not be allowed on populations that are below 54% of their estimated carrying capacity, so that the risk of extirpation of the population is not seriously increased by exploitation; and 3) the highest possible continuing yield should be obtained from a population. The IWC agreed that the second objective had the highest priority. The final version of the RMP was accepted in 1994 (IWC 1994b).

The RMP (Figure 1) includes a catch limit algorithm (CLA) and rules addressing some potential uncertainties and variabilities. The CLA applies to a single stock when stock identity is certain. The rules address uncertainties in stock identity, recent abundance, and stock status, and variability in the sex ratio of the catch.

A "Revised Management Scheme" (RMS), which has not been adopted by the IWC, is designed to ensure that the RMP is implemented as planned (Figure 1). An important issue in implementation is the monitoring of harvests. Ideas proposed for the RMS include national and international observers aboard whaling vessels, DNA profiles of all whales killed, and a vessel monitoring system for satellite tracking of all whaling vessels. Besides monitoring harvests, the proposed RMS might provide data for retrospectively evaluating the efficiency and safety of the RMP, testing assumptions of the robustness trials, improving accuracy of the population models, monitoring the status of harvested stocks, and monitoring the humaneness of the harvest.

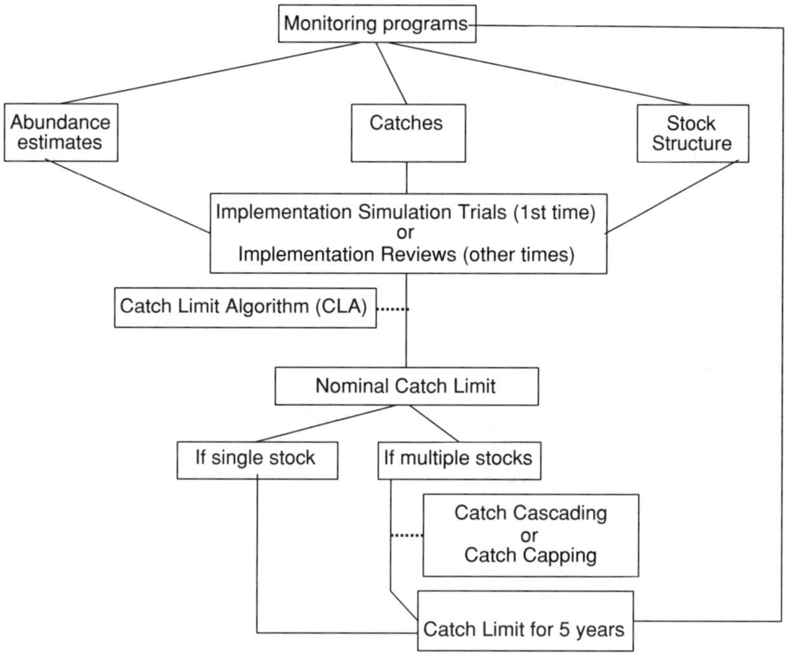

FIGURE 1. Schematic of management scheme proposed by the IWC. Adopted RMP includes items below, and not including, the monitoring programs. Proposed RMS would include RMP, several types of monitoring programs, and perhaps more.

Catch limit algorithm

To account for uncertainties, the IWC designed the CLA to calculate a catch limit, satisfy the three conflicting objectives, and be robust to past, present, and future uncertainties in abundance and harvest estimates, stock identity, and population dynamics. The IWC developed the CLA by examining the performance of five potential catch laws using simulations that cover a wide range of plausible scenarios (Kirkwood 1992).

The basic steps of a simulation were as follows. An initially unexploited whale population was created and historical catches were removed from the population. A full age-structured model governed dynamics of the simulated population and its response to catches. Abundance data were generated to have the same properties as data from abundance sighting surveys. For 100 simulated years, a control law was applied to the population. At the end of 100 years, population status and total catches were determined.

Initially, the control law was subjected to a set of mild tests, called base case trials. If these trials were 'passed', then it was subjected to a set of stronger tests for robustness to departures from assumptions, called robustness trials. Five statistics were used to evaluate performance of the control law: 1) distributions (median, upper, and lower percentiles) of the total catch over 100 years; 2) final population size after 100 years; 3) lowest population size during the 100 years; 4) continuing average catch over the final 10 years (i.e., a surrogate for the realized long-term yield); and 5) inter-annual variability in catch limits.

CLA trials.—Base case trials examined all combinations of the following assumptions in which harvest came from a single population with no uncertainty in stock structure: 1) at the onset of management, the population was either unexploited (i.e., a "development" case), reduced to 30% of its unexploited abundance (i.e., a "rehabilitation" case), or reduced to 60% of its unexploited abundance (i.e., a "sustainable" case); 2) MSY rates (i.e., MSY as a percentage of MSYL) for a population were either 1%, 4%, or 7%, with only the 1% MSY rate being used in the "sustainable" case; 3) abundance survey data were unbiased and available for the first year of management and for every fifth year after that.

Robustness trials investigated the effects on the five statistics when a control law was used in situations where the assumptions were departed from. Initially, the following seven trials were conducted, one at a time: 1) population dynamic parameters varied, with MSYL actually being 40–80% of the estimate, ages at maturity differing due to delayed density dependent effects, carrying capacity showing trends, and the MSY rate showing cyclic changes; 2) initial abundance was different than that examined in base case trials; 3) abundance survey data had upward and downward biases, and were collected at different frequencies; 4) relationships between CPUE data and true abundance differed, including no relationship at all; 5) catch histories prior to the onset of management were uncertain or inaccurate (including underestimation by half), and long periods of protection occurred before management started; 6) carrying capacity fluctuated (including reduction by half); and 7) episodic events (e.g., epidemics) occurred irregularly at which time the population was abruptly halved.

Subsequently, simultaneous departures of multiple assumptions were investigated. Two or more of the following parameters varied: 1) MSY rates, 2) initial depletion levels, 3) periods of protection before management, 4) biased sighting surveys, and 5) biased catch histories (Smith et al., in press). To maximize the power of detecting nonrobustness, a partial factorial design and the same set of random numbers were used to simulate 810 cases of 10 underlying population model scenarios. The performance statistics were the median final depletion and the difference between the 5th and 96th percentiles of the final depletion distribution.

Uncertainty in the relationship between IWC management areas and single biological populations was also a concern to the IWC. To investigate the robustness of a control law to multiple stock situations, the control law was subjected to trials that simulated coastal and pelagic whaling scenarios where stock identity was uncertain.

Coastal whaling trials examined a scenario where whales were taken by a coastal whaling station from a small portion of a management area. The management area was assumed to contain one or two biological populations. If two populations were present, then the boundary between them was assumed to be unknown.

Pelagic whaling trials examined a scenario based on Antarctic whales where the identity of stocks matched a variety of hypotheses, and the MSY rates and initial depletion levels varied. The various stock hypotheses included two stocks being managed as one, one stock being managed as two, stock ranges overlapping or shifting, and stock ranges overlapping and shifting.

CLA specification.—The algorithm that most satisfied all three of the RMP's quantitative objectives in all scenarios was the 'C' procedure; it was chosen as the CLA (IWC 1994a; Cooke 1999). The computer code to implement the 'C' procedure is complex and long; the FORTRAN code is nine pages long (IWC 1994a). As input, the 'C' procedure requires a time series of annual data on catches, absolute abundance estimates for each management area, and the variance–covariance matrix of those absolute abundance estimates. In general, if past catches are large compared with current abundance, then the 'C' procedure calculates a small catch limit, or even no catch. Otherwise, catches are a small fraction of the abundance estimates, typically less than 2%.

The 'C' procedure calculates a catch limit for each management area. Catch limits are based on an assessment where a nonage-structured Pella-Tomlinson type population dynamics model is fit to a time series of absolute abundance estimates using a Bayes-like approach, where at least one abundance estimate exists for an area. If no abundance estimate exists, then the catch limit for that area is zero. The 'C' procedure presumes a wide range of population status could have produced the observed time series of abundance and catches. From a random sample of this range, a distribution of potential catch limits is calculated. To ensure a catch limit certainty to one whale, the final catch limit was defined as approximately the 41st percentile of the distribution. The exact percentile is currently being refined (IWC 2001). By adding data from new abundance surveys (and new catches, if they exist), the distributions of possible population status and catch limit are gradually narrowed down. This generally results in catch limits increasing over time because additional abundance data generally reduce the uncertainty about population status.

CLA constraints.—Cooke (1995) came to three conclusions after the development and testing of the five potential control laws. First, regular direct surveys to estimate absolute abundance were a prerequisite for satisfactory management. Despite considerable efforts, no one was successful in finding a management procedure that worked well without such data. Second, when abundance estimates from regular surveys were available, benefits of other types of data were marginal. The most useful other type of data were total catches. Regular absolute abundance estimates from surveys were, thus, considered both necessary and sufficient for good management. Third, safe management could only be achieved by limiting catches to a small proportion of the absolute abundance. Procedures that allowed higher catches and relied on the detection of trends in abundance did not perform well.

After conducting base case and robustness trials, the IWC Scientific Committee concluded that the most important factor affecting the performance of a control law was stock identity (Cooke 1995). Moderately important factors were bias in abundance and catch estimates, and bias in the variance of these estimates. Least important factors were population dynamic parameters (e.g., age structure, reproduction, density dependence), environmental changes, and sudden events (e.g., epidemics and other mass mortalities).

Stock identity was the most important factor. When stock boundaries were poorly known, depleting or even extirpating a stock was possible even when the total population was not depleted. Uncertainty in stock identity was most critical when populations overlapped on their feeding grounds, and when the extent of the overlap varied between years. This is to some extent an artifact of the performance criteria selected. If the criterion was restated as conserve 'whales,' not conserve 'whale populations,' then stock identity would probably not have been as important.

Bias in abundance estimates was considered as only moderately important because it was not until bias was greater than 50% that the consequences to management became severe. A persistent bias was more serious than a temporary bias that affected only a few estimates. In addition, underestimated variances of abundance were more detrimental than overestimated variances. Cooke (1995) concluded that as survey methods improve, it is more important to use the improved methods to increase precision and decrease bias of current abundance estimates, than to maintain comparability with previous estimates.

Environmental changes and sudden events emerged as least important because catch limits were adjusted with each new abundance estimate, which would reflect effects of such environmental changes and sudden events. In addition, because the catch limit was a small proportion of the abundance estimate, the relative impact of harvesting, as a proportion of the population size, was always small. Even when harvesting in very poor environmental conditions, the impact due solely to harvesting was small, relative to the impact due to the environmental changes and sudden events.

When multiple assumptions were violated simultaneously in single-stock simulation trials, Smith et al. (in press) found that the 'C' procedure was robust, with two exceptions. When severely negatively biased historical catches and low estimated abundance variance were combined, the final whale population was lower than expected. When this scenario was combined with high levels of inter-annual abundance variance, the variability of the final depletion level was unacceptably high. Smith et al. (in press) concluded that because scenarios remain where the CLA may not be robust, and hence not achieve the desired objectives, it is essential that the CLA be tested on each whale stock to which the CLA might be applied.

Rules on uncertainty and variability

As noted earlier, the RMP includes the CLA and the rules addressing some potential uncertainties and variabilities. This section provides details on the rules.

Uncertainty in stock identity.—The rules for addressing uncertainty in stock identity are based on the nature and extent of the area in which the whales are to be harvested. A "Small Area" is defined as an area (possible disjoint areas) small enough to contain whales from only one population, or, if whales from two or more populations are present, then the relative catches from those populations would be in proportion to the relative abundance of the populations. A "Combination Area" is defined as a disjoint union of two or more Small Areas. A "Medium Area" is defined as the area that encompasses all of the known or suspected range of the population(s) of interest, and thus, all of the Small Areas for that population. A "Large Area" is defined as a major ocean area that excludes whales from other population of the same species.

When whales are to be harvested from only a single Small Area, then the calculated (i.e., nominal) catch limit for the Small Area becomes the realized catch limit for that area. When whales are to be harvested from a Combination Area, then the nominal catch limit for the Combination Area may be modified by "catch-cascading," where the nominal catch limit for the Combination Area becomes

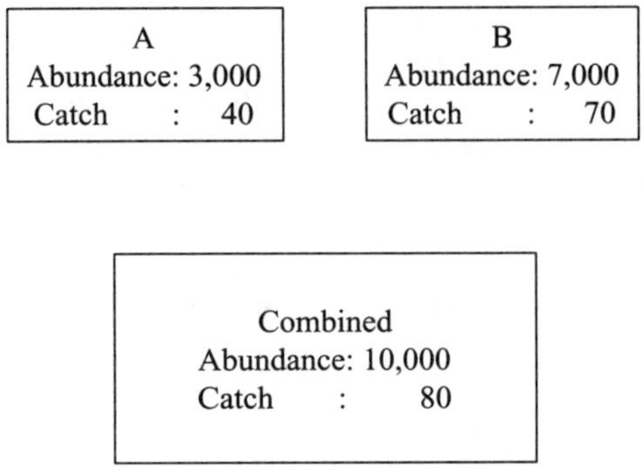

	Catch Limit		
Rule	A	B	Combined
Cascading	24=(3000/10000) . 80	56	80
Capping	39=(40/110) . 80	51	80

FIGURE 2. Example illustrating catch-cascading and catch-capping rules to two Small Areas, A and B. Population size for Small Area A is 3,000, and for Small Area B is 7,000. Catch limit for Small Area A is 40 and for Small Area B is 70, when the CLA is applies to each Small Area separately. Catch limit for the combined area is 80, when the CLA is applied to the combined area.

the realized catch limit, but is distributed among the Combination Area's component Small Areas in proportion to the relative abundance estimates of population in the Small Areas (Figure 2). When whales are to be harvested from either a Medium or Large Area, then the nominal catch limit for the Medium or Large Area may be modified by "Catch-capping," where the nominal catch limit of the Medium or Large Area may become the realized catch limit, but is distributed among the Medium or Large Area's component Small Areas in proportion to the relative nominal catch limits for the Small Areas (Figure 2).

If the sum of the nominal catch limits calculated for the Small Areas that make up a larger area (Combination, Medium, or Large Area) exceeds the nominal catch limit calculated for the Combination, Medium, or Large Area, then the realized catch limit in each Small Area is the nominal catch limit from the larger area that is scaled so that the sum of the Small Area catch limits is equal to the nominal catch limit for the Combination, Medium, or Large Area.

Variability in sex ratio of catches.—The CLA was developed without consideration of the sex of harvested whales. When the proportion, P_f of female whales in the catch from the most recent five years exceeds 50% of the catch taken from a Small Area, then the catch limit for that Small Area should be adjusted downwards by the ratio $0.5/P_f$.

This adjustment should be waived if it is agreed that the data from the most recent five year period are too limited to provide a useful indication of the expected sex ratio of future catches.

Uncertainty in recent abundance.—The catch limit for a Small Area is reduced when the time series of absolute abundance estimates used in the CLA does not include an estimate within the past eight years. In each succeeding year beyond eight years in which there is no absolute abundance estimates, the catch limit is reduced by 20% of the unadjusted nominal catch limit. This rule ensures that catch limits are reduced linearly to zero after 13 years without an absolute abundance estimate.

An eight-year period was selected to trigger the phase-out because after about eight years, the population status cannot be predicted reliably. However, the IWC realized that the eight-year criterion was arbitrary because the performance of the CLA did not degrade excessively for intervals longer than eight years.

Uncertainty in stock status.—Before the CLA can be applied to a stock (or stocks), "Implementation Simulation Trials" must be conducted, and after the CLA is applied, "Implementation Reviews" are required. Implementation Simulation Trials involve the delineation of Small Areas and, as appropriate, Medium and Large Areas, as well as the selection of Catch-cascading or Catch-capping, and should consider all available biological, stock identification, and operational data for the particular stock(s) of interest.

Implementation reviews investigate all prior data from the stock(s), and incorporate new data as appropriate to decide if management should be changed; for example, to redefine Small Areas or change the selection of Catch-cascading or Catch-capping. These reviews are scheduled for generally not more than five years after the completion of the previous Implementation Simulation trial or review. Such reviews can be carried out more frequently if new important information becomes available (e.g., new stock structure data).

U.S. Management Scheme

Background and objectives

Under the 1972 U.S. Marine Mammal Protection Act (MMPA), the U.S. National Marine Fisheries Service (NMFS) is responsible for the management and conservation of populations of whales, dolphins, porpoises, seals, sea lions and fur seals that reside in U.S. waters, and the U.S. Fish and Wildlife Service (FWS) is responsible for manatees, walruses, sea otters, polar bears, and dugongs that reside in U.S. waters. The primary objective of the MMPA is to maintain the health and stability of the marine ecosystem by accomplishing the following: 1) maintain a marine mammal population level above its optimum sustainable population (OSP) level, and, at or above the level at which the population ceases to be a significant, functioning element in its ecosystem; 2) restore populations that have been reduced below these levels; 3) approach a zero mortality incidental catch rate of marine mammals by commercial fisheries; and 4) minimize unnecessary disruption of fishing activities. OSP is defined as the number of animals at the maximum net productivity level (MNPL) of the population, keeping in mind the carrying capacity of the habitat and the health of the ecosystem.

After the MMPA was authorized, various attempts were made to define and estimate OSP for cetacean populations (i.e., whales, dolphins, and porpoises). However, just as the IWC could not parameterize the biological models in the NMP due to lack of data, the U.S. scientists could not define and estimate the biological parameter OSP due to lack of data. This failure resulted in an inadequate enforcement of the MMPA. In an attempt to rectify this situation, the U.S. implemented a five-year moratorium on the MMPA's regulations to allow collection of necessary data, and to gain a better understanding of fishery-marine mammal interactions, at the same time that fishing activities continued.

In 1994, amendments to the MMPA defined a more practical management scheme (Figure 3) based on the principle that assessments should precede the use of resources, and that managers should consider the possible consequences of uncertainty, and act accordingly (Mangel et al. 1996). Similar to the IWC's RMS, the MMPA mandates a feedback-loop management scheme. The MMPA management scheme provides checks and balances to address, or at least identify, unforeseen uncertainties, and also address nonscientific issues such as, developing regulations to reduce incidental marine mammal catches in fishing activities to a sustainable level. Specifically, the MMPA management scheme includes the following: 1) periodically assess marine mammal status using the potential biological removal control law, 2) annually review levels of interaction between commercial fisheries and marine mammals, 3) periodically review scientific aspects by independent Scientific Review Groups, and 4) when needed, establish Take Reduction Teams to develop and monitor Take Reduction Plans that reduce fishery incidental catches.

Potential biological removal

Central to this management scheme is marine mammal stock assessment reports. These reports are mandated to contain the following information: descriptions of stock distribution; estimates of abundance, growth rate, and annual human-caused mortality; descriptions of fisheries that interact with the stock; and statements summarizing the status of each stock. Assessments must be updated and reviewed, at least every three years. Assessments of strategic stocks are required to be reviewed annually.

Status is determined by the difference between human-caused mortality levels and potential biological removal (PBR) levels. The PBR level is defined as the maximum number of animals, not including natural mortalities, that may annually be removed from a marine mammal stock, while allowing

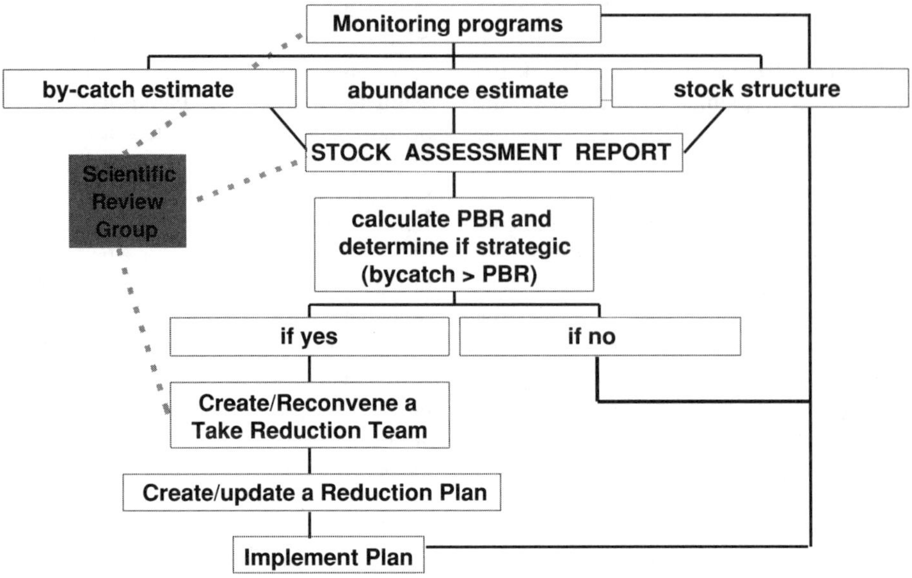

FIGURE 3. Schematic of management scheme mandated by the U.S. Marine Mammal Protection Act (MMPA).

that stock to reach or maintain its OSP level. The PBR is essentially a control law similar to the IWC's CLA. If a stock's human-caused mortality is greater than its PBR, then the stock status is considered 'strategic'. A stock is also considered strategic, if it is listed as threatened or endangered under the Endangered Species Act, as depleted under the MMPA, or is declining, and within the foreseeable future is likely to be listed as a threatened species under the Endangered Species Act.

To achieve the MMPA goal of maintaining populations above OSP, PBR is defined as

$$PBR = N_{\min} \cdot \frac{1}{2} R_{\max} \cdot F_R, \tag{1}$$

where $N_{\min}$ is the minimum population estimate of the stock, $R_{\max}$ is the maximum theoretical or estimated net productivity rate of the stock when the stock size is small, and F_R is a recovery factor between 0.1 and 1.0. The minimum population estimate is defined as the number of animals in a stock, based on the best available scientific information on abundance, and also on the precision and variability of the abundance estimate. This definition provides a reasonable assurance that the true stock size is equal to or greater than the minimum population estimate. Net productivity rate is defined as the annual per capita net rate of increase in the stock size (i.e., additions due to reproduction, less losses due to mortality). The intent of F_R is to compensate for effects of undefined uncertainties on the recovery of a stock to its OSP level, and to ensure the time to recovery is not significantly increased, especially for stocks listed as endangered, threatened, and depleted.

PBR trials.—To design practical definitions of the three components of PBR, and to ensure management objectives would be met when plausible uncertainties exist, U.S. scientists used simulations of population dynamics and robustness trials (Taylor 1993; Taylor and DeMaster 1993; Wade 1998). This process was accomplished in three steps: define objectives, conduct base case trials, and conduct robustness trials.

The first step was to define quantitative objectives and corresponding performance criteria. Objective one was populations should be maintained above MNPL (50% of carrying capacity (K)). To achieve this objective and to measure long-term performance, $N_{\min}$ was specified such that, if a population started at $N_{\min}$, experienced human-caused mortalities equal to PBR, and F_R equaled to 1.0 for 100 years, then the population size at the end of 100 years was above MNPL, with 95% probability. In addition, to measure short-term performance, $N_{\min}$ was specified such that, if a population started at MNPL, then it was still at or above MNPL after 20 years, with 95% probability.

Objective two was a population should recover to a level close to its carrying capacity, or preexploitation population level (90% of K). To achieve this objective, F_R was defined such that, if a population started at 0.05 K and experienced human-caused mortalities equal to PBR for 200 years, then the population size would equilibrate to a level at about K (0.9 K), with 95% probability.

Objective three was a population, known to be at a low level relative to its preexploitation level (0.05 K), should recover at a rate close to its maximum rate. To achieve this objective, F_R was defined such that, if a population started at 0.05 K and experienced human-caused mortalities equal to PBR, then the time it took to recovery to NMPL (0.5 K) was not delayed by more than 10% (with 95% probability), as compared with the recovery time of a population with no human-caused mortalities.

Step two of the PBR trials was to conduct base case simulations to define $N_{\min}$ and F_R to meet the performance criteria. In these simulation trials, marine mammal population dynamics were modeled with a discrete form of a nonage- or sex-specific generalized logistic equation that incorporated population size, maximum net recruitment rate, preexploitation population size (i.e., carrying capacity), and a shape parameter that controlled the amount of nonlinearity in the density-dependent response of the net recruitment rate.

In each base case simulation trial, carrying capacity was set at 10,000, and MNPL at 0.5 K. The basic steps of a simulation trial were as follows. Starting in year zero, the true population size was estimated using the population dynamics model. Next, starting in year one, for every fourth year, an abundance estimate was randomly chosen, and a new PBR value was estimated. Finally, annual human-caused mortalities, equal to PBR, were subtracted from the current population size. This process was repeated for a specified number of years (20, 100 or 200 years) and final population sizes were recorded. Each trial was repeated 2,000 times, generating a distribution of final population sizes.

Abundance estimates were assumed to have a lognormal distribution, with the mean equal to the true population size, and the coefficient of variation equal to either 0.2 or 0.8. Incidental fisheries mortalities were assumed to have a Gaussian random distribution, with the mean equal to PBR, and the coefficient of variation equal to 0.30.

The first component of PBR, $N_{\min}$, was calculated as the lower percentile of a lognormal distribution:

$$N_{\min} = \frac{N_{\text{best}}}{\exp\left(z\sqrt{\ln(1 + \text{CV}(N_{\text{best}})^2)}\right)}, \quad (2)$$

where N_{best} was the best estimate of the current population size, $\text{CV}(N_{\text{best}})$ was the coefficient of variation of N_{best}, and z was the standard normal variate (i.e., equals to 1.96 for the 2.5th percentile, 1.645 for the 5th, 1.282 for the 10th, 0.842 for the 20th, etc.). The choice of z, which affects the level of confidence in $N_{\min}$, was solved to meet objective one.

The second component of PBR, $R_{\max}$, has been directed estimated for only a few marine mammal populations. Survival rates have not been directly estimated because of a lack of biological samples, and a lack of a long series of accurate abundance estimates. Accordingly, plausible default values for R_{max} were defined by examining all available data from all stocks (Wade 1998). The default values were set to 0.04 for cetaceans, and 0.12 for pinnipeds. In base case simulations, the value of the true $R_{\max}$ used in the population dynamics model was defined as the $R_{\max}$ used in the PBR calculation.

The third component of PBR, F_R, was initially assumed to be unity when solving for z in $N_{\min}$. Subsequently, F_R was solved to meet objectives two and three.

Step three in the PBR trials was to conduct robustness trials to evaluate if the definitions defined in the base case trials (step two) were still adequate when assumptions were relaxed and plausible uncertainties were included. Eight robustness trials were conducted (Wade 1998): 1) estimated human-caused mortality was equal to $1/2$ actual human-caused mortality; 2) estimated abundance was twice actual abundance; 3) estimated $R_{\max}$ was twice actual $R_{\max}$; 4) estimated CV of abundance was less than the actual CV (estimated CV of 0.2 when actually 0.8 and estimated CV of 0.8 when actually 1.6); 5) estimated CV of human-caused mortality was $1/4$ actual CV; 6) abundance surveys were conducted every 8 years, instead of the default 4 years; 7) MNPL was set to 0.45 K, rather than the assumed 0.5 K; and 8) bias was added to the human-caused mortality estimate, and MNPL was equal to 0.7 K.

These trials were considered to represent plausible 'worst-case scenarios'. Trial six was later expanded to find the optimum time interval between abundance surveys, given a range of precision in the abundance and human-caused mortality estimates (Wade and DeMaster 1999).

PBR specifications.—After conducting the base case and robustness trials, the MMPA objectives were met by defining the components of PBR as following (Wade 1998):

$N_{\min}$ = 1) the 20th percentile of a lognormal distribution of the abundance estimate (which is equivalent to the lower limit of a 60% 2-tailed confidence interval), or 2) a direct count of animals, such as a count of hauled-out seals.

R_{max} =1) 0.04 for cetaceans, and 0.12 for pinnipeds and sea otters (default values), or 2) a reliable stock specific estimate from a peer-reviewed journal, or from a review group, such as a Scientific Review Group or the Scientific Committee of the IWC.

F_R = 0.1 for stocks listed as endangered, and 0.5 for stocks listed as depleted, threatened, or of unknown status (default values). The value of F_R should be adjusted depending on the coefficient of variation of the mortality estimate (discussed later).

Guidelines on uncertainty and variability

To account for other types of uncertainty and to maintain quality data, additional guidelines were developed (Barlow et al. 1995; Wade and Angliss 1997). These guidelines are discussed below.

Uncertainty in abundance estimates.—Because only one value for N_{min} is required to calculate PBR, information from a time series of abundance estimates would be underutilized if previous abundance estimates were ignored. To use time series information, and to decrease uncertainty in the current abundance estimate, three guidelines were developed.

The first guideline is if appropriate, a weighted average of abundance estimates from an eight-year period should be used for N_{best}, where the weight for each abundance estimate is its variance. This guideline is considered appropriate when there are no obvious trends in abundance, estimates are for the same region, and other known factors have not drastically affected the status of the stock.

The second guideline is if a stock is declining, then F_R should be reduced. The value of F_R should account for the magnitude and duration of the decline, although no specific guidelines have been provided.

The third guideline is, if the current abundance estimate is nine years or older, then PBR should be considered unknown, not zero. In these circumstances, a decision on the stock status should be made on a case-by-case basis.

Uncertainty in incidental catch estimates.—Because only one value of the human-caused mortality estimate is required to be compared with PBR, information from a time series of estimated human-caused mortalities would be underutilized. To use time series information, and account for uncertainties in the mortality estimates, two guidelines were developed.

The first guideline is, if appropriate, a five-year unweighted average of annual mortality estimate should be calculated. This is considered appropriate if the fishery has not changed significantly within the five-year period.

The second guideline is, when the coefficient of variation (CV) of the mortality estimate is high (>0.3), F_R should be reduced. The value of F_R should be 0.48 when the CV of the mortality estimate is 0.3–0.6; 0.45 when the CV is 0.6–0.8; and 0.40 when the CV is greater than 0.8. The reduced value of F_R is necessary because F_R was defined in the simulations when the CV of estimated mortalities was assumed to be less than or equal to 0.3.

Uncertainty in sex composition of human-caused mortalities.—In situations where human-caused mortalities are predominately (>50%) females, F_R should be reduced to compensate for the greater impact of this mortality on the stock. However, the guidelines have not specified how large the reduction should be.

Uncertainty in other factors.—The intent of the recovery factor is to compensate for uncertainties not accounted for in other components of PBR. This provides flexibility to appropriately adjust the recovery factor to meet MMPA objectives. For example, if human-caused mortality estimates are unbiased because of high observer coverage, then it may be appropriate to increase the recovery factor to reflect this greater certainty. However, no specific guidelines exist on the appropriate value of F_R. For stocks of unknown status, recovery factors of one are suggested to be reserved for cases where estimates of N_{min}, R_{max}, and human-caused mortality are all unbiased, and where stock identity is unequivocal.

Uncertainty in stock identity.—Stock identity for most marine mammals in U.S. waters is uncertain. The robustness trials did not explicitly include uncertainty in stock identity or multi-stock scenarios. To address potential problems due to uncertain stock identity, a general guideline is: "For the purposes of management under the MMPA, a stock is recognized as being a management unit that identifies a demographically isolated biological population. In the absence of adequate information on stock structure and fisheries mortality, a species' range within an ocean should be divided into stocks that represent defensible management units..."

Classifying interactions between fisheries and marine mammals

The MMPA mandates annual classification of all U.S. commercial fisheries into category I, II or III, based on whether the fishery has a frequent, occasional, or remote likelihood, respectively, of marine mammal incidental mortality and serious injury. Categorization is built on a two-tiered, stock-specific approach that first addresses the total impact of all fisheries on each marine mammal stock, and then addresses the impact of individual fisheries on each stock. The tiers are:

Tier 1: If total annual mortality and serious injury across all fisheries that interact with a stock is less than or equal to 10% of PBR for the stock, then all fisheries interacting with this stock should be placed in Category III. Otherwise, the fisheries should be subject to Tier 2 classification rules.

Tier 2—Category I: Annual mortality and serious injury of a stock in a given fishery is greater than or equal to 50% of the PBR level.

Tier 2—Category II: Annual mortality and serious injury in a given fishery is greater than 1% and less than 50% of the PBR level.

Tier 2—Category III: Annual mortality and serious injury in a given fishery is less than or equal to 1% of the PBR level.

The MMPA mandates that vessels in fisheries with frequent or occasional marine mammal incidental mortality and serious injury levels (Categories I or II) must register, and must carry an observer, if requested by NMFS. Observers collect data used to estimate total incidental catch, update the classification of the fishery, and design/monitor a Take Reduction Plan.

Independent scientific reviews

The MMPA mandates the establishment of three independent regional Scientific Review groups representing Alaska, the Pacific Coast (including Hawaii), and the Atlantic Coast (including the Gulf of Mexico). These groups periodically review scientific aspects. Responsibilities for these groups include: providing advice on stock assessment reports and methods to reduce incidental catches in fishing operations; identifying uncertainties and research needs; and making recommendations on scientific issues.

Regulations to reduce incidental catches

When a stock is classified as strategic because fishery-caused mortality estimates are greater than PBR, the MMPA mandates a Take Reduction team be established to prepare a draft Take Reduction plan. The Team is composed of experts in marine mammal biology, fishery practices, management, and conservation of marine mammals. Membership may include representatives of all stakeholders, such as representatives from state and federal agencies, regional fishery management councils, interstate fisheries commissions, academic and scientific organizations, environmental groups, fishery groups that incidentally catch the stock, Alaska native organizations, and Indian tribal organizations.

Ideally, the team develops a Take Reduction plan that is a consensus agreement. However, even if the plan is not a consensus, NMFS must publish in the *U.S. Federal Register* a plan that reduces fishery-caused mortality to a level below PBR, and a discussion on the differences between the published plan and the plan developed by the team. Following public review and comments, a final plan is published and implemented. After that, the team(s) that developed the plan periodically meets to monitor the success of the plan and recommend modifications, if needed.

Comparison of the Two Management Schemes

In the past, scientists and managers managed marine mammal stocks using detailed biological assessments. The IWC implemented the NMP, which depended on estimating parameters in detailed biological models, and the United States implemented the 1972 version of the MMPA, which depended on estimating the biological parameter OSP. These attempts were not successful because for most stocks, the necessary data to estimate the biological parameters were not available. For stocks where assessments were completed, a wide range of opinions often ensued concerning the interpretation of the assessments and the need for management measures. As a result, in some cases, no management actions were taken. In other cases, such as those dealt with by the IWC, large inter-annual changes in the allowable catch were calculated. And, in other cases where management actions were taken, long-term consequences of the management actions were unknown.

Because of this history, both the IWC and United States undertook another approach to define their management schemes. These approaches included specifying quantitative objectives and performance statistics; conducting simulations to define and test control laws, using only data and types of uncertainties that were reliably collected; investigating the robustness of control laws to different types of uncertainties; determining long-term effects of the control law; periodically reviewing and updating data to monitor control law performance and identify uncertainties (a feedback-loop management scheme); and, finally, developing guidelines for data collection and analysis methods.

There are numerous advantages to the simulation approach. Simulations allow many tests to be conducted quickly, and long-term consequences of control laws to be evaluated. To determine precisely how a control law is performing, the status of a simulated population can, after any time period, be precisely determined (Cooke 1999). In addition, and most importantly to the marine mammal populations, extinction of a simulated population is of no consequence. Disadvantages of the simulation approach are objectives, plausible parameter estimates, scenarios to test, and performance statistics should represent reality, and should be agreed upon by all stakeholders. In addition, interpretation of results may be difficult, especially, when there are conflicting objectives.

The feedback-loop strategy is used at two levels in both the IWC and U.S. management schemes. On one level, within the control law, new data can be incorporated to reduce the uncertainty in the catch limits. On another level, within the management scheme, a system of checks and balances to identify unforeseen uncertainties include periodically reviewing the assessment, collecting data other than that required by the control law, monitoring catches, and educating the parties catching the marine mammals.

Both management schemes require only periodically collecting new data. This allows data collection to be logistically, and economically feasible. Simulations were used to determine the most appropriate schedule for data collection. It was found that annual updates were not necessary, updates every four or five years were sufficient.

Both management schemes also contain guidelines on data collection and analyses. These guidelines are expected to evolve as methods are improved, and new questions and uncertainties are encountered.

The general approach of incorporating uncertainties via simulations was similar in both cases. However, the approach taken by the IWC was more comparative and comprehensive. The IWC developed and compared five control laws that used different input data and algorithms to calculate a catch limit. This comparison allowed the IWC to define appropriate input data and methods to incorporate some types of uncertainties, and yet still satisfy management objectives. Concerns were raised that the assumptions made within the simulation model dictated the success of a control law. To address these concerns, hundreds of base case and robustness trials were conducted, which modified model assumptions, and incorporated many different scenarios. Developers of the MMPA did not use such an exhaustive approach, perhaps because the IWC results were already available, and the IWC's conclusions could be adapted without repeating all the work.

Both schemes incorporate some types of uncertainty into the control law, and other types into associated rules and guidelines, though details differed. One obvious difference between the control laws developed by the IWC and the United States is the level of complexity, as illustrated by comparing the length of the IWC's CLA (nine pages of FORTRAN code) to the MMPA's PBR (equations 1 and 2). The PBR equation is easy to understand, while the CLA's algorithm is not. The CLA is complex because it is designed to use information from a series of abundance and catch estimates, and to explicitly incorporate the uncertainties of these estimates. Thus, as the time series grows, catch limits became more precise. This is because, as the time series grows, the stock status is known more precisely; and thus, the catch limit is also known more precisely. The PBR equation does not explicitly use information and uncertainty from a series of abundance and mortality estimates. However, because time series of marine mammals in U.S. waters are now becoming sufficiently lengthy, guidelines were developed to utilize this information.

Uncertainty in stock structure is handled differently within the two schemes. The most important factor affecting performance of the IWC's CLA is stock identity (Cooke 1995). Because of this, many multispecies robustness trials were conducted, and complex rules were developed to allocate catch limits in such way to compensate for potential unknown multispecies relationships. This level of testing was not conducted for PBR, and rules regarding multi-stocks were not specified in the MMPA guidelines.

The IWC and MMPA control laws also differed with respect to flexibility. The CLA is rigid, and like a black-box, when abundance and catch data of baleen whales are entered, a catch limit is outputted. The IWC does not recommend the CLA algorithm be redefined or modified. This is deemed not necessary because the CLA was demonstrated to be robust to many types of uncertainties. The MMPA does not follow this philosophy. The MMPA is specifically written to allow values of any or all of the three components of PBR to be changed, if scientifically justifiable. This flexibility appears to be necessary for two reasons. One, PBR was not exhaustively tested for robustness to many types, and combinations of uncertainties and invalid assumptions; thus flexibility may be needed in some specific cases. And, two, perhaps more importantly, PBR must be applied to all stocks of whales, dolphins, porpoises, seals, sea lions, sea otters, polar bears, and manatees that reside in U.S. waters. It does not seem possible for all those species to be managed safely by one rigid control law, especially without more extensive testing.

Conclusion

Using simulations to develop and test control laws to estimate catch limits appears to be an appropriate approach to incorporate some types of uncertainty, and determine robustness to other types of uncertainty. Moreover, to account for known uncertainties not already addressed, the IWC and MMPA management schemes include guidelines and rules on data collection and analyses, and to identify unforeseen uncertainties, the management schemes include periodic reviews and monitoring plans.

Acknowledgments

I would like to thank Tim Smith and Philip Hammond for the many hours of discussions about these management schemes. In addition, I appreciate the comments on previous drafts that Fred Serchuk and Jon Gibson provided.

References

Barlow, J., S. L. Swartz, T. C. Eagle, and P. R. Wade. 1995. U.S. marine mammal stock assessments: guidelines for preparation, background, and a summary of the 1995 assessments. U.S. Department of Commerce, NOAA Technical Memorandum NMFS-OPR-6.

Cooke, J. G. 1995. The International Whaling Commission's revised management procedure as an example of a new approach to fishery management. Pages 647–658 *in* A. S. Blix, L. Walløe, and Ø. Ulltang, editors. Whales, seals, fish and man. Elsevier B.V., Amsterdam, The Netherlands.

Cooke, J. G. 1999. Improvement of fishery-management advice through simulation testing of harvest algorithms. ICES Journal of Marine Science 56:797–810.

International Whaling Commission (IWC). 1977. Chairman's Report of 27th Meeting. Report of the International Whaling Commission 27:6–15.

IWC. 1994a. A programme to implement the Catch Limit Algorithm (Annex I). Report of the International Whaling Commission 44:153–167.

IWC. 1994b. The Revised Management Procedure (RMP) for baleen whales (Annex H). Report of the International Whaling Commission 44:145–152.

IWC. 2001. Chairman's Report of 52nd Meeting. The Journal of Cetacean Research, and Management 3 (supplement): 1–76.

Kirkwood, G. P. 1992. Background to the development of Revised Management Procedures (Annex I). Report of the International Whaling Commission 42:236–243.

Mangel, M., L. M. Talbot, G. K. Meffe, M. .T. Agardy, D. L. Alverson, J. Barlow, D. B. Botkin, G. Budowski, T. Clark, J. Cooke, R. H. Crozier, P. K. Dayton, D. L. Elder, C. W. Fowler, S. Funtowicz, J. Giske, R. J. Hofman, S. J. Holt, S. R. Kellert, L. A. Kimball, D. Ludwig, K. Magnusson, B. S. Malayang III, C. Mann, E. A. Norse, S. P. Northridge, W. F. Perrin, C. Perrings, R. M. Peterman, G. B. Rabb, H. A, Regier, J. E. Reynolds III, K. Sherman, M. P. Sissenwine, T. D. Smith, A. Starfield, R. J. Taylor, M. F. Tillman, C. Toft, J. R. Twiss, Jr., J. Wilen, and T. P. Young. 1996. Principles for the conservation of wild living resources. Ecological Applications 6:338–362.

Smith, T., T. Schweder, C. Allison, D. Palka, A. Punt, and D. Butterworth. In press. Effects of combinations of uncertainties on the risk performance of the IWC's commercial whaling management procedure. Special Issue of Cetacean Research and Management Journal.

Taylor, B. L. 1993. "Best" abundance estimates and best management: why they are not the same. U.S. Department of Commerce, NOAA Technical Memorandum NMFS-SWFSC-188. Available from SWFSC, P.O. Box 271, La Jolla, California 92038.

Taylor, B. L., and D. P. DeMaster. 1993. Implications of non-linear density dependence. Marine Mammal Science 9: 360–371.

Wade, P. R., and D. P. DeMaster. 1999. Determining the optimum interval for abundance surveys. Pages 53–66 *in* G. W. Garner, S. C. Amstrup, J. L. Laake, B. F. J. Manly, L. L. McDonald, and D. G. Robertson, editors. Marine mammal survey and assessment methods. A. A. Balkema, Rotterdam, The Netherlands.

Wade, P. R. 1998. Calculating limits to the allowable human-caused mortality of cetaceans and pinnipeds. Marine Mammal Science 14(1):1–37.

Wade, P. R., and R. P. Angliss. 1997. Guidelines for assessing marine mammal stocks: Report of the GAMMS workshop April 3–5, 1996, Seattle, Washington. U.S. Department of Commerce, NOAA Technical Memorandum NMFS-OPR-12.

American Fisheries Society Symposium 27:171–196, 2002

Uncertainty, Risk, and Decision: The PVA Example

DANIEL GOODMAN
Environmental Statistics Group, Ecology Department, Montana State University
Bozeman, Montana 59717, USA

Introduction

Uncertainty is pervasive in natural resource management; attainment of the management goals often depends both on present quantities that cannot be measured accurately and on future environmental disturbances that cannot be predicted accurately. When we make management decisions under such circumstances, we recognize that the outcome may be quite different from our intentions, and we perceive this possibility as risk. But, decisions must be made all the same, so we attempt to take the uncertainty, and hence the risk, into account in our decisions, through bet hedging and setting margins of safety. The point of this chapter will be to show that the decision process can still be, in its own way, an exact science, not withstanding the uncertainty.

Given that we cannot eradicate uncertainty, the sense in which decision making under uncertainty can be an exact science is a matter of coping with the uncertainty in a way that is recognizably optimal. That is, we have to learn to view natural resource management as a kind of gambling and consider how we might conduct that gambling as intelligently as possible. In this perspective, each management decision is equivalent to the placing of a bet, and the measure of success isn't so much winning every bet, which is impossible, but rather, success lies in evaluating the odds well enough to place our bets so that in the long run our gains exceed our losses. To this end, we need a means to recognize what is a good bet, what is a bad bet, and what is a bet that would best be deferred until more information becomes available, even if there is some concrete cost to the delay or some cost to the additional information. The mathematical discipline of statistical decision theory is concerned with understanding this sort of decision making as a formal optimization problem (Berger 1985).

There are two key components to the evaluation of uncertain management decisions. These are the costs of each possible outcome and the probabilities of each outcome. Generally the costs (or benefits) are either measured in economic terms, such as yield, or in policy terms, having to do with the social value placed on such things as biodiversity. The probabilities are a more technical issue.

The present chapter will be devoted first to showing how the probabilities can be calculated in a rigorous, objective, and fairly precise way, despite all the uncertainty; second, this chapter will show how the probabilities can be used for cost–benefit analysis, to optimize the decisions; and third, this chapter will show how the probabilities can be used for cost–benefit analysis to optimize the investment in more data. The exposition will be developed around an idealized application of these ideas to population viability analysis, for management of endangered populations.

Population Viability Analysis

Population viability analysis (PVA) is the enterprise of calculating the probability of population or species extinction. PVA attempts to take into account everything that is known and everything that is not known about the dynamics of the population in question. Factors that do not affect dynamics are irrelevant to the population viability analysis. Factors that do affect dynamics should be incorporated via their effects on dynamics. Where the effect on dynamics is uncertain, as with many genetic factors, the uncertainty in this effect should be quantified explicitly and carried through the analysis as an additional parameter that is represented by a distribution.

Everything that is known about the dynamics of a population should be captured in a statistical inference procedure that combines different kinds of information. Some information may be in the form of case specific data on the population in question; other relevant information may be in the form of comparative data on other "similar" populations. Bayes' formula accomplishes the mathematical merging of case specific and comparative information.

Everything that is *not* known should also be captured in the statistical inference procedure, by using a model that is broad enough to cover all possibilities and carrying out the inference so as to obtain probability distributions for the parameters. The uncertainty in the parameters is quantified, simply by the spread of the distributions (for example, in terms of the variances).

Because of the various uncertainties, the resulting calculation of the time to population extinction is itself represented as a probability distribution. The *risk* of population extinction is a combined function of the distribution of time to extinction and our assigning of costs to extinctions at various times; extinctions in the more distant future are assigned a lower cost. If we adopt a simple cost function that assigns costs according to whether the extinction occurs before or after a particular time horizon, set by policy, then the risk is the fraction of the distribution of time to extinction that falls in extinction times that occur sooner than the specified time horizon. Obviously, for correct calculation of this tail area of the distribution to the left of this time horizon, we need to know the whole distribution—not just simple summaries such as the mean time to extinction. And, shortcuts that distort the apparent spread of the distribution of time to extinction can greatly distort the calculated risk.

Even with hypothetically-known parameter values for the population dynamic model, the prediction of time to extinction would still be uncertain, because the model is *stochastic*; some components of the population dynamics process are inherently random processes whose collective outcomes may have known distributions, but whose specific outcome in a given realization cannot be predicted in any further detail, so the resulting time to extinction will itself be predicted only as a probability distribution. The parameter uncertainty merely adds to the inherent uncertainty in the prediction of time to extinction, by representing the uncertain parameters as distributions (just as the inherent stochasticity of the process gives rise to distributions). The only fundamental difference between inherent stochasticity and parameter uncertainty, in this light, is that parameter uncertainty can, in principle, be reduced, by obtaining more data. The formulations developed in this chapter can, in fact, be used to evaluate the likely benefits of collecting more data, by calculating the probability that the additional data will change the inference by a large enough amount to correctly modify the decision about the population. If that probability is too small, relative to the costs of obtaining the data and the costs of the consequences of the decision, then the proposed data collection is not worthwhile.

There is a temptation, when conducting population viability analysis as a computer modeling exercise, just to replay the recent past as a stochastic simulation and to record the distribution of time to extinction in these simulations. There are two main flaws in this approach. One is that the indefinite extrapolation of trends, such as those connected with habitat loss, may not be justified. For example, the point of the population viability analysis may be to identify an amount of remaining habitat that, if protected, would be sufficient to confer an acceptable survival probability. So, the population viability analysis must model realistic future scenarios with respect to habitat, harvests, incidental mortality, and similar factors that really are the results of decisions—not just passive trends. A second flaw to the blind simulation approach is that the recent past, for which we have data, is only a sample. The true distribution of environmental variation may be rather different from what occurred in that particular window of time. And, the random noise, affecting our observations in that particular window of time, could distort the apparent values of important dynamic parameters. The formal *inference* approach, outlined in this chapter, addresses these sampling effects automatically and scales the quantified uncertainty in the parameter distribution, according to the influence of the size of the sample of the observed past history.

The logic for correctly-calculated population viability analysis, and for population risk assessment, will be illustrated in explicit detail in this chapter, only for a drastically oversimplified model. The point is to show how the problem is set up: to define an appropriate utility function, to calculate the probabilities of outcomes, to factor in the uncertainty about parameters, and to calculate the value of additional data. The logic extends with no particular difficulty to realistic cases, but the mathematical, computational, and modeling challenges, then, of course, become much greater.

The Role of Population Viability Analysis in Decisions under the Endangered Species Act

There is a growing body of opinion that the term "endangered," under the Endangered Species Act, applies to a population for which the probability of extinction within some specified time interval exceeds some maximum tolerable level. The literature contains mention, for example, of the criterion "more than 5% probability of extinction within 100 years," as an example of a rigorous definition of this form.

The choice of 100 years as the time horizon, and of 5% as the maximum tolerable probability, is a matter of policy whose merits will not be discussed here. The form of the criterion, however, will be accepted in this chapter as *the* way to give objective, technical meaning to the regulatory definition of "endangered."

If, accordingly, the decision for listing under the Endangered Species Act depends on a determination whether the probability of extinction, by time horizon H, exceeds some critical probability C, then consistency might require related technical definitions for other related decisions mandated in this legislation. The criterion for delisting might reasonably be a determination that recovery actions have achieved conditions under which the probability of extinction, by time horizon H, is less than C. The criterion for adequacy of a *recovery* plan (or *habitat conservation* plan) might reasonably be that the actions proposed in the plan will, if implemented, achieve conditions under which the probability of extinction, by time horizon H, is less than C. And, the criterion for *jeopardy* might be that a proposed action will, if implemented, increase the probability of extinction, by time horizon H, to a level above C.

In this view, a determination of the probability of extinction, within time H, is *the* bottom line for the technical analysis bearing on the decisions. Various contributing factors—such as genetics, captive breeding, incidental take, habitat protection, habitat loss, metapopulation structure, density dependence, trends in population size, trends in habitat quality, and so on—matter to the decision only insofar as they affect the calculated probability of extinction within time H. By the same token, a *correct* calculation of the probability of extinction within time horizon H is obliged to take all such contributing factors into account, insofar as they apply.

Correct Calculations: When Is a Probability True?

The previous section has emphasized correct calculation of the probability of extinction. The stress on correctness of the calculation, rather than on truth of the result, may seem odd, but it is deliberate—and telling. A given population can only become extinct once, and this will happen at a particular future time. Population viability analysis does not pretend to predict *that* particular time, in that instance, with much precision. Indeed, if we could really predict the actual time of extinction, we would not label the analysis a *risk* analysis—we would call it a prediction.

We resort to the vocabulary of *risk*, because we know that the prediction of time to extinction embodies a great deal of uncertainty. We also know that, from a practical standpoint, appreciable uncertainty in the prediction of time to extinction is inevitable. Notwithstanding the uncertainty, we feel a need to make management decisions. So, we hope for a decision-making procedure that can make rational use of uncertain predictions. And, we hope for a risk assessment procedure that makes these uncertain predictions in an objective, rigorous, and reproducible way.

The uncertain prediction is stated as a *probability distribution* of the time to extinction. What does it mean for such a probability distribution to be true?

There are many schools of thought concerning the meaning of probability. There is a school of thought, usually labeled "subjectivist" or "subjective Bayesian," which holds that probability statements are basically a kind of psychological reporting that expresses subjective strength of belief. This school offers no reason why one person should accept another person's statement of probability as a guide to intervening in the world, so this, obviously, is an unsatisfactory basis for public decision making.

An alternative school of thought, conventionally called "frequentist," gives meaning to probability statements, by giving them a frequency interpretation. This school imagines a possibility of unlimited replication of the conditions in the case under consideration to constitute a "universe" of cases that are indistinguishable, before the event, with respect to anticipated outcome, but which lead to a variety of different outcomes, as each unfolds in time. The frequency distribution of outcomes for the universe of cases is what gives meaning to a probability statement about any one case. Saying that, for *this* case, the probability of outcome A is P, means that for the universe of cases that are indistinguishable before the event, the frequency of outcome A is P.

Now, the frequency interpretation could be persuasive for public decision making, provided the frequencies of outcomes in the imagined universe of indistinguishable-cases-before-the-fact corresponded to actual frequencies in "the world." The dominant hypothesis-testing theory of statistical inference is variously called "frequentist," "classical," or "Fisherian," for reasons which are only of historical interest. What does concern us about this conventional theory of inference is that it constructs its imagined universe of indistinguishable cases, by imagining repeated sampling of the observations from a hypothesized reality, in order to consider the distribution of data that might result. This fails to correspond to actual frequencies, "in the world," of values of the property of interest, because, in practice, we already have our data when we begin the inference process. Our interest lies in the distribution of the uncertain parameter being inferred, not in the distribution of imaginary data that we "could" have obtained but didn't.

A third theory of statistical inference constructs its initial universe of indistinguishable-cases-before-the-fact as an empirical histogram of comparative information obtained from actual experience, and then uses Bayesian calculations to narrow the distribution, according to the data that were actually observed in the case of interest. Provided the experiential base is rich enough for the before-the-fact histogram to correspond well enough to actual frequencies in the world, this approach will give results that are persuasive for public decision making. Essentially, this combines the best elements of the frequentist approach to probability (an objective frequency interpretation of probability) and the best elements of the Bayesian inference approach (computing the probability that really is of interest, treating the data as given, and putting a distribution on the unknown parameter). We will adopt this definition of probability, and this approach to inference, for the present discussion.

In this view, then, we stressed correct calculation, rather than *truth*, of a probability distribution of time to extinction, because the correctly inferred probabilities (which in this sense are invariably "true") depend as much on our information, which define the set of cases that are indistinguishable-before-the-fact and which include the case-specific data, as they depend on the eventual reality of when extinction actually takes place, for *this* case. Correctly calculated probabilities inferred from data are *conditional*, on the data and on the extent of our knowledge, according to which sets of cases are distinguishable or indistinguishable. With different data, or different general knowledge, the correctly calculated probability will be different, and this different probability will be true, even though the eventual time of actual extinction for the population in question is unaffected by the data or the general knowledge.

The logic of this approach, advocated in this chapter, combining a frequency interpretation of probability with a Bayesian machinery for statistical inference, can be traced directly at least as far back as a paper by von Mises (1942), and the approach is implicit in the original paper of Bayes (1763). Unfortunately, in recent practice, the terminology has become somewhat confused with the emergence of a family of techniques called "Empirical Bayes," which, in fact, embody a variety of different concepts of probability and methods for obtaining the prior, some of which are consistent with the approach outlined here, and some of which are not. More details of the history and the mathematical basis are reviewed in Goodman (2002b).

Theoretically, the conceptual framework advocated in this chapter is very "clean" and rigorous. In actual application, it is quite unproblematic when there is sufficient comparative information to define the empirical histogram of indistinguishable-cases-before-the-fact as a convincing distribution, which is called the prior distribution. But, when the comparative information is (or is thought to be) too limited, or too indirect, for direct construction of a convincing histogram, there is a problem of inference in constructing the prior itself. The solution to this problem has not become standardized. Several previously published procedures, and one new one, are discussed in Goodman (2002b). For purposes of this chapter, we will not be concerned with the details of estimating the prior. Here, we merely note that, from a practical standpoint, the differences in procedure for determining the empirical prior are unimportant when the amount of comparative data is large, and we note further that, by defining a broad enough class for comparison, the amount of comparative information can often be made "large."

Risk Assessment: Probability and Utility

A probability distribution of outcomes does not describe a *risk*. We need to put some measure of value, cost, or, more generally, utility, on each outcome, in order to calculate the risk as a sum of the utility of each possible outcome, weighted by the probability of that outcome.

For the determination of endangered status under the Endangered Species Act, acceptance of a criterion such as "greater than 5% probability of extinction within 100 years" would imply a binary utility function for extinction events, depending on the time of their occurrence. An extinction more than 100 years in the future does not count against an evaluation under this criterion, but an extinction any sooner is scored as an undesired outcome. The *risk*, from this standpoint, is the convolution of the step function utility (as a function of the time of the extinction) and the probability of each time to extinction. In other words, the risk is the probability of extinction within 100 years, and the risk tolerance embodied in this particular interpretation of policy would be a 5% probability of extinction within 100 years.

With this understanding, that the risk is a combined effect of a probability distribution (of time to extinction) and a utility function (depending on whether the time to extinction is before or after 100 years), we see that *uncertainty* is part of the risk. Uncertainty is just a measure of the spread in the distribution of outcomes of concern (time to extinction). In this respect, there is no fundamental difference between uncertainty owing to an inherently unpredictable phenomenon and uncertainty owing to sampling error affecting a parameter estimate. Both come into play in a calculation of the probability distribution of time to extinction, and both contribute to the risk, calculated as the probability of extinction within 100 years, only through their influence on the tail area of the distribution to the left of the 100-year time horizon.

If the mode of the distribution of time to extinction is beyond 100 years, both forms of uncertainty increase the risk, because an increased spread in the distribution increases the fraction of the distribution that spills over to the left of the 100-year boundary. From a management or planning standpoint, the important difference between the two forms of uncertainty lies in what we can do about them—parameter uncertainty can be reduced by collecting more data, whereas inherent uncertainty in the process cannot. To this extent, we can change the risk by collecting more data.

In this sense, in a *correctly* conducted population viability analysis, there is no uncertainty about the risk, even though there is bound to be great uncertainty about the time to extinction. The uncertainty about the time to extinction is fully represented in the correctly-calculated distribution of time to extinction. But, if this distribution itself is correct, any tail area of the distribution—including the one that we define to be the risk—will also be correct.

From this perspective, we see immediately that some of the popular shortcuts employed, in order to domesticate the uncertainty in a risk assessment, are not valid. One of the most glaring of such errors arises when a "best estimate of the risk" is incorrectly put forward, by substituting a central estimate for some uncertain parameter in place of the distribution that correctly represents that uncertain parameter. Another version of this error arises when a "conservative risk estimate" is put forward, by substituting a pessimistic confidence bound—usually the 5% tail—for some uncertain parameter (or parameters) in place of the distribution(s) that correctly represents that uncertain parameter(s). In our PVA example, the risk *is* a tail area, so any tampering with the variance of the distribution of time to extinction can distort the critical tail area, thereby distorting the calculated risk.

For example, if the mode of the distribution of time to extinction is well beyond 100 years, replacing the distribution of an uncertain population dynamic parameter with its central estimate will shrink the distribution of time to extinction toward its mode, shrinking the tails of the distribution, and thereby underestimating the risk. Conversely, if the mode of the distribution of time to extinction is well below 100 years, replacing the distribution of an uncertain parameter with its central estimate will overestimate the risk.

Replacing the distribution of an uncertain parameter with an upper or lower confidence limit can produce similarly large distortions of the risk estimate, with no particular justification. Why the lower 95% confidence limit, for example? Why not 90% or 99%? And, the mapping of the percentile of the uncertain parameter to the quantitative amount of distortion of the risk estimate can vary wildly from analysis to analysis, depending on how many parameters are pessimistically "fixed" (the effects compound with multiple parameters), depending also on where the "fixed" parameter lies in the chain of calculation leading to the bottom line risk calculation, and depending further on what fraction of the distribution of time to extinction constitutes the tail area before "fixing" the selected parameter or parameters. If there is a desire to achieve some particular degree of "conservatism" in a decision process, this should be built into the utility function and the error tolerance; attempting to achieve conservatism, by tampering with the statistical inference, confuses the issue and will yield erratic results.

Dynamic Models and Parameter Estimation

Population viability analysis has two intellectual roots. One root is in evolutionary and island biogeographic theory, where there is motivation to model, in a fairly idealized, mathematical way, the dynamics of local and global extinction processes, in order to understand, in a general way, geographic patterns of species occurrence, large scale patterns of species abundance, and the nature of adaptation to environmental fluctuation. The second root is in environmental management, where there is motivation to calculate—for actual, particular species and populations—the probabilities of population or species extinction, in order hopefully to devise concrete management plans that control these extinction rates to feasible acceptable levels.

The two traditions, one theoretical and the other applied, have in common the calculation of extinction probabilities, but there are also important differences between the two traditions.

The theoretical tradition is concerned with the creation of mathematical models to explore the nature of random population growth and decline processes in the abstract. The interest, here, has been in finding mathematically tractable representations that do, at least impressionistically, include some of the features that we think operate in reality. The sorts of analyses that are considered satisfactory, in this tradition, generate results, such as a sensitivity analysis, to examine the relative importance of various factors in determining average time to extinction or determining the mathematical form (linear, logarithmic, saturating, etc.) of the dependence of mean time to extinction on a particular factor. These analyses follow a hypothetical format; they state that *if* a population's dynamics are governed by such and such a function with such and such stochasticity, *then* the resulting distribution of time to extinction will have such and such properties. In other words the focus is on forward projection from an assumed starting point.

The theoretical tradition is deficient, from the standpoint of concrete applications, in two main regards; the theoretical tradition neglects messy details in dynamics and functional form that unfortunately can have a large effect in any particular case, and the theoretical tradition is not concerned with parameter estimation.

The notion of uncertainty arising from parameter estimation is foreign to the theoretical tradition. In this tradition, the parameters are not estimated; they are either left as symbolic variables with unspecified numerical values, or their values are just made up, for purposes of illustration. But, the parameter uncertainty in a real risk assessment is a substantial part of the risk, and if the parameters are made up, so is the risk assessment.

The probability communicated by a real risk assessment is a *conditional* probability. It is a probability conditional on the data. In a population viability analysis, the risk is stated as the tail area of a distribution of time to extinction, where that distribution is based on inferences from data. So, there needs to be a formal mechanism connecting data to the population viability analysis. Four relevant treatments in the literature are Dennis et al. (1991), which uses a rigorous estimation procedure, at least for point estimates of the parameters, and Taylor et al. (1996), Ludwig (1996), and Wade (2002), which use a Bayesian inference procedure for propagating parameter uncertainty through the analysis. Goodman (2002a) reviews the literature on PVA from the particular perspective of uncertainty analysis.

In the discussion that follows, we will present a recipe for connecting data to a population viability analysis for a decision process in a formal way. The conceptual framework will be that of statistical decision theory (Berger 1985). The literature treating endangered species management from the perspective of statistical decision theory is not yet extensive. There is a more-developed literature on applications of statistical decision theory to broader fisheries management questions (Walters and Hilborn 1976; Ludwig and Walters 1982; Thompson 1992; Punt and Hilborn 1997; McAllister and Kirkwood 1998) and applications of Bayesian methods to the estimation of parameters needed for management decisions (Wade 2000, in press; Walters and Ludwig 1994).

Much of the discussion in this chapter will be developed in the vocabulary of the theoretical tradition of PVA, in order to build a bridge between the mathematical rigor of the theoretical tradition and the data dependence that ought to be characteristic of the applied tradition. This recipe will constitute a method for addressing one of the deficiencies of the theoretical tradition—its neglect of parameter estimation—as a template for applied population viability analysis. The second deficiency—reliance on oversimplified, mathematically tractable representations—will not be treated in this discussion. Indeed, for purposes of clarity in exposition, this discussion will perpetuate that deficiency.

Illustration with a Simple Metaphor: Dynamics Represented by a Poisson Process and "Extinction" Defined as the Occurence of *M* Further Events

The full schedule of analysis that we need to formalize for a rigorous PVA proceeds stepwise from *observation* to *inference of parameter distributions* to *probabilistic prediction* with *uncertainty propagation*. The theoretical literature provides many models for probabilistic prediction in a hypothetical PVA, but without substantive guidance on the connection to data via inference of parameters in a real PVA.

For purposes of illustration, we may walk through the full schedule of analysis by considering a simple physical stochastic process: the approximate dynamics of decay events in a given mass of radioactive material. We may take the time sequence of decays to be a metaphor for the sequence of catastrophic events in the population dynamics: for example, the net rate of winking out of subpopulations in a metapopulation with insufficient connectivity for recolonization, or the cumulative downward spiral of status of a small population owing to effects such as inbreeding depression, density depensation, and disease outbreak. We may take the number of decays, observed during a sample time interval, to be a metaphor for the data from which the parameters governing the population dynamics are to be estimated. And, we will take "extinction" to occur when a specified number of further decays have taken place after some specified future time. For example, this critical number of decays may correspond to the number of subpopulations in the metapopulation, and extinction of the metapopulation would occur when local extinctions had claimed all the subpopulations. This metaphor is, of course, chosen for mathematical simplicity.

This metaphor is about at the limit of complexity of PVA-like models, for which the inference can be presented as a closed-form mathematical expression. The simple mathematics is helpful for clarity of this introductory exposition. The model is sufficiently realistic to illustrate the logic-connecting data, inference, population viability analysis, risk assessment, the cost–benefit analysis of intervention decisions, and the cost–benefit analysis of decisions to collect more data. However, this model definitely is not realistic enough for reasonable use in an actual PVA application. A more realistic example would require more complicated (and less transparent) numerical techniques to carry out the statistical inference.

Probablistic Forward Projection in the Illustration

Where the number of items with potential for decay is large compared with the number of decays we are concerned with, the time sequence of decay events approximates a textbook Poisson point process (see, for example, Karlin and Taylor 1975). The occurrence of decays is "random," but there is a probability distribution governing the time sequence. In particular, under this approximation, the probability of a decay, at any instant, is constant and is independent of decays in the previous instant.

For a Poisson point process with fixed rate parameter, the number of decays occurring, during a time interval of given duration, will be Poisson distributed. That is, where the rate parameter is r decays per unit time, the number of decays n that will occur in an interval of duration t is distributed according to the probability density function

$$p(n|r) = \frac{e^{-rt}(rt)^n}{n!}. \quad (1)$$

Some summary statistics of this distribution are the average

$$E(n|r) = rt, \quad (2)$$

and the variance

$$Var(n|r) = rt. \quad (3)$$

Biologically, of course, this is not, in detail, a realistic representation of the local extinction process. So, our interest will be confined to the formal relationships among data, inference, uncertainty, and risk, and not with the particular extinction rates in this model.

Calculation of the Risk, If the Rate Parameter Were Known

For a Poisson point process with fixed parameter, the time elapsed, from some arbitrary starting time until the occurrence of a given number of decays, will be gamma distributed. That is, where the rate parameter is r decays per unit time, the time elapsed t, till the occurrence of m decays, is distributed according to the probability density function

$$p(T|r) = \frac{re^{-rT}(rT)^{m-1}}{(m-1)!}. \quad (4)$$

Some summary statistics of this distribution are the average

$$E(T|r) = \frac{m}{r}, \quad (5)$$

and the variance

$$Var(T|r) = \frac{m}{r^2}. \quad (6)$$

So, even with a fixed rate parameter r, the "time to extinction" in this physical analog is a random variable, but with a known distribution. Accordingly, we cannot predict the time to extinction in this model, but we can state the mean time to extinction, the variance in the time to extinction, or any particular percentile of the distribution of time to extinction.

For purposes of a "regulatory" determination of the probability that extinction will occur before a critical time horizon, we would need to compute the area of the tail of the distribution that is to the left of the critical time horizon H. For the gamma distribution of equation (4), this tail area is

$$p(T < H|r) = \frac{r^m}{(m-1)!} \int_0^H e^{-rT} T^{m-1} dT. \quad (7)$$

Calculation of the Risk with Uncertainty in the Parameter

The "result," in equation (7), follows the form of a traditional, theoretical PVA. It computes the desired quantity, the probability of extinction before time H, by modeling the future dynamics as a stochastic forward projection, where the parameters (in this case, r) are assumed known. But, of course, the parameter r, or its equivalent in a real case, would not be known; it would be estimated. In the present example, r would be estimated from observations on the number of decays occurring during some past sample interval.

Parameter estimation is largely ignored by the theoretical tradition. It would be conventional, within this tradition, to carry out a *sensitivity analysis*, to quantify how much a given change in the value of the parameter r will change the value of the result. But, this knowledge of sensitivity does not allow us to qualify, correct, or adjust our PVA result, $p(T < H)$, until we can quantify the uncertainty in the parameter r, itself. A "small" sensitivity could still have a large effect, if the uncertainty in the parameter is large, for example.

A correct PVA will propagate the parameter uncertainty, as quantified by a correct parameter estimation procedure, through equation (7), to obtain a result that reflects both the inherent uncertainty in the dynamical process and the uncertainty in the parameters of that process. In the model we are considering here, the process uncertainty has its origin in the random nature of the Poisson process, resulting, for example, in a variance in the time to extinction, as given by equation (6), even when the parameter r is known. And, the parameter uncertainty arises from the uncertainty in r, given that we only have the observation n upon which to base our estimate of r. We capture the effects of both uncertainties as

$$p(T < H|n) = \frac{1}{(m-1)!} \int_0^\infty \int_0^H p(r|n) r e^{-rT} \times (rT)^{m-1} dT dr \quad (8)$$

where n is the case specific data, from which r is inferred, and $p(r|n)$ is the distribution that represents what we know and what we don't know about r, based on whatever data we have.

This correct answer is definitely not the same as using the "best estimate" of the parameter as the value for r in equation (7), nor is it the same as using a so-called "worst case" value of the parameter for r in equation (7). Note that the correct result for applied decision making, equation [8], is conditional on the data, unlike the theoretical result, equation (7), which was conditional on a *hypothetical value* of a parameter that in practice is not known.

Inference of the Parameter Distribution from Observations

To begin the real task of parameter estimation, for this analysis, we must define the distribution of values of the parameter r, from the "universe" of masses of decayable material (or metapopulations) that we would consider indistinguishable, in this respect, from the one at hand, in the absence of case specific observational data bearing on rates governing the dynamics. We construct the distribution from previous comparative data on such masses or populations. For the moment, since we are, in any case, dealing with a metaphor, we will not concern ourselves with the details of how such a distribution is specified from comparative data, nor will we concern ourselves with the implications of the obvious possibility that the construction of this distribution is itself an estimation process subject to its own sources of error.

For mathematical convenience, we will specify the distribution of comparative information on the decay rate parameter r, as a gamma distribution with shape parameter a and scale parameter b:

$$p(r) = \frac{be^{-br}(br)^{a-1}}{\Gamma(a)}, \quad (9)$$

where $\Gamma(a)$ is the standard gamma function and where both parameters a and b are treated as known from the comparative information.

We can imagine that the mass of items with potential for decay (or the metapopulation), in the case at hand, came to our attention, via random selection, from a "universe" of similar-looking masses (or metapopulations), whose rate parameters had the distribution given by equation (9). Accordingly, in the absence of any data specific to the case at hand, equation (9) describes our knowledge of the rate parameter for this case; repeated sampling from this universe would give rise to cases that would be indistinguishable in the absence of case specific data, and the repeated sampling would give rise to a physical collection of cases, where the frequency of actual rate parameter values would be as in equation (9).

Coincidentally, with r distributed according to equation (9), the average of the distribution for the value of the rate parameter is

$$E(r) = \frac{a}{b}, \tag{10}$$

and the variance is

$$Var(r) = \frac{a}{b^2}. \tag{11}$$

Thus, a primitive way of specifying the distribution of r, in the universe of indistinguishable-cases-before-the-fact, would be to invert equations (10) and (11) and, then, determine a and b from observations on the mean and variance of the values of r in a sample of cases from this universe, for which data on r were available.

The case-specific data, for the case at hand, for which we are conducting a PVA, would consist of the observation of n decays, during a sampling interval of duration t. The inference procedure for associating a probability distribution with values of the parameter r, given the data n, uses Bayes' formula:

$$p(r|n) = \frac{p(n|r)p(r)}{\int_0^\infty p(n|r)p(r)dr}. \tag{12}$$

Substituting into equation (12) the particular functions for $p(n|r)$ and $p(r)$, from equations (1) and (9), respectively, gives

$$p(r|n) = \frac{(b+t)e^{-(b+t)r}((b+t)r)^{a+n-1}}{\Gamma(a+n)}, \tag{13}$$

which is itself a gamma distribution.

The mean and variance of this distribution of r, given the data, are

$$E(r|n) = \frac{a+n}{(b+t)} \tag{14}$$

and

$$Var(r|n) = \frac{a+n}{(b+t)^2}. \tag{15}$$

The physical frequency interpretation of equation (13) is that, if we repeatedly sampled masses of items with potential for decay (or metapopulations with potential for deteriorating status) from a "universe" of similar looking cases, whose decay rate parameters had the distribution given by equation (9), and, if, for each case, we observed the number of decays occurring in an interval of duration t (the same duration for each), then the subset of cases, for which the number of observed decays was n (the actual number observed for the one mass in question), would have had a distribution of values for r, as given by equation (13).

In other words, the probability distribution for r, based on inference from observing the particular data n, has meaning as a literal frequency histogram for values of r from a pool of similar cases that, from the information available to us, are indistinguishable both before and after the collection of the data n. Because this distribution for r has a physical interpretation as a literal frequency histogram, the distribution may be used for forward propagation in calculations that depend on the value of r.

Linking the Inference to the Risk Calculation

Propagating the distribution of the parameter r through the forward projection process, predicting the distribution of times to extinction (substituting equation [13] into equation [4]), gives a compound gamma (Pearson Type VI) distribution of time to extinction:

$$p(T|n) = \frac{\Gamma(a+n+m)}{\Gamma(a+n)\Gamma(m)}(b+t)^{a+n}T^{m-1} \times (b+t+T)^{-(m+a+n)}, \tag{16}$$

which is to say that $(b+t)/(b+t+T)$ is beta distributed, with parameters $(a+n)$ and m.

The mean and variance of the distribution of the time to extinction, given the data, are

$$E(T|n) = \frac{m(b+t)}{a+n-1}, \tag{17}$$

provided $(a + n) > 1$, and

$$Var(T|n) = \frac{m(b+t)^2(m+a+n-1)}{(a+n-1)^2(a+n-2)}, \quad (18)$$

provided $(a + n) > 2$.

The probability of extinction prior to the stated time horizon is

$$p(T < H|n) = \frac{\Gamma(a+n+m)}{\Gamma(a+n)\Gamma(m)}(b+t)^{a+n}\int_0^H T^{m-1} \times (b+t+T)^{-(m+a+n)}dT. \quad (19)$$

Assuming that that we are correct about our choice of model (in this case the Poisson), equation (19) is *the* risk, taking all the uncertainty into account. We note that the calculated risk is not, itself, uncertain—it simply *is* the risk, including the uncertainty of the information upon which it is based. If we had more information, or better information, or different information, the uncertainty would be different, and the risk would be different. This dependence of the risk on the available information is not a weakness of the method—it is common sense. After all, if we had complete and perfect information, there would be no risk—there would only be a perfect prediction of whatever the outcome will be.

The tortuous mathematics in equation (19), connecting the risk to the information available to us at the time, may seem a nuisance. But, this math is necessary to get the risk calculation right, and the math also opens the door to further calculations that are at the heart of rational planning, for investment in future data collection to improve the PVA. Explicit dependence of the risk calculation on the available information allows useful calculations of how much of the risk is attributable to the various sources of uncertainty, and it allows useful calculations of the probability that the risk will be revised by a specified amount, if more data are collected. This latter calculation is essential for weighing the merits of collecting more data.

Comparison to Risk Calculations Based on "Best Estimates" of Parameters

We may compare the correct calculations of risk, culminating in equation (19), with some attractive, but incorrect, shortcuts. The maximum likelihood estimate for the decay rate parameter, based on observing n decays in an interval of duration t, is simply

$$\hat{r} = \frac{n}{t}. \quad (20)$$

Substituting this value in equation (4) for the conditional distribution of the time to extinction gives

$$p(T|\hat{r}) = \frac{ne^{-\frac{nT}{t}}\left(\frac{nT}{t}\right)^{m-1}}{t(m-1)!}. \quad (21)$$

Compare this to the correct distribution, equation [16], based on a distribution, rather than an "estimate," for r.

Substituting $\hat{r}$ in equation (5) for the conditional mean time to extinction gives

$$E(T|\hat{r}) = \frac{mt}{n}. \quad (22)$$

Compare this to the correct value, equation (17), based on a distribution, rather than an estimate, for r.

Substituting $\hat{r}$ in equation (6) for the conditional variance in the time to extinction gives

$$Var(T|\hat{r}) = \frac{mt^2}{n^2}. \quad (23)$$

Compare this to the correct value, equation (18), based on a distribution, rather than an estimate, for r.

Substituting $\hat{r}$ in equation (7) to calculate the conditional risk, the conditional probability of extinction before time horizon H, gives

$$p(T < H|\hat{r}) = \frac{(n/t)^m}{(m-1)!}\int_0^H e^{-\frac{nT}{t}}T^{m-1}dT. \quad (24)$$

Compare this to the correct value, equation (19), which propagates the uncertainty in r by using a distribution for r.

The only consolation to be drawn from these comparisons is that, as the amount of data becomes overwhelming, n and t both become very large, and the correct equations obviously converge on the maximum likelihood "best estimate." But, this is cold comfort, for it only confirms the correctness of the correct equations and does not, in any way, justify the shortcut, when, as is inevitable in a real extinction risk assessment, the data do not overwhelm the parameter estimation procedure, and the parameter uncertainty remains substantial.

Partitioning Uncertainty in Time to Extinction Attributable to the Parameter Uncertainty

Equation (18) quantifies the total uncertainty in the time to extinction as the variance in time to extinction, given the information available to us at the time. From equation (6), which gives the variance in time to extinction as a function of the unknown parameter, and equation (13), which gives the distribution of the unknown parameter given the information available to us, the currently expected uncertainty in time to extinction, quantified as a variance in time to extinction, if the uncertainty in the rate parameter were entirely resolved in the future, is

$$E(Var(T|r))|n = \int_0^\infty Var(T|r)p(r|n)dr = \frac{m(b+t)^2}{(a+n-2)(a+n-1)}. \quad (25)$$

This relates to the total current uncertainty in time to extinction as

$$\frac{E(Var(T|r))|n}{Var(T|n)} = \frac{(a+n-1)}{(a+n+m-1)}. \quad (26)$$

Partitioning of Risk Attributable to the Various Uncertainties

Equation (7) gives the magnitude of the risk as a function of the unknown rate parameter. Averaging over values of the rate parameter weighted according to their probability, as given by equation (13), to find the expected value of the risk, if the parameter were known, merely recapitulates the result we already have in equation (19) because the *expected* risk is *the* risk. Nevertheless, there are other informative calculations we can perform to anticipate what the magnitude of the conditional risk would be, if the parameter uncertainty were resolved.

For example, we may solve equation (7) numerically, to find the value of the rate parameter for which the risk (with rate parameter known) just equals the risk corresponding to the risk tolerance set by policy. Let C designate the risk tolerance (e.g., 5% probability of extinction within 100 yrs), and let the associated critical value of the rate parameter be r_c. From equation (13), we have the probability distribution of the rate parameter r, given the information available to us, so the tail area of this distribution to the left of r_c gives the probability, from the perspective of the current information, that resolution of the parameter uncertainty in r will result in a calculated risk below the policy tolerance:

$$p(p(T < H|r) < C)|n = p(r < r_c|n) = \frac{(b+t)^{a+n}}{\Gamma(a+n)} \int_0^{r_c} e^{-(b+t)r} r^{a+n-1} dr \quad (27)$$

where r_c is defined implicitly from

$$p(T < H|r_c) = \frac{r_c^m}{(m-1)!} \int_0^H e^{-r_c T} T^{m-1} dt = C. \quad (28)$$

However interesting it may be to know the probability that resolution of *all* the parameter uncertainty will reduce the risk to an acceptable level, this information is largely academic because we don't know (and probably couldn't afford) the amount of data-collection activities that would have to be carried out in order to eliminate the parameter uncertainty. A calculation, that would be more to the point, is quantification of the probable risk reduction from a specified investment in additional observations that, of course, will not resolve the parameter uncertainty *entirely* but will reduce the uncertainty by a probable amount that can be calculated.

Probability of Reducing Risk by Obtaining More Data

Imagine that there is the option of conducting observations for an additional sampling interval of duration s before the countdown begins for the m events that define extinction. This option would carry management costs, since the decay events that occur during this sampling interval would have to be compensated, somehow, by management interventions, so that, at the end of the additional sampling interval, the actual status of the population is no worse than before. This option would also confer potential information benefits, since, at the end of the sampling interval, we would probably have a better estimate for the parameter r, and then, our ensuing decision would probably have a lower error rate.

We begin the analysis of comparing the cost and benefits of the additional data by asking what is the probability that the specified additional sampling effort will reduce the risk by a specified amount. From equation (1), substituting ν for n and s for t, we have the probability of ν events during the new sampling interval of duration s, as a function of the rate parameter r; and from equation (13), we have the probability distribution of r, given the existing data n. Combining, we get

$$p(\nu|n) = \int_0^\infty p(\nu|r)p(r|n)dr = \frac{(b+t)^{a+n}s^\nu}{(b+t+s)^{a+n+\nu}\nu!}\prod_{j=0}^{\nu-1}(a+n+j). \quad (29)$$

If ν events are observed in the new sampling interval, the revised inference for the distribution of the rate parameter r, following the logic of equation (13), will be

$$p(r|\nu,n) = \frac{p(r|n)p(\nu|r)}{\int_0^\infty p(r|n)p(\nu|r)dr} = \frac{(b+t+s)e^{-(b+t+s)r}((b+t+s)r)^{a+n+\nu-1}}{\Gamma(a+n+\nu)}. \quad (30)$$

So, if ν events are observed in the new sampling interval, the new distribution of time to extinction, following the logic of equation (16), will be

$$p(T|\nu,n) = \frac{\Gamma(a+n+m+\nu)}{\Gamma(a+n+\nu)\Gamma(m)}(b+t+s)^{a+n+\nu} \times T^{m-1}(b+t+s+T)^{-(m+a+n+\nu)}. \quad (31)$$

The new uncertainty in the time to extinction, quantified as a variance, will be

$$Var(T|\nu,n) = \frac{m(b+t+s)^2(m+a+n+\nu-1)}{(a+n+\nu-1)(a+n+\nu-2)}. \quad (32)$$

And, the new risk will be

$$p(T < H|\nu,n) = \frac{\Gamma(a+n+m+\nu)}{\Gamma(a+n+\nu)\Gamma(m)} \times (b+t+s)^{a+n+\nu}\int_0^H T^{m-1} \times (b+t+s+T)^{-(m+a+n+\nu)}dt. \quad (33)$$

Accordingly, to find the probability, before the fact, that an additional sampling interval of duration s will result in a revised conclusion that the risk is *less* than the critical value C, if the conclusion based on existing data (equation [19]) is that the risk currently is greater than C, we would use equation (33) to find ν_c, the largest value of ν for which $p(T < H|\nu,n) < C$, and then, we would use equation (29) to find the probability of a value of ν that small or smaller. These substitutions give

$$p(\nu \le \nu_c|n) = \left(\frac{b+t}{b+t+s}\right)^{a+n}\sum_{\nu=0}^{\nu_c}\left(\frac{s}{b+t+s}\right)^\nu \times \frac{1}{\nu!}\prod_{j=0}^{\nu-1}(a+n+j), \quad (34)$$

where the product series is defined to equal one for $\nu = 0$.

This directly addresses one aspect of the expected value of the additional investment in data collection: the probability, before the fact, that this investment in data collection will lead to a revised decision that the population is actually in a lower-risk category than was concluded before the collection of the new data.

Probability of Increasing Risk by Obtaining More Data

Imagine a situation where analysis of the risk, based on the comparative information and the data in hand (equation [19]), showed that the risk was above the level C set by policy as the tolerance, while analysis of the probability of revising the decision, if more data were obtained (equation [34]) showed a trend of increasing and substantial probability of concluding that the risk is below the tolerance as the amount of future sampling increased. Under these circumstances, it might be tempting, but mistaken, to anticipate the results of future sampling and decide, in advance, that the risk "really" is below our tolerance threshold. The superficial, and incorrect, justification for jumping to this conclusion, notwithstanding that fact that with the present data the risk was above the tolerance, would be the fact of the presently computable (equation [34]) probability that a conclusion of lower risk would be borne out by future data. What this premature, and incorrect, line of argument neglects is that there is a complementary probability that the future data will, in the actual event, show that the risk is higher than the risk presently calculated with the data now in hand. The present risk calculation (equation [19]) already accounts for both possible outcomes of future sampling, and no further anticipation of the results of future sampling is legitimate in the evaluation of the total risk from the perspective of the present.

To gain an understanding of the complementary probability to equation (34)—the probability that future sampling will lead to a conclusion that the risk is higher than the present calculation—let us consider a situation where there are three categories for management action (rather than the two categories, risk above or below C, that we have been exploring, thus far). Let us consider, reasonably enough, that there might be a policy commitment to trigger additional (and more expensive) protection and recovery actions, if the risk were above a secondary, "very" critical level C_2. Imagine that, with the data in hand, equation (19) showed a risk above C and below C_2. Equation (34) gives the probability that additional sampling will revise the decision and assign the population to the category with risk below C. What is the probability that the future sampling will lead to a revised decision that assigns the population to the category with the risk above C_2?

From equation (33), we can solve for ν_{c_2}, the smallest value of ν for which $p(T < H|\nu, n) > C_2$ for a given duration s, of future sampling. Then, our interest would be in the probability of observing a future number of decays that large or larger:

$$p(\nu \geq C_2) = 1 - p(\nu < \nu_{c_2}) = 1 - p(\nu \leq \nu_{c_2} - 1). \tag{35}$$

We can use equation (34) to solve for $p(\nu \leq \nu_{c_2} - 1)$, and the complementary probability that we are interested in follows immediately from equation (35).

As information increases, probabilities of various outcomes diverge, from their initial intermediate values, and move towards larger and smaller values. After the fact of extinction, for example, the probability of extinction before a specified time (that may, by then, be in the past) is either 0 or 1. Where there is process uncertainty, as well as parameter uncertainty, the probabilities diverge toward two asymptotes less extreme than 0 and 1, before the fact, since, before the fact, the process uncertainty remains, even if all the parameter uncertainty has been removed by availability of unlimited sampling data bearing on the parameter.

Though we cannot do better than equation (19) for calculating the total risk from the perspective of the present, we can legitimately "decompose" the total risk into a spectrum of possible outcomes of future sampling, and we can weigh the merits of an investment in more sampling, by considering the probabilities and costs of those respective outcomes versus the cost of the additional sampling. As part of that evaluation, we might, for example, wish to calculate the probability that the future sampling will lead to a revised decision that is a *better* decision.

Probability of a Better Decision After Obtaining More Data

The probability of a "better" decision will be explored in this section, by comparing the decision based on inference from the combined existing and future data to the decision that would be appropriate, if we knew the actual value of the rate parameter r. In other words, we compare with the decision that we would make, if we had infinite data bearing on r. This is a clearly defined standard, but it is somewhat contrived, since r, itself, is not an observable, when we have limited data. In a real application, we might consider endangered species management decision rules that respond to observables, and we would seek to compute the probability that various subsequent management decisions (that incur various costs) will be triggered, based on observables that signal whether previous decisions did or did not have the intended effect. But, that analysis depends on defining a more elaborate—in fact, adaptive—system of management decision rules than is now commonplace. This would be a good idea (Goodman 2002a) but is beyond the scope of the present chapter. Here, we consider the more abstract property—the probability that our risk classification agrees with the classification we would make, if we knew the value of the rate parameter r.

We are concerned with a total of four possible outcomes of additional future sampling, where the current decision, based on data in hand, is that the risk is above C. The four outcomes are

1. Revise the decision, to risk is less than C, and this revision agrees with the standard (i.e., this is the same decision we would make, if we knew the true r).
2. Revise the decision, to risk is less than C, and this revision conflicts with the standard (i.e., this is the opposite of the decision we would make, if we knew the true r).
3. Let the conclusion stand, that risk is above C, and this conclusion agrees with the standard (i.e., this is the same conclusion we would draw make, if we knew the true r).

4. Let the conclusion stand, that risk is above C, and this conclusion conflicts with the standard (i.e., this is the opposite of the conclusion we would draw, if we knew the true r).

Conventionally, the four respective decision outcomes are classified, in order, as true negative, false negative, true positive, and false positive. Generally, each will have a different utility, if it materializes. The utility structure depends on the costs of the various management options, the speed with which mistaken management can be recognized, and the ease with which mistakes can be corrected. The costs may include social or political costs, as well as the economic costs of implementation.

The cost of a true negative is usually the lowest, since it requires the least ongoing management, and events will prove that the minimal management was sufficient. The cost of a true positive is usually higher, since it requires more management intervention. The cost of a false negative usually is higher still because of the attendant political embarrassment and emergency management required when events show that the initial decision to treat the case as a negative was incorrect. The cost of a false positive is usually higher than that of a true negative (since, at least initially, the positive decision will trigger the higher level of management intervention) and usually lower than that of a true positive (since, eventually, it will probably be recognized that the higher level of management is unnecessary, and then, management costs will be reduced).

Which one of these outcomes materializes depends on the future data we get and on the underlying true, but incompletely known, value of the rate parameter r. The probabilities of the outcomes, from the perspective of not yet having obtained the new data but having committed to obtain them, depends on the probabilities of the various possible numbers of decays, that might be observed, and on the distribution, based on the data in hand, of the true value of r.

With no additional data, in the scenario we are considering, we are already committed to a conclusion that risk is greater than C, so the only possible consequences of that conclusion are (3) and (4). For the case at hand, where we don't know r, we do have the distribution of r from equation (13), and we have the probability that r is less than r_c, from equation (27). Therefore, the probability of outcome (4), under the scenario of no additional data, is given directly by equation [27], and the probability of outcome (3), with no additional data, is its complement.

With additional data (i.e., with $s > 0$), there is a probability of a revised decision that could lead to outcome (1) or (2), and the probabilities of outcomes (3) and (4) will also be different from what they were, with no additional data. We wish to determine the amount by which the new set of probabilities of the respective outcomes is more favorable, when there is a commitment to new data, compared with the set of probabilities, when the decision is based only on the data currently in hand. For this, we need to compute the probability of each outcome, when there is a commitment to a future observation period of nonzero duration s.

From equation (33), we already have ν_c, the greatest number of new events that could be observed in a future sampling interval of duration s, compatible with revising the conclusion, based on current data, that risk is greater than C, to a new conclusion that risk is less than C. From equation (34), we have the probability of a revised decision, the probability that ν_c or fewer decays will be observed, during the future sampling interval of duration s. That probability of a revised decision is the sum of the probabilities of outcomes (1) and (2).

The probability of outcome (1), itself, is the probability of r being less than r_c times the probability of ν_c being less than, or equal to, ν_c, conditional on r being less than r_c. Thus,

$$p(outcome(1)) = p(r < r_c)p(\nu \leq \nu_c|r < r_c), \tag{36}$$

and similarly,

$$p(outcome(2)) = p(r > r_c)p(\nu \leq \nu_c|r > r_c), \tag{37}$$

$$p(outcome(3)) = p(r > r_c)p(\nu > \nu_c|r > r_c), \tag{38}$$

and,

$$p(outcome(4)) = p(r < r_c)p(\nu > \nu_c|r < r_c). \tag{39}$$

Returning to equation (36) and the term $p(\nu \leq \nu_c|r < r_c)$, we can compute the probability of observing ν_c or fewer events, conditional on r_c, by using equations (1), (13), and (27):

$$p(\nu \le \nu_c | r < r_c) = \sum_{\nu=0}^{\nu_c} p(\nu | r < r_c)$$
$$= \sum_{\nu=0}^{\nu_c} \int_0^{r_c} p(\nu|r) \frac{p(r|n)}{p(r < r_c|n)} dr$$
$$= \frac{1}{\int_0^{r_c} e^{-(b+t)r} r^{a+n-1} dr} \sum_{\nu=0}^{\nu_c} \frac{s^\nu}{\nu!}$$
$$\times \int_0^{r_c} e^{-(b+t+s)r} r^{a+n+\nu-1} dr. \quad (40)$$

So,

$$p(outcome(1)) = \frac{(b+t)^{a+n}}{\Gamma(a+n)} \sum_{\nu=0}^{\nu_c} \frac{s^\nu}{\nu!}$$
$$\times \int_0^{r_c} e^{-(b+t+s)r} r^{a+n+\nu-1} dr. \quad (41)$$

Similar derivations give the probabilities of outcomes (2), (3), and (4), or alternatively, to save some ink, those probabilities may be obtained with a little algebra from the probability of outcome (1) and the solutions for $p(r < r_c)$ (equation [27]) and $p(\nu \le \nu_c|n)$ (equation [34]) and the relationships

$$p(outcome(1)) + p(outcome(2)) = p(\nu \le \nu_c), \quad (42)$$

$$p(outcome(1)) + p(outcome(4)) = p(r < r_c), \quad (43)$$

and,

$$p(outcome(1)) + p(outcome(2)) + p(outcome(3)) + p(outcome(4)) = 1. \quad (44)$$

With a commitment to a longer and longer period of future observation (i.e., as s increases), the probabilities of the outcomes of "mistaken" decisions, outcome (2) and outcome (4), will decline asymptotically to zero, and the probabilities of the outcomes of "correct" decisions, outcome (1) and outcome (3), will increase asymptotically to their equivalents with perfect knowledge of r. In that limit, the probability of outcome (1) becomes simply the probability that r is less than r_c, as given by equation (27), and the probability of outcome (3) becomes its complement.

Similar calculations can be done for the outcomes of further observation, when the conclusion, with the existing data, is that the risk is below the critical threshold.

Cost–Benefit Analysis for the Collection of Further Data

The evaluation of the merits of collecting future data can be conducted as a cost–benefit analysis. Each of the four outcomes (true negative, false negative, true positive, false positive) has a distinct utility. The expected net cost, computed in advance, for a commitment to a future data collection episode of duration s_i, is a sum of the direct cost of the data collection activities, the cost of the management required to compensate for the decays that occur during the observation period, and the costs of each of the four outcomes, weighted by their probabilities as calculated by equations (41–44) for that value of s. The optimal commitment to collecting further data is the s_i, which has the lowest expected net cost.

Numerical Example

Consider an example where the comparative information on similar populations shows that, over that set of populations, the distribution of the rate parameter r is a gamma distribution with shape parameter $a = 2.5$ (dimensionless), and scale parameter $b = 6$ (in units of time, which we will specify in years for this example). This distribution has a mean $E(r) = 0.417$ decays per year (equation [10]) and a standard deviation $\sigma_r = 0.264$ (equation [11]). This distribution, capturing the comparative information on r, is plotted as the dashed line in Figure 1.

The general shape of the distribution for r is reasonable. Values lower than zero are not physically possible. There is a mode to the distribution around $r = 0.22$ (so, the mean time for a subpopulation to wink out for this modal population would be about 5 years). Considerably higher decay rates are possible but less probable, giving rise to a long right tail to the distribution. The breadth of the distribution reflects the range of values for r in the collection of populations that are comparable with respect to properties that bear on r. Since our interest in assembling the comparative information is only in the values of r, the "comparable" populations may be quite different with respect to their respective metapopulation structure (e.g., very different values of m, the number of subpopulations), and they may, accordingly, have very different times to extinction. The

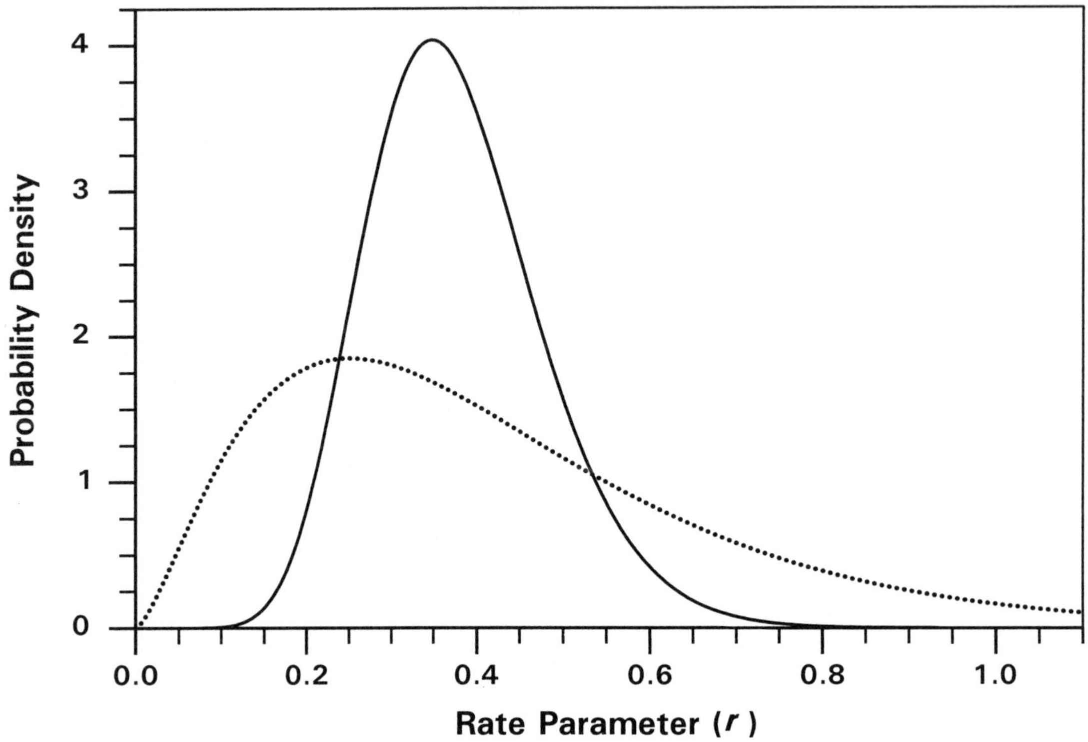

FIGURE 1. Distribution of the rate parameter r, before and after the inclusion of case-specific data in the inference. (a) Dotted line: the distribution of the rate parameter, based only on comparative information from similar populations. (b) Solid line: the distribution of the rate parameter, with the case specific data (11 decays observed, in a sampling interval of duration 30) taken into account.

defining property of members of this collection is solely that, without case-specific data on the decay rate for the population in question, we would classify these populations as similar to the population in question, with respect to r.

Now, imagine that, for the population in question, we had collected case-specific data and observed 11 decays, over a sampling interval of 30 years duration. These case-specific data, combined with the comparative information, lead to the inference that the distribution of the rate parameter, for this case, is a gamma distribution with shape parameter $a = 13.5$ and scale parameter $b = 36$ (equation [13]). This distribution has a mean $E(r) = 0.375$ (equation [14]) and a standard deviation $\sigma_r = 0.102$ (equation [15]). This distribution, the inference for r, is plotted as the solid line in Figure 1.

The general shape for this new, inferred distribution for r, for the population in question, is reasonable, and its relation to the distribution from the comparative information is also reasonable. The mode of the inferred distribution is at a larger value for r (about 0.27), corresponding to a higher volatility than the mode of the distribution of r for the collection of populations that were deemed similar, without regard for observations on the number of decays in a sampling period, but the distribution is narrower. This is to say that, for the same metapopulation structure (same m), the modal population from the comparable set would have a longer persistence time than would a population with the modal r, from our inferred distribution for the population of interest.

Not surprisingly, the addition of the case-specific data, in this instance, has markedly reduced the uncertainty about the parameter r, but, perhaps, not as much as we might like. The remaining uncertainty in r is a reflection of the strength of the data. Evidently, a sampling interval of 30 years still leaves noticeable uncertainty. For comparison, we might consider the inferred distribution, if, instead of our actual sampling, we had hypothetically conducted our observations for a sampling interval of duration 120 years and had observed 44 decays (the same ratio of decays to duration of observation as in

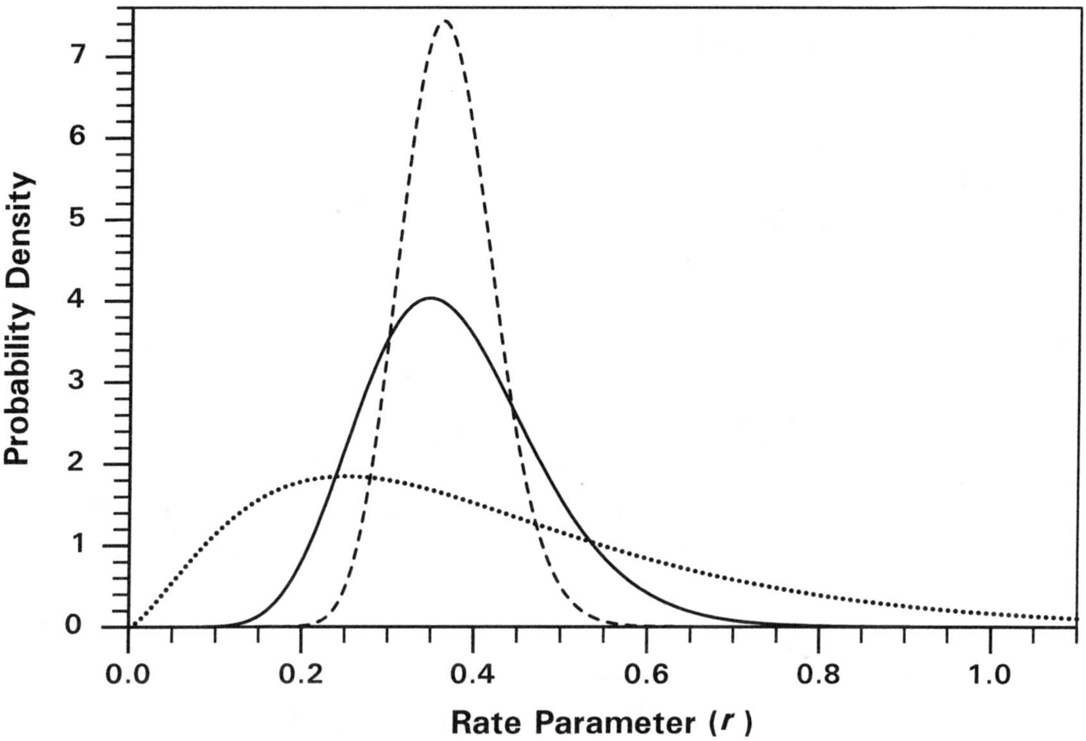

FIGURE 2. Distribution of the rate parameter r that would be inferred, if, hypothetically, we had observed 44 decays in a sampling interval of 120 years, rather than the actual data, 11 decays in 30 years. (a) Dashed line: the inferred distribution of the rate parameter that would have resulted, if we had hypothetically observed 44 decays in 120 years. (b) Solid line: the distribution of the rate parameter with the actual case specific data (11 decays observed, in a sampling interval of duration 30, taken into account. (c) Dotted line: the distribution of the rate parameter, based only on comparative information from similar populations.

the actual sampling). This would give rise to a much narrower inferred distribution for r, plotted as the dashed line in Figure 2, where the inference from the actual data is still plotted as a solid line, and the distribution, based on comparative information, is a dotted line (as in Figure 1).

But, if the only data we have, in hand, are the observation of 11 decays in a 30-year interval, we are not justified in jumping to the conclusion that the narrower distribution in Figure 2 is correct. After all, we drew it based on an *assumption* about the number of decays that would be observed in a hypothetical sampling interval of 120-years duration—we don't really know that we would observe this number of decays, if we actually conducted observations for 120 years. There is a *probability*, which we can compute before the fact, that a sampling over a 120-year interval would yield 44 decays, but there are also probabilities of the sampling yielding other numbers of decays, which, in turn, would lead to different inferences. In a later calculation, we will consider this predicted distribution of future data, in order to evaluate the worth of further sampling.

Our decision about the value of investing in more data (longer sampling interval duration) will be deferred for the moment, so that we can consider the implications of the uncertainty for the bottom line—the risk calculation—rather than basing the decision to collect more data only on our impression of the magnitude of the uncertainty in an intermediate parameter. Our real goal does not hinge on obtaining an inferred distribution of r that we feel good about just because it looks good. The real issue is whether the uncertainty, with our present data, is committing us to a more expensive decision that we might, with a reasonably high probability, reverse, on the basis of collecting a bit more data, whose cost is small compared with the probable cost savings attributable to the reversed decision; if so, this would constitute a compelling argument for collecting more data. After

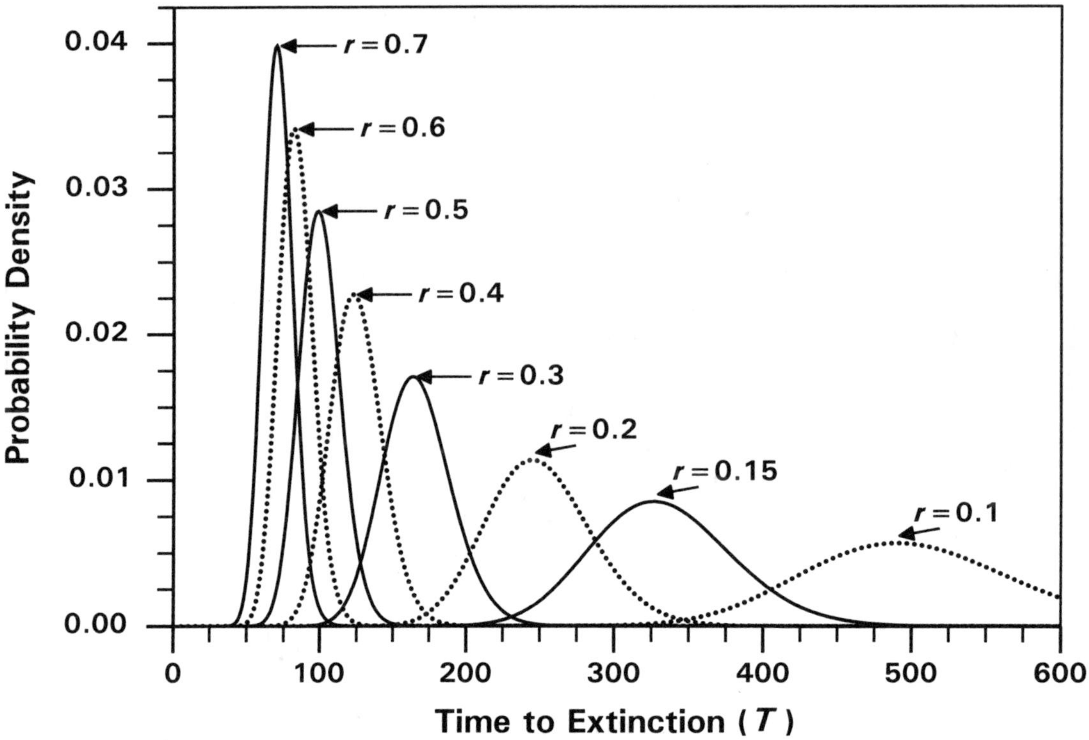

FIGURE 3. Distributions of the time to extinction T, for different fixed values (as labeled) of the rate parameter r.

a few more preliminaries, we will take up the matter of future data in that light. For the moment, we must continue with the calculation of the risk based on the present data.

Let extinction be defined as occurring after 50 further future decays. For any particular value of the rate parameter r, the resulting distribution of time to extinction T is given by equation (4). The uncertainty in T, encompassed in that distribution, is entirely owing to uncertainty in the process and is not due to parameter uncertainty. The distribution of time to extinction is plotted in Figure 3, for a variety of values of r. We see that the process uncertainty in the time to extinction increases as r decreases. This is because the mean time to extinction increases for lower r (smaller decay rate), and the longer the time to extinction, the more time the process uncertainty has to operate on the population, giving rise to the greater spread in the distribution of T.

In accordance with our inferred distribution for r, each value of r has a probability, given by equation (13) (and plotted as the solid line in Figure 1). Therefore, some of the values of r, for which we plotted the resulting distribution of T in Figure 3, are more probable than others. In Figure 4, the respective distributions of T are weighted by the probability of the r on which each was based—now, the area under each curve is proportional to the probability of the r that gave rise to it. The sum of all such weighted distributions, for all possible values of r, gives the distribution of time to extinction, with the uncertainty in the rate parameter taken into account. This weighted summation of the component distributions is accomplished mathematically in equation (16); Figure 4 gives a preview of the shape that we will expect for the result of equation (16), with this example.

For this example, the distribution of the projected time to extinction T, based on the inferred distribution for r given the data and given the comparative information, is the compound gamma distribution described by equation (16), with a mean of $E(T) = 144$ years (equation [17]) and a standard deviation $\sigma_T = 47.5$ (equation [18]). The risk, given these data, is 0.154 (equation [19]). This distribution of T is plotted as the solid line in Figure 5, and the risk is the shaded area of that curve. If we adopt the risk tolerance of $C = 5\%$, the risk in this case would be above our tolerance level.

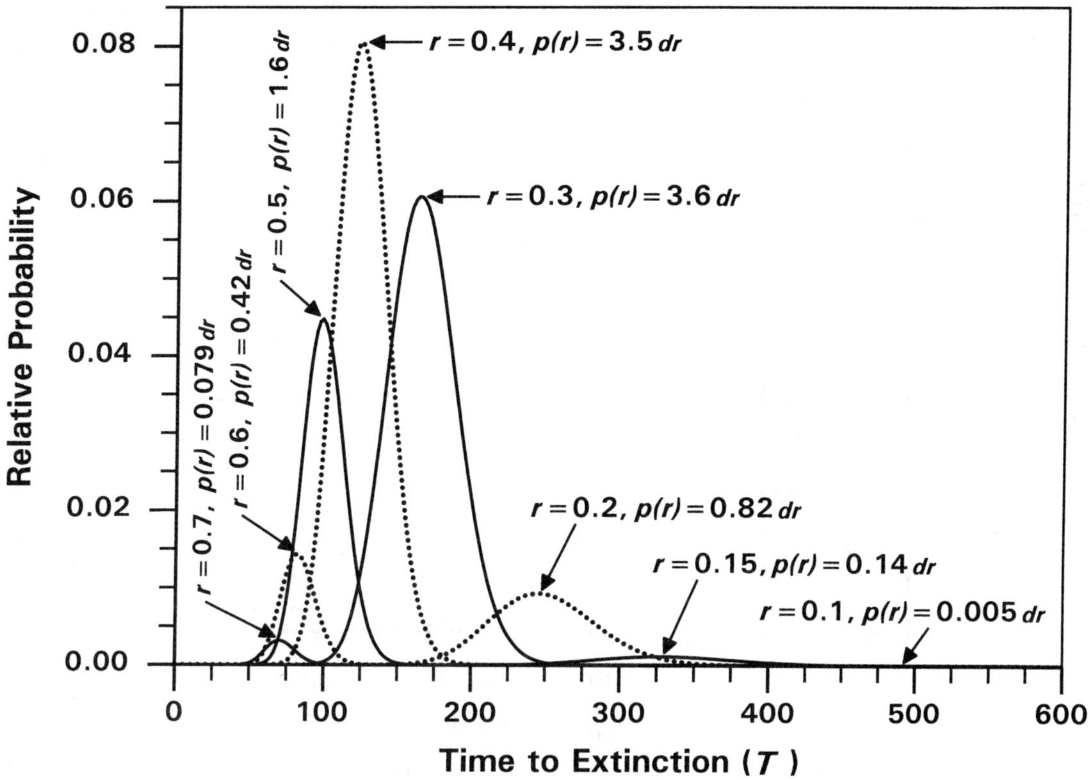

FIGURE 4. Distributions of the time to extinction T, for different fixed values (as labeled) of the rate parameter r, where each distribution is weighted according to the probability (as labeled) of that value of r. The y-axis is scaled so that the area under the curve for any individual distribution of T would be equal to one, if the relative probability of the associated value of r were one.

By contrast, the distribution of the projected time to extinction, based on the comparative information alone, is a compound gamma distribution (also from equation (16), but with t and n set to zero) with a mean of $E(T) = 200$ years (equation [17]) and a standard deviation $\sigma_T = 287$ (equation [18]). The risk, based only on the comparative information, is 0.315 (equation [19]). The distribution of T is plotted as the dotted line in Figure 5, and the associated risk is the shaded area of that curve. We see that the greater uncertainty, in the absence of case specific data, results in greater uncertainty in the rate parameter r (Figure 1), greater uncertainty in the time to extinction T (Figure 5), and higher risk (Figure 5).

It is interesting to note, in Figure 5, that the modal time to extinction, after inclusion of the case specific data (solid line), is greater than the modal time to extinction before their inclusion (dotted line), even though the modal decay rate after inclusion of the data is larger (solid line in Figure 1) than before (dotted line in Figure 1). Evidently, in the transformation of decay rate to probability of time to extinction (equation [4]) and the convolution with probabilities of decay rates (equation [16]), the spread of the inferred distribution for r can overwhelm the effect of the mode of the distribution for r. Increased uncertainty in r moves the mode of the distribution of times to extinction to the left (shorter times to extinction become more probable), in this example.

To compare the correct calculation with one of the many illegitimate shortcuts in risk calculation, the maximum likelihood estimate of r, given the data, is $\hat{r} = 0.367$ (equation [20]). Using this single estimated value of r, and ignoring the uncertainty in the parameter, leads to a predicted distribution of time to extinction that is a gamma distribution, with shape parameter 50 and scale parameter 0.367 (equation [21]), corresponding to a mean time to extinction of $E(T) = 136$ (equation [22]), a standard deviation of the time to extinction $\sigma_T = 19.3$ (equation [23]), and

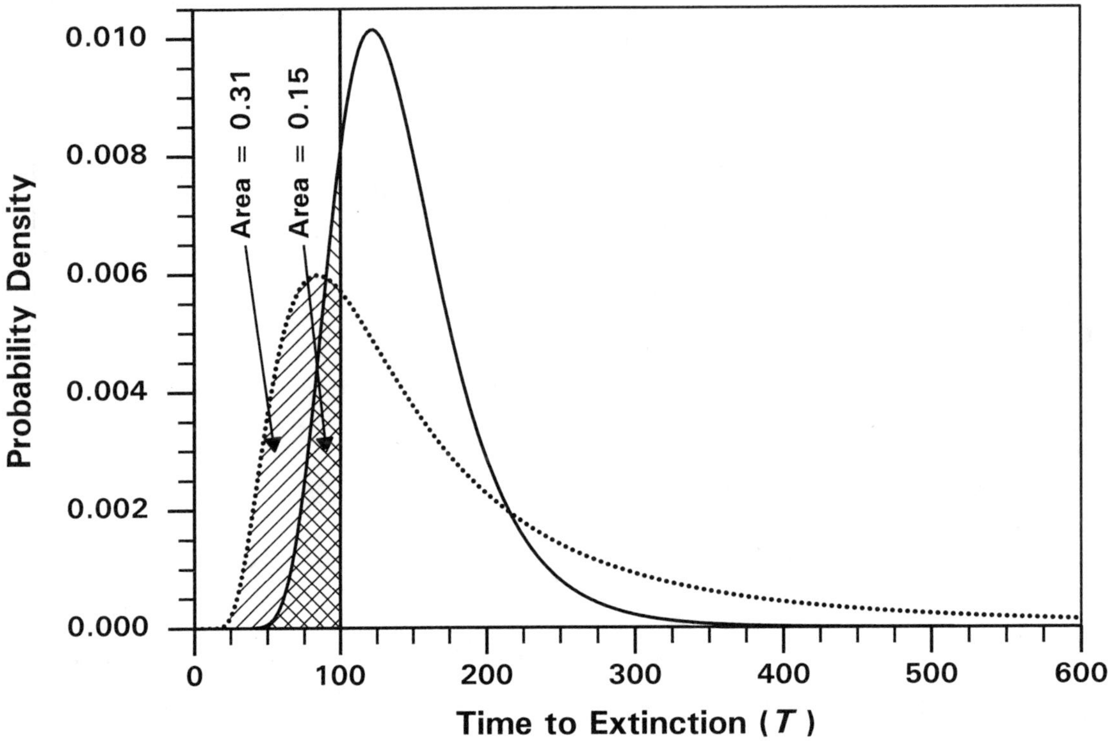

FIGURE 5. Distribution, of the projected time to extinction T, before and after the inclusion of case-specific data in the inference. In both distributions, the area to the left of the cutoff at 100 years defines the risk. (a) Dotted line: the distribution of the time to extinction, based only on comparative information. (b) Solid line: the distribution of time to extinction, with the case specific data taken into account.

a risk of 0.02 (equation [24]). Compare this with the correct values, $E(T) = 144$, $\sigma_T = 47.5$, and a risk of 0.154. The neglect of parameter uncertainty leads to a substantial underestimate of the uncertainty in time to extinction and a gross underestimate of the risk. The distribution of time to extinction, based on the "best estimate" of r, and ignoring uncertainty in r, is plotted as the dashed line in Figure 6, and the associated incorrect estimate of risk is the shaded area of that curve. It is apparent how errors, in representing the uncertainty in r, propagate errors in calculating the uncertainty in T, which greatly distorts the calculated risk corresponding to the tail area of the distribution.

If all uncertainty in r were somehow resolved, the expected uncertainty in the time to extinction (expressed as a variance in T) would be reduced to 20% of its value, with the data now in hand (equation [26]). The probability that the risk, calculated after all uncertainty in r were somehow resolved, would be below the risk tolerance of 5% is 0.592 (equation [27]).

Properties of the future recalculation of the risk that would be expected before the fact, if additional data collection (before the countdown for m decays to extinction) were proposed, are plotted in Figure 7, as a function of the duration of the future sampling interval. This calculation is based on merging the predicted future data with the sampling data already in hand (11 decays observed in an interval of 30 years). The double line shows the maximum number of future decays in the future sampling interval that would be consistent with a conclusion that the risk is below the tolerance of 5% (equation [33]). The solid line shows the probability of observing this few or fewer decays (equation [34]), which, therefore, is the probability that a future sampling of that duration will lead to a future conclusion, that the risk is less than the policy threshold of 5% probability of extinction within 100 years.

The peculiar sawtooth appearance to the curve, describing the probability of a revised decision, in Figure 7, is owing to the necessarily integer nature of the data in this example. The data are counts of events

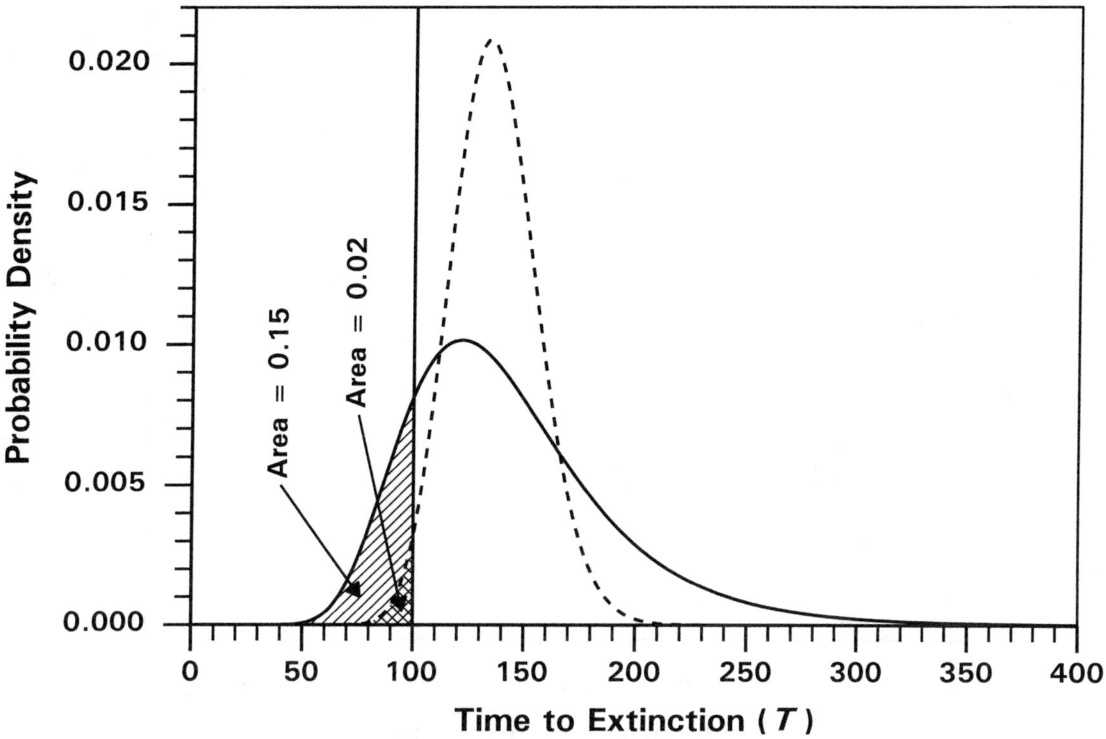

FIGURE 6. Distribution of the projected time to extinction T, computed by an incorrect shortcut. (a) Dashed line: the distribution of time to extinction, computed using the maximum likelihood estimate of the decay rate in place of the inferred distribution for the decay rate. (b) Solid line: the correct distribution of time to extinction, using the correctly inferred distribution of the decay rate. In both distributions, the area to the left of the cutoff at 100 years defines the risk. The incorrectly predicted distribution of time to extinction (dashed line), based only on a "best estimate" of the rate parameter, ignoring the uncertainty in the parameter estimate and ignoring the comparative information, greatly underestimates the risk in this case.

(decays), and there is no definition for a fractional decay, so the double line, showing the threshold number of decays for a revised decision, is necessarily a step function. Over any interval on the sampling interval axis (x-axis), where the step function is flat, the probability of observing this (fixed) number of decays, or fewer, declines, as the sampling interval increases in length, because the longer interval provides more opportunity for observing decays.

The calculations show that, for this example, there would be no point to investing in a future sampling interval of less than 7.2 years duration, because those future data could not possibly change our conclusion based on the current data, in this example, that the risk is above the tolerance level of 5%. Future data collection from a longer sampling interval might well be worthwhile, since it could lead to a revised conclusion that the risk is below the tolerance level. The probabilities of a revised decision could be used in a cost–benefit analysis, for identification of an optimal duration for the future sampling interval. The logic of the optimization would be to compare the cost of conducting that future sampling against the product of the probability of a revised decision and the cost saving that would be realized by the revised decision, if it occurred. Examples of revised decisions that might save money include delisting, declining to list, determining no jeopardy, or accepting a less-extensive critical habitat designation. If the direct costs of the proposed future sampling exceed the possible savings weighted by the probability of the savings, the proposed future sampling would be expected to exact a net cost; if the direct costs of the proposed sampling are less than the probability-weighted savings, then the proposed sampling would be expected to confer a net benefit in cost reduction.

The trend in the single-lined curve, in Figure 7, shows a diminishing return as the length of sampling

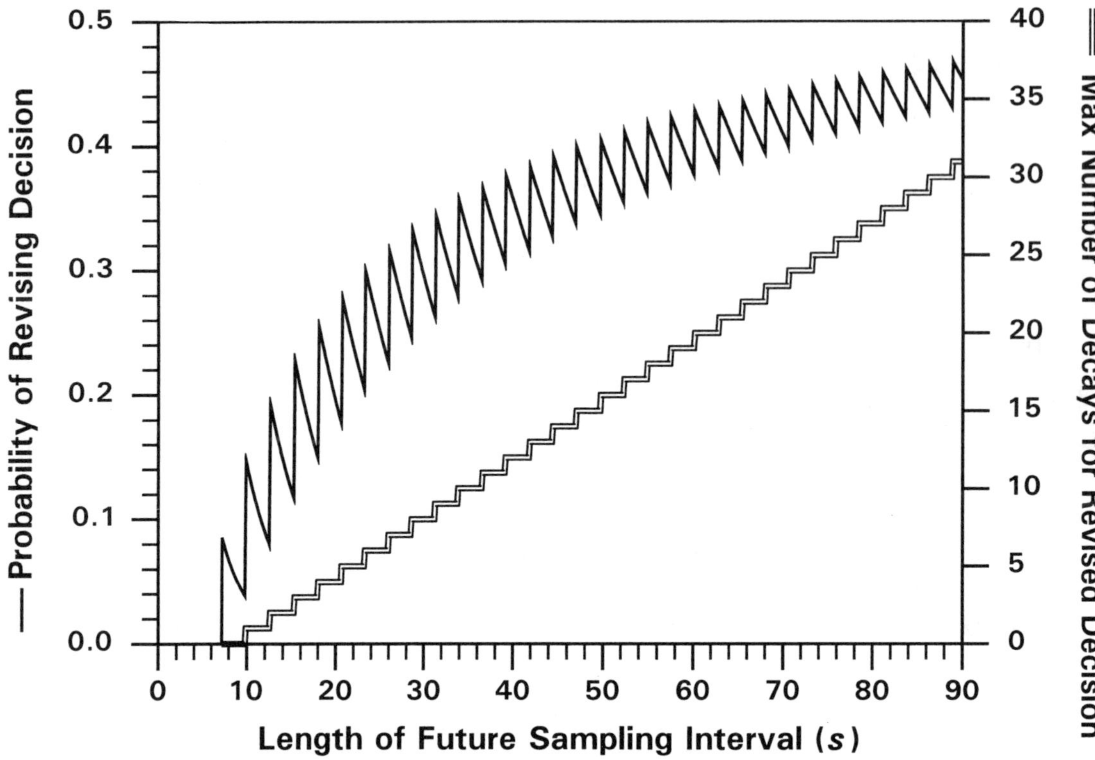

FIGURE 7. Analysis of the probabilities, before the fact, that future sampling will yield new data that will result in a revised calculated risk that is below the tolerance, defined by policy as a function of the length of sampling interval for the future data collection. (a) Double line: the maximum number of observed decays in the future sampling interval that, when included with the comparative information and the sampling data already in hand, will lead to a conclusion that the risk is below the tolerance level. (b) Single line: the probability that the future sampling will lead to a conclusion that the risk is below the tolerance level.

interval increases. Indeed, we already have the result from the calculation on elimination of all uncertainty in r (equation [27]), that the curve should asymptote at a probability of 0.592, for an infinite sample interval. At large sampling intervals, the remaining parameter uncertainty in r is negligible, whereas the remaining uncertainty in T is process uncertainty that is not affected by increasing sampling to determine r. Because of the diminishing returns, we are assured that the optimal commitment to future sampling will either be no future sampling (because no amount of future sampling has a high enough probability of revising the conclusion about the risk) or an intermediate amount of future sampling.

To compare the probability that future data will lead to a conclusion that the risk is larger than we thought, consider a secondary "very" critical risk level of $C_2 = 0.50$. With the present data, in our present example, equation (19) showed a risk of 15%, so the determination with the current data would be that the risk is above the 5% threshold and below the 50% threshold. Equation (35) gives the probability that future data will lead to a revised conclusion that the risk is above 50%. The results are graphed in Figure 8.The double line shows the smallest number of future decays observed in a future sampling episode of duration s that allows the conclusion that the risk is greater than 50%, and the single line shows the probability of observing this many decays or more.

Just as Figure 7 shows a trend of increasing probability of concluding that the risk is below the 5% tolerance, as the amount of future sampling increases, Figure 8 shows a trend of increasing probability of concluding that the risk is above the 50% tolerance, as the amount of future sampling increases. This probability of exceeding the 50% risk threshold also shows a pattern of diminishing "return," as it,

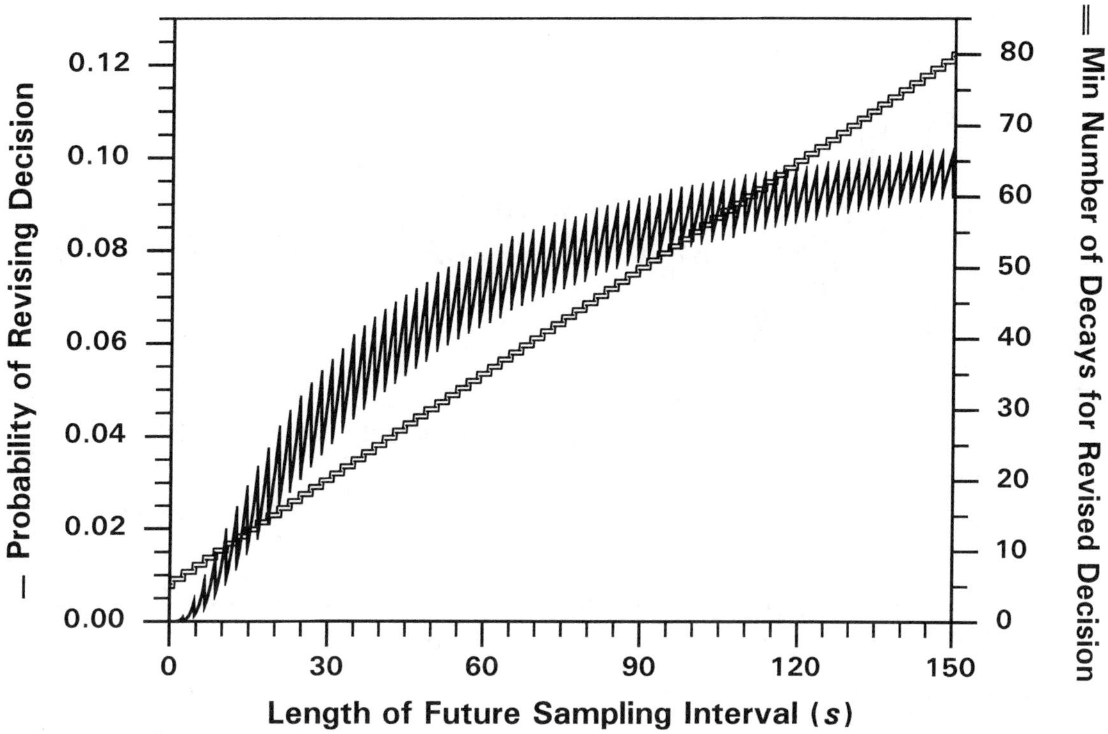

FIGURE 8. Analysis of probabilities, before the fact, that future sampling will yield new data that will result in a revised calculated risk that is above a secondary tolerance level, defined by policy as a function of the length of sampling interval for the future data collection. (a) Double line: the minimum number of observed decays in the future sampling interval that, when included with the comparative information and the sampling data already in hand, will lead to a conclusion that the risk is above the secondary tolerance level. (b) Single line: the probability that the future sampling will lead to a conclusion that the risk is above the secondary tolerance level.

too, asymptotes at a constant value, when the length of the sampling interval is sufficient to eliminate all practical uncertainty in r. We may implicitly solve equation (28) to find r_{c_2}, the value of the decay rate parameter at which the risk just equals $C_2 = 0.5$, for a known constant decay rate. Then, we may use equation (27) to find the complement of the probability, before the fact, of a decay rate that high, or higher, if all uncertainty in r were removed. For this example, these calculations show that the probability of concluding that the risk is above 50% asymptotes at 12.1%, with infinite amounts of future data.

As the amount of future sampling increases, so does the probability that use of the future data will lead to a decision that agrees with the decision we would make, if we had perfect knowledge of the rate parameter r_c. In the present example, where the decision with the data in hand is risk $>C$, Figure 9 shows probabilities of the four possible outcomes of future data as a function of the duration of the period of observation: (1) the probability that the future data will *correctly* revise that decision to risk $<C$ (equation [41]), (2) the probability that the future data will *mistakenly* revise that decision to risk $<C$ (equation [37]), (3) the probability that the future data will *correctly* let the conclusion stand that risk $>C$ (equation [38]), and (4) the probability that the future data will *mistakenly* let the conclusion stand that risk $>C$ (equation [39]).

In the limit of increasing amount of future data, the probabilities of the outcomes corresponding to *correct* decisions (outcomes 1 and 3) asymptote at the probabilities of $p(r < r_c|n)$ and $p(r > r_c|)$, respectively, and the probabilities of the outcomes corresponding to *mistaken* decisions (outcomes 2 and 4) go to zero. Figure 10 shows the probabilities of the four possible outcomes for a future sampling interval of duration up to 1,500.

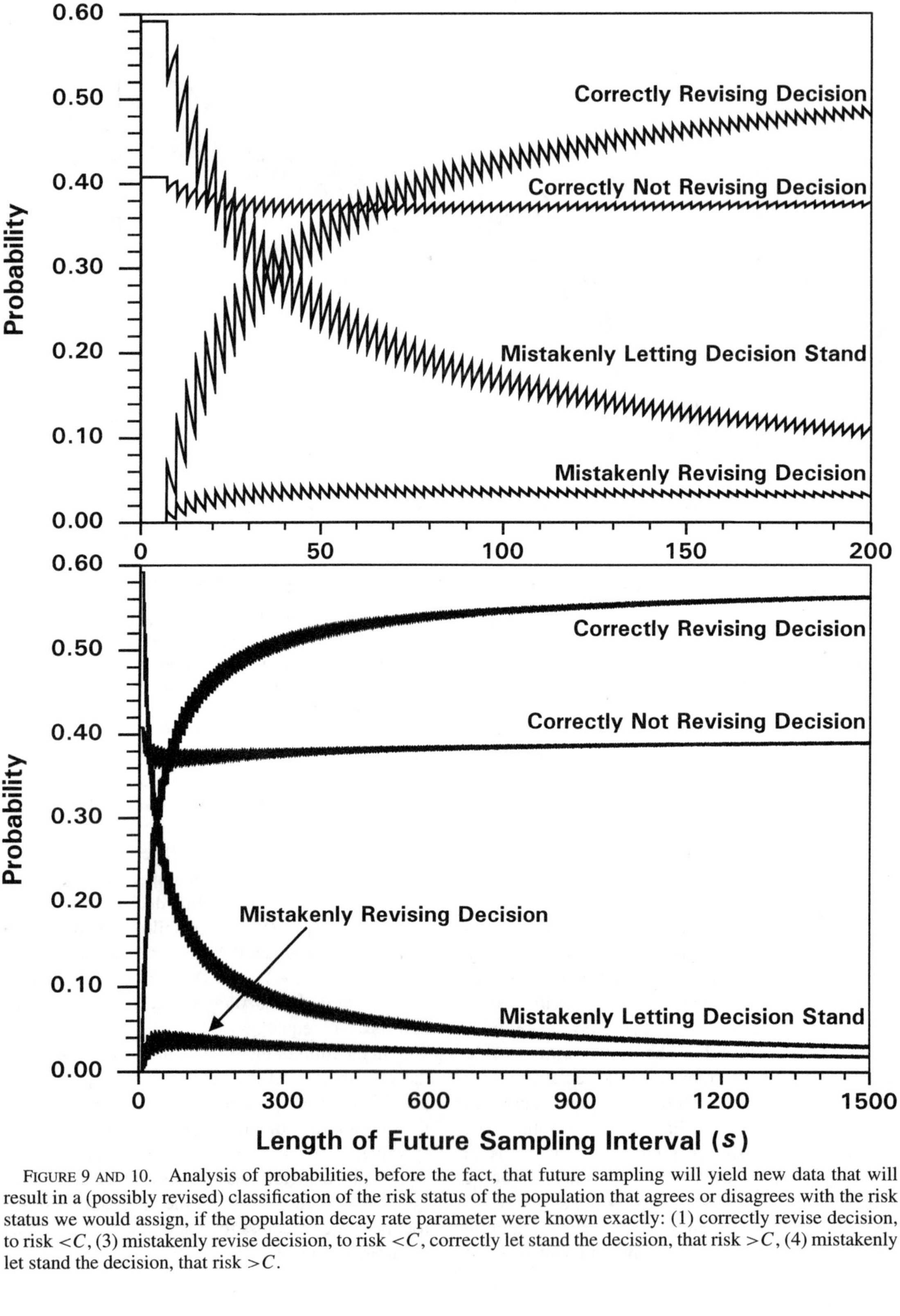

FIGURE 9 AND 10. Analysis of probabilities, before the fact, that future sampling will yield new data that will result in a (possibly revised) classification of the risk status of the population that agrees or disagrees with the risk status we would assign, if the population decay rate parameter were known exactly: (1) correctly revise decision, to risk $<C$, (3) mistakenly revise decision, to risk $<C$, correctly let stand the decision, that risk $>C$, (4) mistakenly let stand the decision, that risk $>C$.

Summary

With extinction risk defined as the probability of extinction before time horizon H, the problem of risk assessment becomes one of conducting a population viability analysis, to calculate the probability distribution of the time to extinction T, and then calculating the tail area of the distribution to the left of H. The decision for intervention (or for regulatory categorization) can then depend on determining whether the risk, so calculated, is greater or less than some risk tolerance level C, set by policy.

A population viability analysis that calculates the probability distribution of time to extinction T necessarily involves some parameters whose numerical values are uncertain. A Bayesian inference procedure will yield a joint probability distribution for the parameters conditional on the data, thus capturing a quantification of the uncertainty. The distribution for the uncertain parameters is propagated forward in a probabilistic calculation of time to extinction, where uncertainty in the parameters and inherent unpredictability of the random population dynamic process are folded together, so that both contribute to the spread of the distribution in time to extinction, and therefore, both contribute to the tail area of that distribution that constitutes the risk. The uncertainty is part of the risk. More disconcertingly, the risk so calculated is conditional on the data, but this is as it should be; the true risk includes the effects of our ignorance and the effects of our dependence on the vagaries of the information that is available to us.

Shortcuts that substitute a parameter "estimate," such as the mean or some confidence bound on the parameter estimate, will distort the calculated risk and are not valid.

The population viability analysis model employed, to generate the distribution of time to extinction, must include all the relevant factors that affect time to extinction. Factors that are insufficiently known must not be omitted for that reason; factors that are insufficiently known must be included as probability distributions on their parameters, and the greater the ignorance about a factor, the broader the distribution(s) representing its parameter(s).

This approach offers the advantage of mathematical rigor and empirical objectivity. It also creates a cleaner separation between science and policy, allowing both to function better. The decision theory procedure outlined here does not "test" any hypothesis, so questions of Type I and Type II error are moot. The result of the analysis is simply a calculation of the risk. Whether that risk is tolerable is a direct matter of policy. Considerations analogous to "power," in a hypothesis testing context, are dealt with in this decision theory procedure, by calculating the probability that additional sampling will reverse the current decision. The fact that evaluation of the merits of additional sampling can then proceed as a simple cost–benefit analysis should come as a relief, compared with the contortions necessary in a hypothesis testing framework for justifying the choice of a particular power level and a particular alternative hypothesis. Finally, the explicit calculation of actual probabilities of outcomes in this decision theory framework encourages development of rigorous adaptive techniques for more efficient and effective management of endangered species. Managers should not be satisfied with less.

Acknowledgements

Portions of this research were supported by DOE cooperative agreement DE-FC07-94ID13317 to Montana State University, EPA cooperative agreement CR-820086-03 to Montana State University, and NMFS contract 40JHNF600188 to the author. The author is indebted to Jim Bence and an anonymous reviewer for helpful comments.

Literature Cited

Berger, J. O. 1985. Statistical decision theory and Bayesian analysis. Springer-Verlag. New York.

Dennis, B., P. L. Munholland, and J. M. Scott. 1991. Estimation of growth and extinction parameters for endangered species. Ecological Monographs 61:115–143.

Goodman, D. 2002a. Predictive Bayesian PVA: a logic for listing criteria, delisting criteria, and recovery plans. In D. R. McCullough and S. R. Beissinger, editors. Population viability analysis. University of Chicago Press.

Goodman, D. 2002b. Taking the prior seriously: Bayesian analysis without subjective probability. In M. Taper and S. Lele, editors. The nature of scientific evidence. University of Chicago Press, Chicago.

Karlin, S., and H. M. Taylor. 1975. A first course in stochastic processes. Academic Press, San Diego, California.

Ludwig, D. 1996. Uncertainty and the assessment of extinction probabilities. Ecological Applications 6:1067–1076.

Ludwig, D., and C. J. Walters. 1982. Optimal harvesting with imprecise parameter estimates. Ecological Modelling 14:273–292.

McAllister, M. K., and G. P. Kirkwood. 1998. Using Bayesian decision analysis to help achieve a precautionary approach for managing developing fisheries. Canadian Journal of Fisheries and Aquatic Science 55:2642–2661.

Punt, A., and R. Hilborn. 1997. Fisheries stock assessment and decision analysis: the Bayesian approach. Reviews in Fish Biology and Fisheries 7:35–63.

Taylor, B. L., P. R. Wade, R. A. Stehn, and J. F. Cochrane. 1996. A Bayesian approach to classification criteria for spectacled eiders. Ecological Applications 6:1077–1089.

Thompson, G. G. 1992. A Bayesian approach to management advice when stock-recruitment parameters are uncertain. Fisheries Bulletin 90:561–573.

Wade, P. R. 2000. Bayesian methods in conservation biology. Conservation Biology 14:1308–1316.

Wade, P. R. In press. The conservation of exploited species in an uncertain world: novel methods and the failure of traditional techniques. Pages 110–143 in J. D. Reynolds and G. Mace, editors. Conservation of exploited species.

Wade, P. R. 2002. Bayesian population viability analysis. In S. R. Beissinger and D. R. McCullough, editors. Population viability analysis. University of Chicago Press.

Walters, C., and D. Ludwig. 1994. Calculation of Bayes posterior probability distributions for key population parameters. Canadian Journal of Fisheries and Aquatic Science 51:713–722.

Walters, C. J., and R. Hilborn. 1976. Adaptive control of fishing systems. J. Fish. Research Bd. Canada 33:145–159.

von Mises, R. 1942. On the correct use of Bayes' formula. Ann. Math. Statist. 13:156–165.